THE DESIGN OF SAMPLE SURVEYS

McGRAW-HILL SERIES IN PROBABILITY AND STATISTICS

David Blackwell and Herbert Solomon, *Consulting Editors*

Bharucha-Reid Elements of the Theory of Markov Processes and Their Applications
Drake Fundamentals of Applied Probability Theory
Dubins and Savage How to Gamble If You Must
Ehrenfeld and Littauer Introduction to Statistical Methods
Gibbons Nonparametric Statistical Inference
Graybill Introduction to Linear Statistical Models, Volume I
Jeffrey The Logic of Decision
Li Introduction to Experimental Statistics
Miller Simultaneous Statistical Inference
Mood and Graybill Introduction to the Theory of Statistics
Morrison Multivariate Statistical Methods
Pearce Biological Statistics: An Introduction
Pfeiffer Concepts of Probability Theory
Raj The Design of Sample Surveys
Raj Sampling Theory
Thomasian The Structure of Probability Theory with Applications
Wadsworth and Bryan Introduction to Probability and Random Variables
Wasan Parametric Estimation
Weiss Statistical Decision Theory
Wolf Elements of Probability and Statistics

THE DESIGN OF SAMPLE SURVEYS

Des Raj
Sampling Expert, United Nations Development Program
Professor of Statistics, University of Ibadan

McGraw-Hill Book Company
New York St. Louis San Francisco Düsseldorf Johannesburg
Kuala Lumpur London Mexico Montreal New Delhi
Panama Rio de Janeiro Singapore Sydney Toronto

The Design of Sample Surveys

Library of Congress Catalog Card number 70-135306
07-051155-1

56789 KPKP 79876

To the memory
of my parents

Contents

Part 3. Planning and Execution of Surveys

Part 4. Data Collection in Selected Fields

Foreword

We are living in a world in which most countries are making strenuous efforts to raise the living standards of their people. In order to achieve balanced development, carefully worked-out plans are drawn up and executed as far as possible. To formulate these plans in a scientific manner, it is essential to have basic facts in numerical terms for the various regions in the country and for the country as a whole.

It is beyond the resources of smaller countries to collect facts year after year from each person, establishment, or farm in the country. Fortunately, as we know now, it is not essential to enumerate each unit in the universe in order to arrive at an acceptable figure for the total. A carefully designed sample can provide the necessary information for guidelines that a country needs, at a cost the country may well afford.

The Statistical Office of the United Nations has been deeply concerned, since its organization in 1946, with ways and means of assisting national governments in obtaining the statistical data so indispensable for planning economic and social development, for checking on current implementation of programs, and for assessing results.

In this work the United Nations has been assisted in two important ways. First, it has been aided by the United Nations Statistical Commission, which initiated and maintained a strong impetus toward the promotion and elaboration of sampling methods and toward the establishment of sampling offices in national governments. The commission was greatly assisted in its efforts by the work of its Subcommission on Statistical Sampling, which recommended principles that could guide the development of suitable methodology in the developing countries. The composition of the subcommission was such as to ensure the highest professional level of recommendations. Sir Ronald Fisher, Prof. P. C. Mahalanobis, Dr. W. E. Deming, Dr. F. Yates, and Prof. G. Darmois gave unstintingly of their time in elaborating the basic principles. In this work they were frequently joined by other distinguished experts.

Second, and this has been equally important in the development of national statistical systems, is the application of sampling methods to the practical problems encountered by developing countries. These countries generally had no long tradition of censuses, or similar periodic compilations, to use as a framework for sampling. The application of sampling methods that would produce acceptable results in such situations therefore required the utmost ingenuity.

The United Nations was fortunate in having the services of such experts as Dr. Des Raj under the auspices of the United Nations Technical Assistance Programs for practical field assignments. Dr. Raj has performed distinguished service in adapting theoretical principles to the practical conditions he found in one country in Southern Europe and one in Africa. At this writing he is continuing his service in another African country, especially oriented to training.

This book deals very competently with the design of sample surveys under the operational conditions prevalent in many developing countries today. It discusses all aspects of sample surveys in a realistic manner. In view of the fact that most countries are now relying increasingly on sampling methods for the collection of information, this book is a very welcome addition to a literature which is far too scanty.

W. R. LEONARD, *former Director*
Statistical Office
United Nations

P. J. LOFTUS, *Director*
Statistical Office
United Nations

Preface

Sample survey, the new technology of our times, has now come to be recognized as an organized instrument of fact finding. Its importance to modern civilization lies in the fact that it can be used to summarize, for the guidance of administration, facts which would otherwise be inaccessible owing to the remoteness or obscurity of the units involved or their numerousness. As a fact-finding agency, a sample survey is concerned with the accurate ascertainment of facts recorded and with their compilation and summarization. This is a specialized task with which this book is wholly concerned.

I have given the phrase "the design of surveys" its widest possible meaning. It includes all preparations needed to launch the survey, the actual conduct of the operations, and the subsequent processing of the data and writing of the report. This book is intended to cover all phases of survey work. Thus the design of the questionnaire, administration of the survey, and methods of processing and tabulating are all included. The treatment is simple and concise and avoids complicated mathematics. Good use has been made of the basic ideas of sampling without proving any formulas. The reader interested in proofs is referred to the author's companion volume "Sampling Theory."

The book consists of four parts. The first part is introductory. Its purpose is to explain the nature and role of sampling methods and the risks involved in making an inference about the population on the basis of a sample. The concepts of sampling error and response error are introduced, and the role of a sample survey versus a complete enumeration is discussed. The reader is introduced to different techniques of sampling and the basic principles are brought home intuitively with the help of numerical examples.

The second part is devoted to a more detailed study of the different techniques of sampling. The procedures of sample selection and estimation of population characteristics and the associated sampling errors are demonstrated through illustrations gathered from diverse sources. Many of the examples are taken from live surveys with which I have been associated from time to time during the last 20 years. The discussion is presented without recourse to advanced mathematics in the hope that it will permit a larger group of readers to understand the methods of data analysis. To give the reader practice and confidence, there are at the end of some chapters a number of problems with answers.

The problems that arise in actually designing a sample survey and putting it into operation form the subject matter of the third part of the book. A thorough discussion is given of the initial steps involving identification of the objectives of the survey and decisions regarding the population to be covered and the methods of collecting the information. An attempt has been made to point out the different defects to which a frame can be subject and the procedures to be devised by which these defects are not allowed to vitiate the results of the survey. The framing of questionnaires, design of the sample, selection and training of interviewers, processing of the data, and writing of the report are the topics that are discussed in considerable detail.

The fourth part of the book is a progress report and a description of the present state of the complexities encountered when data are collected in diverse fields such as population, agriculture, and labor. The range and flexibility of the use of sampling methods for collecting information is brought out. The illustrations have been selected from a variety of sources so as to convey a sense of perspective. A careful examination is made, wherever relevant, of the effects of conditioning, telescoping, and fatigue on the quality of the data collected. The rationale of the operational definitions used to throw light on complex phenomena such as ill health or unemployment is explained. Some of the important sample surveys being conducted all over the world are described at their proper place.

The book has been written to serve a wide variety of tastes. Administrators who are responsible for commissioning surveys and interpreting the results will find the first part a sufficient guide to the basic principles. The second part of the book will be useful to those whose job is to analyze the data collected from surveys. Included in this category are young professional statisticians and students of statistics. Sampling statisticians entrusted with the task of designing the sample will find Chaps. 8 and 9 particularly useful. Those in charge of fieldwork or the processing of data will find in Chap. 10 some food for thought. A subject-matter specialist such as a demographer can refer to the relevant chapters in Part 4 in order to get a glimpse of the data collection problems encountered in his field. Finally, the book will be of interest to users of the results of surveys who wish to know the circumstances under which information based on samples can be considered to be trustworthy.

Some repetition is inevitable when a book is intended to be used by persons having different backgrounds and tastes. The present effort is no exception. The advantage of the arrangement chosen, however, is that the different parts of the book are more-or-less self-contained. Knowledge of a particular part of the book is not essential for understanding the chapters that follow. The reader can begin almost anywhere.

The author of a book is grateful to almost anyone who has touched the field. I take pleasure in acknowledging my debt to all those persons, named or unnamed, whose results I have drawn upon during the writing of this book. The book could not have been written without the opportunity, made possible by the United Nations and the governments of Greece, Ethiopia, and Nigeria, to use sampling methods on a variety of populations, for which I shall ever remain grateful. Thanks are also due to Mr. Cornelius Ogunsile for his patient help in typing.

DES RAJ

part one

Introduction

1
The Nature of Sampling Methods

1.1 INTRODUCTION

This is the age of statistics. We learn from the daily newspapers that the production of industry in the country is up by 4 percent over the previous year; the number of unemployed now stands at 3.5 million; the area under food crops has shrunk by 2 percent. A demographer claims that the population of the country has risen by 1.5 percent during the year while the number of suicides has more than doubled. It is broadcast on the radio that as many as 70 percent of the adults smoke more than two packets of cigarettes a day and so on. Confronted with this mass of statistics, the common man becomes bewildered and starts wondering how these figures are arrived at and at what cost.

It should be conceded at the outset that we do need figures to make the right type of decision. Government, business, and the professions all seek the broadest possible factual basis for decision making. In the absence of data on the subject a decision taken is just like leaping into the dark. We do need statistics, in fact better and better statistics. Many of the statistics we find in newspapers are the by-product of day-to-day administration. There are some other basic facts about the nation which are ordinarily collected through periodic censuses. And then there are statistics collected through sample surveys in which just a fraction of the universe is used to provide information for the whole.

1.2 THE SAMPLING PROCEDURE

The basic idea is simple. Information is needed about a group or population of objects such as persons, farms, or firms. We examine only some of the objects and extend our findings to the whole group. Thus a few blocks may be taken from a city and lists of persons made in each sample block. This information is used for estimating the total number of persons in the city. As another example, a number of articles is taken from the manufactured product, the number of defective articles in the sample is determined by visual inspection or otherwise, and an estimate is made of the proportion of defective articles in the whole product. In both cases, a sample of objects is taken to make an inference about the whole group.

There are three elements in the process: selecting the sample, collecting the information, and making an inference about the population. The three elements cannot generally be considered in isolation from one another. Sample selection, data collection, and estimation are all interwoven and each has an impact on the others. Sampling is not haphazard selection; it embodies definite rules for selecting the sample. But having followed a set of rules for sample selection, we cannot consider the estimation process independent of it; estimation is guided by the manner in which the sample has been selected.

1.3 SAMPLING ERRORS

It appears at first sight that sampling is a risky proposition. If a sample of blocks is used to estimate the total number of persons in the city, the blocks in the sample may be larger than the average. This will overstate the true population of the city. In sampling from a product, the sample can be free of defects; yet as much as 10 percent of the manufactured product may be defective. Thus we concede that the figure obtained from the sample may not be exactly equal to the true value in the population. The reason is that the estimate is based on a part and not on the whole. Another way of saying it is that the sample estimate is subject to sampling errors or sampling fluctuations. But, as the reader will find in later chapters, these errors can be controlled. Modern sampling theory helps in designing the survey in such a manner that the sampling errors can be made small.

Some writers suggest the use of a "representative" sample as a safeguard against the hazards of sampling. This is an undefined term which appears to convey a great deal but which is unhelpful. If it means that the sample should be a miniature of the population in every respect, we do not know how to select such a sample. Actually, we can deliberately choose a distorted sample in order to make a better estimate. Consequently this term will not be employed in this book.

1.4 RESPONSE ERRORS

There are many kinds of errors involved when data are collected from a sample of objects or from the whole group. Suppose we want to estimate the average age of a group by asking a sample of persons. Some persons may not know their age exactly; some others may overstate it as a matter of prestige; a few others may refuse to give their age; in a few cases the enumerator may record the age wrongly; and so on. Thus errors of various types may creep into the results. These errors are present whether you take a sample or canvass every unit in the population.

Then there are errors arising from unclear definitions. Suppose we want to estimate the number of persons in an area. What do we mean by the term "the number of persons" in the area? Do we want to take in those who normally live there, or those who spent the previous night in the area? What about a guest in the house or the father on a trip to a foreign country or the boy at school in the neighboring commune? If you want to count only those who normally live there, how do you define the usual residents? You cannot define all your terms with mathematical exactitude, and so you do not know exactly what you want to measure. This brings about errors.

Some of the errors in the data are of the random type: that is, they average to zero over the sample. This happens when the errors are not deliberate or intentional. Some units will overstate, and others will understate, resulting in a net difference of zero. There are other errors of a systematic type which are more serious. In an inquiry conducted to estimate the average number of parcels operated by farmers, the farmers may deliberately understate the number for fear of taxation. This type of error will not cancel out over the sample but will rather persist. Such errors are called systematic errors (4).

1.5 ATTITUDE TOWARD DATA COLLECTION

We should reconcile ourselves to the situation that errors are always present when data are collected. We have to live with errors. A change of attitude is needed for a proper appreciation of this subject. To pursue the example of estimating the number of persons in an area, we cannot even define our terms rigorously. Even if we could, the population is changing at every moment. On the other hand, the survey or inquiry must cover a period of time. The population must have changed by the time the results become available. And then we are not so much concerned with what the position was at the time of the inquiry but with what it would be when action is to be taken on the basis of the inquiry. There is thus a strong case for reorienting ourselves toward the philosophy of data collection.

Of course the errors of a survey depend considerably on the essential conditions under which it is conducted. If resources are meager and skilled personnel are not available, the errors will be large; these errors can be reduced if money and the right kind of personnel are available.

1.6 DOMAIN OF SAMPLING

Sampling methods are being increasingly used in almost all spheres of human activity. There are population studies for finding the number of persons in an area, their distribution by sex and age, the number of births and deaths, and the amount of internal migration. Then there are studies of labor problems, the number of hours worked and wages paid, whether women receive lower wages than men in the same industry, the occupational structure of the population, and the number of those actively seeking work. Another field is that of agriculture. The total area under cultivation, the pattern of cropping, the size and quality of livestock, and the number and area of holdings and their tenure are some of the problems in which sampling methods are being used. In the industrial sector we want to know the number of establishments by kind, the number of persons engaged, the cost of raw materials, and the value of production. Similarly, sampling methods are used for determining the sales of retail stores and the value of their inventories. Sampling is commonly used for assessing attitudes and opinions of the population. And nowadays a good deal of auditing of accounts is conducted on a sampling basis. The range of applications of the sampling method is enormous.

1.7 CHARACTERISTICS OF INTEREST

Although sampling methods can be employed for any conceivable purpose, there are four characteristics of the population in which we are usually interested: population total (such as the total number of beggars in a city), population mean (the average number of persons in a household), population proportion (the percent of cultivated area devoted to corn), and population ratio (the ratio of expenditure on recreation to that on food). The populations encountered in practice are finite in that the number of objects contained in them is limited.

In addition to summary figures such as the mean or proportion, the entire distribution of a variate may be of interest, such as the distribution of broken homes in a community or the response to a political crisis on a university campus. In all these cases the goal is *description* of the population.

There are also situations in which the goal is *explanation*—to find out why a distribution takes the form it does. What accounts for broken homes? What accounts for the support and opposition which a political movement on campus generates? There may also be cases in which both description and explanation are of interest. For example, the purpose may be to find out both how students react to a situation and why they react the way they do.

Most surveys are descriptive in nature. Description is of two types: simple and differentiated. The distribution of responses to a question is an example of simple description. But when these responses are broken down by, say, age, sex, income, or education, we have an example of differentiated description. Differentiated description is used to see how the distribution varies among subdivisions of the population. Quite often differentiated description is an initial step to explanatory analysis.

1.8 CENSUS VERSUS SURVEY

It might be felt that sampling is not a dependable method and that nothing short of a complete enumeration of the whole population can ever be contemplated. This is an argument that has now lost its ground for the following reasons (1).

In the first place, the data obtained from a census are not entirely without errors. It needs a huge organization to conduct a countrywide census. An army of enumerators, tabulators, and supervisors must be employed to collect the data and process it. Since there is an upper limit to the number of such qualified persons in a country, the data collected may be subject to serious errors.

Some of these errors which are of a random character tend to cancel out but there are many others of the nonrandom type which tend to remain more-or-less fixed and constitute a systematic error. Due to the large scale of operations the census usually has a large systematic error. On the other hand, the volume of operations in a sample survey is small. It is possible to recruit more qualified workers, give them better training, and improve the methods of measurement and supervision. By improving the quality of work the systematic error can be reduced. And by using the modern methods of sample design the sampling error, too, can be reduced to the point that it becomes less important than the systematic error. We thus come to the following conclusion. There are no sampling errors in a census but the systematic error is large. A sample survey would be subject to comparatively small systematic error and a sampling error which is even less important. The total error of a sample survey will, therefore, often be much smaller than the systematic error alone of a census.

Second, the census by itself provides no measure of the margin of error to which the data are subject, although considerable error may be present.

This is different from a well-organized sample survey which is capable of providing both the estimates and the random errors to which they are subject. (The systematic errors will pass unnoticed in either procedure.) It becomes difficult to interpret the figures collected from a census without adequate knowledge of the limitations to which they are subject.

And then there is the paramount consideration of cost involved in the collection of data. Because of the small volume of operations, a sample investigation can often be carried out at a fraction (say, one-fifteenth or one-twentieth) of the cost needed for a census when the object is to produce national totals. For the same reason, the results can be produced with considerably greater speed when the sampling method is used. Indeed, quite often the results of a census have to be obtained on the basis of a sample selected from it in order to be able to use them before they become outdated.

Finally, the scope of the inquiry can be widened by using the sampling process. Information of a more specialized type or of a more delicate nature cannot be collected in a census, for the number of specialists is limited. Such questions can, however, be asked in a sample survey, and reasonably accurate data can be collected. You cannot, for example, subject every one in the country to a medical examination for collecting statistics on health. But you can possibly do it in a smaller inquiry based on a sample.

But this is not the whole story. There are occasions when the sample survey will not do the job. Suppose you want information for every village in the country and there are 586,000 villages in the population. You cannot possibly devise a sample of a reasonable size which can provide you with this information for each village. You will have to take a census in this case. A census is the only possibility when local data are needed for each subdivision of the country.

And then there is the question of coverage. It is difficult to have complete coverage of the population in a sample survey. Even some of the best surveys have been found to cover only 95 percent of the population. It becomes difficult to check whether all those intended to be taken in the sample have actually been taken. It is easier to check on this in the census since every one is supposed to be included.

1.9 SAMPLE SURVEY IN CONJUNCTION WITH CENSUS

It should be pointed out that the roles of the sample survey and the census are not always competing. It is extremely useful to have, say, a census of population every 10 years. This will provide bench-mark basic statistics on the number of persons, their age and sex distribution, marital status and household composition. The census provides an opportunity to collect data on a sampling basis on items of a more delicate character, such as employment, the number unemployed and their characteristics, and so on. The two procedures can work together. For the intercensal period data on the population will naturally be collected through carefully designed sample surveys. These surveys, conducted at low cost, can give up-to-date and reliable information about changing conditions and trends. And the results of the previous census can be used with advantage to get firm estimates of population totals. In addition to broadening the scope of the census, the sampling method can be used for producing advance estimates from the census and for providing postenumeration checks on quality and coverage. Although it is not essential, the precensus test of a census may also be planned on a sample basis (3).

1.10 DEPENDABILITY OF SAMPLING

When the inquiry is based on a sample, the results obtained will ordinarily differ from the true values aimed at. In fact different samples will produce different results. But we cannot judge the validity of a single sample by finding how far it differs from the true value, which is unknown. It is the entire apparatus used—the sampling procedure, the data collection, and the manner in which estimates are made—which will have to be judged as satisfactory or not. Fortunately the mathematical theory of sampling can be used for this purpose provided the sample is selected properly. The sample itself can give guidance as to how different samples will differ from each other and provide a measure of sampling errors. (The sampling errors will depend on the variability in the population, the size of the sample, the method of selection, and the method of calculation.) From the same sample we can make different types of calculations in order to determine the best way of making the estimate. It is possible to provide bounds on the magnitudes of the errors in the results and to devise means of lowering the error bounds as required. It is these aspects of the sampling method which have earned it an important place in scientific inference. Sampling is now considered a reliable and organized instrument of fact finding.

1.11 JUDGMENT SAMPLING

It might be thought that the quickest way of selecting a sample is by judgment. Thus if 20 books are to be selected from a total of 200 to estimate the average number of pages in a book, someone might suggest picking out those books which appear to be of average size. The difficulty with such a procedure is that consciously or unconsciously, the sampler will tend to make errors of judgment in the same direction by selecting most of the books which are either bigger than the average or otherwise. Such systematic errors lead to what are called *biases*. A second disadvantage is that the range of variation as observed in such a sample does not give a good idea of the variability in the population. This happens because the sampler is unlikely to select units which are too small or too large, although such units do exist in the population. Furthermore, the situation does not improve by asking a larger number of persons to select the samples by judgment. And there is no objective method of preferring the judgment of one person to that of another. We cannot predict the type or the distribution of the results produced by a large number of samplers, nor can we predict the manner in which these will differ from the true value aimed for.

The basic reason for our inability to handle judgment samples is that we cannot calculate the probability that a specified unit is selected in the sample. We are therefore unable to determine the frequency distribution of the estimates produced by this process. In the absence of this information the sampling error cannot be objectively determined.

1.12 PROBABILITY SAMPLING

The situation is completely different when we use probability sampling methods in which each unit in the population has a known and nonzero chance of being selected in the sample. Such samples are usually selected with the help of random numbers. (These numbers, now available in millions, are constructed from the set 0 to 9, in which each of the 10 digits is equally likely to be taken with a probability of 1/10.) Having selected a probability sample, we can use the theory of probability to determine the frequency distribution of the estimates derivable from the sampling and estimation procedure used. In this manner different procedures can be compared objectively. And, what is very important, a measure of the sampling variation can be obtained objectively from the sample itself. Valid inferences can be derived from the sample by making use of the statistical theory of inference (2).

1.13 ACCURACY AND PRECISION

An estimate calculated from the sample is said to be precise if it is near the expected value, that is, census count taken under identical conditions. It may not necessarily be near the true value aimed at; that is, it need not be accurate. Precision refers to closeness to the expected value while accuracy refers to closeness to the true value. The expected value will be different from the true value when the errors present in the data do not average to zero. These errors arise from faulty measurement techniques, defective questionnaires, ill-defined concepts, and so on. Such errors may be no less important than the sampling errors. Although our aim is to be as accurate as possible, the sample can only give guidance as to how precise we are. If nonsampling errors are unimportant, precision and accuracy are not different, and the sample can throw light on the accuracy of the results.

1.14 THE QUESTION OF COST

The resources for conducting a sample survey are always limited. We have thus to look for procedures which are simple and efficient, procedures which can be completed within the time schedules and which take into account all administrative requirements. Only those procedures are ordinarily considered from which an objective estimate of the precision attained can be made from the sample itself. Furthermore, it should be possible to carry through the procedures according to desired specifications. If the cost of the survey is specified, the best procedure to be chosen is that which gives the highest accuracy. When the level of accuracy is predetermined, the best procedure is that for which the cost of the survey is minimum. This is the guiding principle of sample design.

1.15 SAMPLING TERMINOLOGY

Some of the terms used in sampling work will now be defined.

A *population* is a group of objects such as persons, establishments, farms. Thus we speak of a population of villages, of fields, of fish in a pond, and so on. A population is *finite* if the number of objects or units contained in it is limited. Most of the populations we deal with are finite.

We are often interested in estimating four characteristics of populations. They are

Population mean or average such as the average number of persons per household, the average earnings of factory workers, the average yield of wheat per acre

Population total such as total number of persons in an area, total number of industrial establishments located in a district, total amount of food grains produced in a state, total number of fish in a lake

Population proportion such as proportion of unemployed persons aged 14 or over, proportion of establishments engaged in manufacturing foods, proportion of fields growing coffee, proportion of factory workers earning less than $300 per month

Population ratio such as ratio of expenditure on foods to that on rent, sex ratio (ratio of males to females), ratio of the volume of goods carried by road to that carried by other means of transport

A population is built up of *elementary units* which cannot be further decomposed. Thus the elementary unit is an individual person in a human population, a card in a file of cards, an establishment in a directory of establishments, and so on.

A group of elementary units is called a *cluster*. Thus we have a cluster of persons (a household), a cluster of farms, etc.

If a human population is divided up into villages for purposes of selection of a sample of villages, the village is the *sampling unit*. In a manufacturing survey the sampling unit may be an establishment or the company.

The fraction of the population selected in the sample is called the *sampling fraction*. Thus the sampling fraction is 1/10 if 15 households are taken in the sample from a population of 150.

The reciprocal of the sampling fraction is called the *raising factor*.

The list or map or any other acceptable material from which the sample is selected is called a *frame*.

When the sample is selected by taking every kth unit, it is called *systematic* sampling.

A sample in which every unit has the same probability of selection is called a *random* sample. If no repetitions are allowed, it is a *simple random* sample selected without replacement. If repetitions are permitted, the sample is selected *with replacement*.

If the units appear in the sample with different probabilities, sampling is said to be with *unequal probabilities*. When the probabilities are based on some measures of size of the units, it is sampling with *probability proportionate to size* (pps).

If the object is to estimate the total production of industrial establishments, the main *variate of interest* is production. We may, however, possess information on an *auxiliary variate* such as employment and use it for making an improved estimate of production.

When the population is divided up into groups and a sample is selected from each group, the groups are called *strata*. If the same fraction is taken into the sample from each stratum, we have used a *fixed sampling fraction* (or proportionate allocation). Otherwise it is a *variable sampling fraction* plan.

If a sample of villages is taken from the totality in the population and a further sample of households is selected from each village in this sample, it is a case of *two-stage* sampling. If a further sample of persons is taken from each selected household, we have used a *three-stage* design, in which the villages are primary sampling units, households are second-stage units, and persons are third-stage units.

Sometimes the first phase of the inquiry is limited to the collection of auxiliary information. This information is used in the second phase for stratification, ratio, or regression estimation. This is called the *double sampling* method.

REFERENCES

1. Cochran, W. G. (1963), "Sampling Techniques," John Wiley & Sons, Inc., New York.
2. Raj, D. (1968), "Sampling Theory," McGraw-Hill Book Company, New York.
3. Yates, F. (1965), "Sampling Methods in Censuses and Surveys," Charles Griffin & Co., Ltd., London.
4. Zarkovich, S. S. (1966), "Quality of Statistical Data," Food and Agriculture Organization, Rome.

2
Basic Principles of Sampling

2.1 INTRODUCTION

We shall present in this chapter the basic principles of sampling and illustrate how sampling works in practice. A small hypothetical population will be used as the testing ground.

2.2 SAMPLING FROM A HYPOTHETICAL POPULATION

Consider the following hypothetical population of 10 manufacturing establishments along with the number of paid employees in each (see Table 2.1).

The purpose is to estimate the average employment per establishment from a random sample of two establishments. There are in all 45 samples each containing two establishments. We can calculate the average employment from each sample and use it as an estimate of the population average. Table 2.2 gives the totality of samples and the average employment given by each. It is clear that the sample estimates lie within the range of 11.0 to 57.5, the true value being 27.0. Some samples give a very low figure while some others give a high estimate. But the average of all the sample estimates is 27.0, which is the true average in the population. We shall express this fact by saying that the sample mean is an unbiased estimate of the population mean; that is, the expected value of the sample mean is the population mean. But, although unbiased, the sample mean varies considerably around the population mean.

Table 2.1 Hypothetical population of 10 establishments

Establishment number	*Number of paid employees y*	*Establishment number*	*Number of paid employees y*
0	31	5	18
1	15	6	9
2	67	7	22
3	20	8	48
4	13	9	27
Total employment 270		Average employment 27.0	

Table 2.2 All possible samples of size 2

Sample	*Average*	*Sample*	*Average*	*Sample*	*Average*
0,1	23.0	1,8	31.5	4,5	15.5
0,2	49.0	1,9	21.0	4,6	11.0
0,3	25.5	2,3	43.5	4,7	17.5
0,4	22.0	2,4	40.0	4,8	30.5
0,5	24.5	2,5	42.5	4,9	20.0
0,6	20.0	2,6	38.0	5,6	13.5
0,7	26.5	2,7	44.5	5,7	20.0
0,8	39.5	2,8	57.5	5,8	33.0
0,9	29.0	2,9	47.0	5,9	22.5
1,2	41.0	3,4	16.5	6,7	15.5
1,3	17.5	3,5	19.0	6,8	28.5
1,4	14.0	3,6	14.5	6,9	18.0
1,5	16.5	3,7	21.0	7,8	35.0
1,6	12.0	3,8	34.0	7,9	24.5
1,7	18.5	3,9	23.5	8,9	37.5

If we increase the sample size to 3, there are 120 samples in all. We can calculate the average employment for each sample and prepare a frequency distribution showing the number of sample averages falling in the different classes. This procedure can be repeated with samples of size $n = 4, 5, \ldots, 9$. The results obtained are given in Table 2.3.

Table 2.3 Proportion of sample means falling in different classes

Employment class	Percent of sample means falling in class Sample size n								
	1	2	3	4	5	6	7	8	9
9–13	10	4	1						
13–17	20	16	10	5	2				
17–21	20	18	20	16	16	12	8	2	
21–25	10	17	15	19	19	23	23	22	20
25–29	10	7	13	19	27	28	29	45	70
29–33	10	6	15	20	18	26	38	31	10
33–37	0	7	12	10	16	11	1		
37–41	0	9	7	8	2		1		
41–45	0	9	4	3					
45–49	10	3	3						
49 or more	10	4							
Total	100	100	100	100	100	100	100	100	100

It is noted that the concentration of sample estimates around the true mean increases as the sample size is increased. While only 30 percent of the samples produced a mean between 21 and 33 for sample size 2, the corresponding percentage is 43 for $n = 3$, 90 for $n = 7$, and 98 for $n = 8$. Here we have observed a universal phenomenon. The concentration of the sample average increases as the sample size increases. This fact is expressed by saying that the sample mean is a *consistent* estimate of the population mean. Provided the sample size is large enough, there is no great risk involved in using the sample estimate for the population parameter (Sec. 1.3).

2.3 THE VARIANCE OF SAMPLE ESTIMATES

We now need a measure of the degree of concentration of the sample estimates around the expected value. This measure is provided by the *variance* of estimates. The deviation of each sample estimate from the expected value is squared and the sum of these squares is divided by the number of samples. This is the variance of the estimate. If all the sample estimates are equal, the variance of the estimate is zero. The greater the variance the less the concentration.

Actually it is not necessary to draw all possible samples to get a measure of the extent to which the sample estimates differ from the value aimed at. By using sampling theory it can be shown that, in random samples of size n, the variance of the simple average or the sample mean $\bar{y}$ is given by (1, p. 35)

$$V(\bar{y}) = \frac{1}{n}\left(1 - \frac{n}{N}\right) S_y^2$$

where

$$S_y^2 = \frac{1}{N-1}\sum (y_i - \bar{Y})^2 \qquad \bar{Y} = \frac{1}{N}\sum y_i$$

and N is the number of units in the population. In the present case

$$\bar{Y} = \frac{\sum y_i}{10} = 27.0 \qquad S_y^2 = \frac{\sum (y_i - \bar{Y})^2}{9} = \frac{2{,}876}{9} = 319.55$$

Hence the variance of the sample mean $\bar{y}$ can be calculated for different values of n. When the sample size is unity, the variance of y, the variate of interest, is 2,876/10 = 287.6, which is called the *population variance* and is denoted by $\sigma_y^2 = \sum (y_i - \bar{Y})^2/N$. The positive square root of the variance is called the *standard deviation*. Table 2.4 gives the variance $V(\bar{y})$ and the standard deviation $\sigma(\bar{y})$ of $\bar{y}$ in random samples of different sizes taken

Table 2.4 Variance of the sample mean

Sample size	$V(\bar{y})$	$\sigma(\bar{y})$	$CV(\bar{y})$
1	287.60	16.9	0.63
2	127.82	11.3	0.42
3	74.56	8.6	0.32
4	47.93	6.9	0.26
5	31.95	5.6	0.21
6	21.30	4.6	0.17
7	13.70	3.7	0.14
8	7.99	2.8	0.10
9	3.55	1.9	0.07

from the hypothetical population of Table 2.1. It is obvious from Table 2.4 that the variance of the sample estimate decreases as the sample size increases. This is in line with the results of Table 2.3 in which the concentration of the sample estimates increases as the size of the sample is increased. The variance or the standard deviation is thus a measure of the spread of the sample estimates around their mean value. The reciprocal of the variance is a measure of concentration.

We have shown in the last column of Table 2.4 the relative standard deviation or the *coefficient of variation* of $\bar{y}$. This is obtained by dividing the standard deviation of $\bar{y}$ by its expected value $E(\bar{y}) = 27.0$. It gives a measure of spread relative to the mean value. The relative standard deviation is a pure number, free of the units in which y is measured.

We can go a step further and calculate from Table 2.3 the proportion of sample estimates which differ from the expected value by less than 1 standard deviation, 2 standard deviations, and 3 standard deviations of the estimate $\bar{y}$. These figures are presented in Table 2.5.

Table 2.5 Concentration of estimates around the expected value

Sample size n	*Proportion of estimates differing from the expected value by less than* $\sigma(\bar{y})$	$2\sigma(\bar{y})$	$3\sigma(\bar{y})$
1	70	90	100
2	64	98	100
3	65	96	100
4	62	97	100
5	62	98	100
6	62	97	100
7	68	97	100
8	67	98	100
9	70	100	100

It will be seen that about 68 percent of the estimates are within 1 standard deviation of the mean, 95 percent are within 2 standard deviations, and all are within 3 standard deviations. This again is a universal phenomenon. In large samples the sample mean $\bar{y}$ follows the normal distribution. For this distribution the percentages in the three classes are 68.5, 95.5, and 99.7 respectively.

Here then is a way of using the results from the sample for making an estimate of a population parameter. If we know the standard deviation $\sigma(\bar{y})$ of the sample mean $\bar{y}$ we can say that the probability is about 0.68 that the interval $\bar{y} \pm \sigma(\bar{y})$ contains the true population mean. In the same way the probability is more than 0.95 that the interval $\bar{y} \pm 2\sigma(\bar{y})$ contains the population mean and it is almost certain that the interval $\bar{y} \pm 3\sigma(\bar{y})$ covers the population mean. These intervals are called *confidence intervals* and the probabilities associated with them are called *confidence coefficients*.

One more question needs to be answered at this stage. How do we know the standard deviation $\sigma(\bar{y})$? It is true that we do not know it. But we can make an estimate of it from the sample just as we have estimated the population mean from the sample. Sampling theory helps us in obtaining such an estimate. The estimate will improve as the sample size is made larger. It should be pointed out that the confidence coefficients do not hold exactly but are approximately right when the sample is used for estimating the standard deviation.

The reader might form the impression from Table 2.4 that the decrease in variance arises from the fact that the sampling fraction (proportion of the population taken into the sample) becomes larger as the sample size increases. This is not so. Suppose there is a very large population containing, say,

100,000 units. Let S_y^2 be 500. If the sample size is 100, the variance of the sample mean is

$$\frac{1}{100}\left(1-\frac{100}{100{,}000}\right)500=\frac{1}{100}\times 0.999\times 500=4.995$$

If the sample size is increased to 200, the variance becomes

$$\frac{1}{200}\left(1-\frac{200}{100{,}000}\right)500=\frac{1}{200}\times 0.998\times 500=2.495$$

The reduction in variance is not due to the fact that the multiplying factor comes down from 0.999 to 0.998 when the sample size is doubled. It is the increase in the sample size that has brought about the reduction in variance.

2.4 UNEQUAL PROBABILITY SAMPLING

There is another method by which we can often achieve greater concentration of the sample estimates around the expected value. Take for example the population of Table 2.1. Suppose we have information from a previous inquiry about the number of employees in these establishments at a prior date. Let this auxiliary information x be as given in column 3 of Table 2.6. We

Table 2.6 Sampling with unequal probabilities

Establishment	y	x	y/x	$X(y/x)$
(1)	(2)	(3)	(4)	(5)
0	31	30	1.03	25.83
1	15	15	1.00	25.00
2	67	60	1.12	27.92
3	20	18	1.11	27.78
4	13	12	1.08	27.08
5	18	15	1.20	30.00
6	9	10	0.90	22.50
7	22	20	1.10	27.50
8	48	45	1.07	26.77
9	27	25	1.08	27.00
Total	270	250		

select a sample of $n = 1$ establishment with probability proportionate to the number of employees x. This can be achieved by assigning as many numbers to the establishment as is the value of x and then choosing one establishment by consulting a table of random numbers (Sec. 3.8). Thus the probability that, say, establishment 6 is selected in the sample is $10/250 = 1/25$. The average employment at the previous date is $\bar{X} = 25$. In order to estimate the present employment, we divide y by the probability with which the establishment has been selected. Thus the divisor for a large establishment is large and that for a small establishment is small. This device has the effect of normalizing the contribution of an establishment toward the sample estimate. (In random sampling a large establishment pushes the mean up while a smaller one pushes it down when in the sample.) Column 5 of Table 2.6 gives the 10 estimates of the population mean for different selections. We note that the probability is now $225/250 = 0.90$ that the sample estimate is confined to the interval 25–29. In sampling with equal probabilities this probability was 0.10 only. This shows that considerably greater concentration of the estimates can be obtained by selecting the sample with unequal probabilities.

The same conclusion is reached if we calculate the variance of the estimate. We learn from theory (1, p. 50) that the variance of the estimate $(\bar{X}/n)\,S(y_i/x_i)$ in samples of size n is given by

$$\frac{1}{n}\frac{1}{N^2}\left(X\sum\frac{y_i^2}{x_i} - Y^2\right)$$

In our example,

$$N = 10 \qquad Y = 270 \qquad X = 250$$

$$\sum\frac{y_i^2}{x_i} = 292.41554 \qquad \frac{1}{N^2}\left(X\sum\frac{y_i^2}{x_i} - Y^2\right) = 2.03885$$

Thus the variance of the estimate is given by $2.03885/n$. In samples of size unity, the variance is 2.04, as compared with 287.60 in the case of sampling with equal probabilities (see Table 2.4). The spread of the sample estimates has been made fantastically small. There is very little risk in using the sample estimate for the population parameter.

It may be pointed out that the reason for the considerable reduction in variance in this case is that there is near-proportionality between the main variate y and the auxiliary character x. If the degree of proportionality is not so high, the results will be less spectacular.

2.5 DIFFERENCE ESTIMATION

Another method of achieving higher precision is the following. The sample is selected with equal probabilities. The auxiliary information is used to estimate the change or the difference of the means $\bar{y} - \bar{x}$ and to this is added $\bar{X}$, the population mean for the auxiliary character x. Then the estimator is $(\bar{y} - \bar{x}) + \bar{X}$, which is called a *difference estimator*. When the sample size is 1, the 10 estimates for the mean $\bar{Y}$ of the population of Table 2.6 are 26, 25, 32, 27, 26, 28, 24, 27, 28, and 27. Now as many as 8 out of the 10 estimates are concentrated around the mean in the interval 25–29. The average of all the estimates is the population mean 27.0, which shows that the difference estimator is unbiased for estimating the population mean. When the sample size is 2, there are 45 possible samples. We can calculate the value of the difference estimator for each sample. In columns 4 and 5 of Table 2.7 is shown the concentration of the difference estimate in this case. It is clear from Table 2.7 that, as compared with the simple average $\bar{y}$, the concentration

Table 2.7 Concentration of the difference and ratio estimates and comparison with the simple average

	Percent of samples falling within the class					
	Simple average		*Difference estimate*		*Ratio estimate*	
Class interval	$n=1$	$n=2$	$n=1$	$n=2$	$n=1$	$n=2$
(1)	(2)	(3)	(4)	(5)	(6)	(7)
9–13	10	4				
13–17	20	16				
17–21	20	18				
21–25	10	17	10	2	10	2
25–29	10	7	80	80	80	98
29–33	10	6	10	18	10	
33–37	0	7				
37–41	0	9				
41–45	0	9				
45–49	10	3				
49 or more	10	4				
Total	100	100	100	100	100	100

is very much higher for the difference estimator. The spread of the estimator can also be studied by calculating its variance for different sample sizes. It can be proved (1, p. 99) that this variance is given by

$$\frac{1}{n}\left(1-\frac{n}{N}\right)\frac{\sum[y_i-\bar{Y}-(x_i-\bar{X})]^2}{N-1}=\frac{1}{n}\left(1-\frac{n}{N}\right)S_u^2$$

For our population

$$\sum(y_i-x_i-\bar{Y}+\bar{X})^2=42$$

Denoting $y-x$ by u, $S_u^2=42/9$. Thus the variance of the difference estimator can be calculated for different sample sizes. The variance is shown in Table 2.8. The table tells the same story. As compared with the simple average the difference estimator is considerably more precise.

Table 2.8 Variance of the difference estimator and the simple average

Sample size n	Variance of the estimator	
	Simple average	*Difference*
1	287.60	4.20
2	127.82	1.87
3	74.56	1.09
4	47.93	0.70
5	31.95	0.47
6	21.30	0.31
7	13.70	0.20
8	7.99	0.12
9	3.55	0.05

2.6 RATIO ESTIMATION

Instead of estimating the difference between $\bar{Y}$ and $\bar{X}$ we may estimate the ratio of $\bar{Y}$ to $\bar{X}$ from the sample and multiply it with the known value of $\bar{X}$. This estimator $\bar{X}\bar{y}/\bar{x}$ is called the *ratio estimator.* The ratio estimator will be fairly precise if the ratio of y to x varies much less than the character y in the population. The concentration of this estimator is exhibited in Table 2.7 when the sample size is one or two.

It may be noted that the average (or expected value) of all the sample estimates is not necessarily equal to the population mean when the ratio estimator is used. For example, the 10 possible estimates based on a sample of size unity are 25.83, 25.00, 27.92, 27.78, 27.08, 30.00, 22.50, 27.50, 26.67, and 27.00. The average of these estimates is 26.728 while the true mean is 27.00. The difference between the expected value and the parameter to be estimated is called the *bias* of the estimate. In this case the ratio estimate is subject to a bias of –0.27. Although biased, it gives a high degree of concentration around the true mean of y (see Table 2.7). In order to get a measure of its concentration we will calculate $E(y-27.0)^2$, the expected value of the square of the deviations from the true mean. This is different from $E(y-26.728)^2$, which is the variance of the estimate. When deviations are taken from the true value, the expected value of the square of the deviation is called the *mean-square error* (MSE). It is easy to see that the mean-square error is the sum of the variance and the square of the bias. For unbiased estimates the mean-square error coincides with the variance.

We can calculate the variance of the ratio estimate and the bias by listing all possible samples and finding the value of the estimate for each. The results obtained for $n=1$ and 2 are given in Table 2.9. It follows that the

Table 2.9 Mean-square error of the ratio estimate

n	*Variance*	*Bias*	*Bias*^{2}	*MSE*
1	3.570	−0.272	0.074	3.644
2	0.976	−0.131	0.017	0.993

ratio estimator is more precise than the difference estimator for $n = 1$ or 2. The bias of the estimator is quite small and may as well be neglected. Actually it can be proved that, for large sample sizes, the bias of the ratio estimate will be negligible relative to the standard deviation (1, p. 87).

The reader should be warned at this stage that the difference and the ratio estimators are not always as good as the present example shows them to be. If the auxiliary character x is not near-proportional to y, these estimators may be no better than the simple average.

2.7 STRATIFICATION

There is another method of making use of auxiliary information for improving the precision of the estimate. With this method the population is divided up into groups or strata which are relatively more homogeneous internally. This can be done by placing within the same stratum those units which are similar with respect to x. Then a sample is selected from each stratum. The sample results from different strata are pooled in order to arrive at an estimate for the whole. Greater precision arises from the fact that the strata are homogeneous, so that stratum means can be estimated with smaller error. Take, for example, the population of Table 2.6. We make the following two strata on the basis of x, employment at an earlier date (Table 2.10). The smaller establishments are in the first stratum and the other ones are in the second stratum. We want to estimate the average size of an establishment

Table 2.10 Stratification of establishments

Stratum 1			*Stratum 2*		
Unit	x	y	*Unit*	x	y
6	10	9	7	20	22
4	12	13	9	25	27
1	15	15	0	30	31
5	15	18	8	45	48
3	18	20	2	60	67

by selecting a random sample of one from each stratum. The sample estimate for each of the 25 possible samples is given in Table 2.11. With this procedure 44 percent of the estimates lie in the interval 21–33 while the corresponding figure is 30 percent in the case of unstratified random sampling. The expected value of the estimate is 27.0; that is, the estimate is unbiased. The variance of the estimate is easily found to be 39.89. With this sample size the variance of the simple average (when no stratification is used) is 127.82 (see Table 2.4). This shows that stratification has been beneficial.

Table 2.11 Sample estimates from a stratified design

Sample	*Estimate*	*Sample*	*Estimate*	*Sample*	*Estimate*
6,7	15.5	4,8	30.5	5,9	22.5
6,9	18.0	4,2	40.0	5,0	24.5
6,0	20.0	1,7	18.5	5,8	33.0
6,8	28.5	1,9	21.0	5,2	42.5
6,2	38.0	1,0	23.0	3,7	21.0
4,7	17.5	1,8	31.5	3,9	23.5
4,9	20.0	1,2	41.0	3,0	25.5
4,0	22.0	5,7	20.0	3,8	34.0
				3,2	43.5

Another mode of stratification may be tried. Suppose the first stratum consists of the two largest units numbered 2 and 8 and the second stratum contains the other eight units. The first stratum is sampled with certainty so that it makes no contribution to the variance of the estimate. A random sample of two units is taken from the second stratum. This gives a total sample of $n = 4$ units from the population. Denoting by $\bar{y}_1$ and $\bar{y}_2$ the sample means from the two strata, the estimate of the population mean is $\bar{y} = (2/10)\bar{y}_1 + (8/10)\bar{y}_2$. This estimate is again unbiased and its variance is found to be

$$\frac{16}{25}\left(\frac{1}{2}\right)\left(1 - \frac{2}{8}\right)\frac{369.875}{7} = 12.68$$

This follows from the fact that S_y^2 in the second stratum is 369.875. The comparable variance for unstratified simple random sampling is 47.93 (see Table 2.4). Stratification has been somewhat useful in bringing down the variance and thereby improving the degree of concentration.

It may be mentioned here that in stratified simple random sampling the estimate of the population average is given by $\sum W_h \bar{y}_h$ where W_h is the relative weight of stratum h. The variance of this estimate is (1, p. 63)

$$\sum W_h^2 \frac{1}{n_h}\left(1 - \frac{n_h}{N_h}\right) S_{yh}^2$$

when a random sample of n_h units is selected from the N_h in stratum h, whose variance is S_{yh}^2. It is clear that the variance of the sample estimate depends on the variances S_{yh}^2 within strata. If the strata can be made fairly homogeneous for the variate y on the basis of x, the quantities S_{yh}^2 will be relatively small. This is how stratification can produce a better estimate.

2.8 CLUSTERING

Another technique is to select the sample in clusters or groups rather than to take individual units. Suppose, for instance, that you want to select a sample of households from a city. It is a very expensive job to make a list of all households in the city before selecting a random sample from it. It is far more convenient to get a map of the city, divide it into identifiable blocks, and select a sample of some blocks. All households in the selected blocks are the object of further inquiry. The sample is selected by taking a few blocks which are clusters of households. The main reason for resorting to cluster sampling is the economy involved.

Ordinarily clusters will be formed by putting together units which are physically near to each other. When there is an opportunity to form clusters by picking out units which are not necessarily geographically contiguous, it is best to place in the same cluster dissimilar units. This is exactly the opposite of stratification, in which similar units are assigned to the same stratum. If each cluster is a miniature of the total universe, one can make a good estimate by selecting just a few clusters.

The principle mentioned in the foregoing paragraph will now be explained with the help of the small universe of 10 establishments (see Table 2.6). We shall make five clusters of two units each by using information on x which is known for each unit. Suppose we make clusters by putting similar units in the same group. Then the composition of the clusters is

(6,4) (1,5) (3,7) (0,9) (2,8)

The figures of average employment $\bar{y}$ for the five clusters are 11.0, 16.5, 21.0, 29.0, 57.5. If one cluster is selected at random and the sample mean used as an estimate of the population mean, the variance of the sample mean is

$$\frac{1}{5}\left[(11.0)^2 + (16.5)^2 + \cdots + (57.5)^2 - \frac{(135)^2}{5}\right] = 267.3$$

This is more than double the variance when two units are selected at random from the whole population (see Table 2.4). The variance for cluster sampling is more than that for unclustered random sampling. Let us now form the clusters by assigning dissimilar units to the same group. The new clusters are

(6,2) (4,8) (0,1) (5,9) (3,7)

and the cluster means are 38.0, 30.5, 23.0, 22.5, and 21.0. If a cluster is picked up at random, the variance of the sample mean $\bar{y}$ now drops down to 41.10, which is less than a third of the variance in unclustered random sampling.

There is no intention to give the impression that one can always form clusters by assigning dissimilar units to the same group. For example, when clusters are formed by putting together households in the same block, the resulting clusters are usually not efficient. The reason is that persons living in the same block are generally found to be more similar than persons living in different blocks. But the point is that it is economical to select the sample in blocks. For the same cost of the survey, we can take many more households in the sample by sampling in clusters than with unclustered sampling of households from the city.

We can go a step further. Instead of collecting information from every household in the selected cluster (block), we may take just a sample of households and collect information from them. In that case the sample is selected in two stages: first the blocks and then the households within blocks. This is called two-stage sampling. The reason for using this method is convenience and the reduced costs associated with it. If a sample is selected directly without clustering, the sample is likely to fall everywhere in the city. It will be expensive to collect the information and supervise the survey. If sampling is done in clusters, we need work in the sample blocks only. It becomes cheaper to collect information this way.

2.9 DOUBLE SAMPLING

Suppose we believe that it is useful to make use of auxiliary information on x for improving the precision of the estimate of the population mean for y. But information on x is not available. In that case two courses are open. We may decide to select a sample of n units, collect information on the variate y under study, and use the sample mean $\bar{y}$ as an estimate of the population mean. Alternatively, a part of the total budget is used for collecting information on x for a fairly large sample taken from the universe. Then a smaller sample is taken for obtaining information on y alone. The two samples are used in the best possible manner to arrive at a better estimate of the population mean. The latter procedure is called double or two-phase sampling. This method will be useful when it is considerably cheaper to collect information on x than on y and when there is high correlation between y and x.

Take, for example, the population of Table 2.6. Suppose the total budget for the survey is \$30. Suppose it costs \$10 to send an investigator to an establishment for collecting information on employment for 1970. However, it costs only \$1 per unit to get information on 1968 employment by mail. We may then collect information on x by mail from all 10 establishments. If we use the difference estimator $\bar{y}-\bar{x}+\bar{X}$, the variance of the estimate is 1.87 (see Table 2.8). Alternatively, we may not collect information on x at all but use the budget of \$30 for interviewing a random sample of three establishments. The variance of the sample estimate $\bar{y}$ will be 74.56 (see Table 2.8). This shows that is is wiser to use the double sampling approach in this situation.

2.10 NONSAMPLING ERRORS

It has been assumed all along that there are no reporting errors in the data collected. Actually, errors of measurement or ascertainment are almost always present when information is collected. The establishments may not have proper records to give information on y. They may deliberately understate the employment figures, assuming erroneously that the inquiry may be connected with taxation, and so on. This will bring about errors in the data.

The problem is how to form usable estimates from the sample in the presence of response errors. Let us start with a simple model to make the ideas clear. Suppose the manager of an establishment tosses a die to determine the figure he is going to report. If the true figure is y, we are given $y+e$ where $E(e)=0$, $V(e)=10$. If we take a sample of two establishments, the expected value of the sample mean will continue to be $E(\bar{y})+E(\bar{e})=$ the population mean. But the variance of the estimate is $V(\bar{y})+V(\bar{e})=V(\bar{y})+10/2$, which is larger than the variance obtained when no response errors are present. In this case the sample mean continues to be unbiased but its variance has increased. If, however, response errors are deliberate, $E(e)$ will not be zero and the estimate will be subject to a bias, which is called the *response bias*. Furthermore, there will ordinarily be some correlation between y and e. Perhaps larger establishments will understate and the smaller ones overstate the true employment. And the data collected from the same establishment may differ from interviewer to interviewer since interviewers have their own personalities and whims. Here then is a very difficult situation confronting us. There are many ways of handling this kind of situation. There is one possibility that is promising. We may collect information on the entire sample, say by mail, in an inexpensive way. A random sample of a relatively smaller size is selected and the best interviewers are sent to these establishments. The interviewers have access to the records of the establishments from which they compile almost the true figures on y. The double sampling technique is then used for making a more accurate estimate of the population parameter. Let us denote the information collected in the larger sample by x and that in the smaller sample by y. Based on the larger sample the sample mean is $\bar{x}'$. The smaller sample gives $\bar{y}$ as the mean for y. It also gives $\bar{x}$ as the mean for x. The estimate to use is $(\bar{y}-\bar{x})+\bar{x}'$. This estimator should be subject to much less response bias. If y is the true value of the unit, there is no response bias present in the estimator.

There may be another type of bias present in the results. This happens when some of the units in the sample do not respond and provide no information whatsoever. For example, the two smallest establishments, numbered 6 and 4, in our population of Table 2.1 may decide not to cooperate in the survey. If a sample of two establishments is taken and a random substitution is made for those establishments which do not cooperate, it is clear that the sample refers to the other eight establishments only whose average for y is 31.0. In these circumstances the sample will overstate the true average and is biased in one direction. Some procedure will have to be found for diminishing this bias.

REFERENCE

1. Raj, D. (1968), "Sampling Theory," McGraw-Hill Book Company, New York.

part two

Techniques of Sampling

3

Methods of Sample Selection

3.1 INTRODUCTION

There are many different ways of selecting a probability sample from a population. The basic requirement is that it should be possible to subdivide the population into what are called sampling units. For example, all persons in a commune may be considered to be grouped into households; all land in a village may be assumed to be divided up into fields. The household in the first case and the field in the second case are the sampling units. The existence of a list or a map showing the various sampling units is an essential prerequisite for the selection of the sample. For the same population there may be several kinds of sampling units to which the selection procedure is applied. The reporting unit and the unit of analysis may not be the same as the sampling unit. For example, the head of a household may report for all members of the household while the unit of analysis for the tabulation of the data is a single person.

3.2 SIMPLE RANDOM SAMPLING

A basic method of sample selection is simple random sampling. In this method each unit in the population has the same probability of being selected in the sample. The selection is usually made with the help of random numbers (Sec. 1.12) after the units in the frame have been numbered from 1 to N. All samples have the same probability and are therefore equivalent. The selection procedure is explained in the examples that follow.

EXAMPLE 3.1

Suppose there are $N = 850$ students in a school from which a sample of $n = 10$ students is to be taken. The students are numbered from 1 to 850. Since our population runs into three digits we use random numbers that contain three digits. All numbers exceeding 850 are ignored because they do not correspond to any serial number in the population. In case the same number occurs again, the repetition is skipped. Following these rules the following simple random sample of 10 students is obtained when columns 31 and 32 of the random numbers given in Appendix 2 are used.

251	546	214	495	074
800	407	502	513	628

Remark If repetitions are included, the sample is said to be selected with replacement. In the present example the sample is selected without replacement.

EXAMPLE 3.2

A bookshop has a bundle of sales invoices for the previous year. These invoices are numbered 2,615 to 7,389 and a random sample of 12 invoices is to be taken. In this case it is not necessary to renumber the invoices from 1 upward. We consult four-digit random numbers, rejecting all numbers below 2,615 and those above 7,389. The random numbers in Appendix 2 produce the following sample:

5,461	4,957	4,071	5,024	5,136	6,281
2,840	6,696	4,624	5,569	3,504	4,548

EXAMPLE 3.3

There are 10 classes in a school, the number of students in each class being given in Table 3.1. The students are numbered from 1 to 574 in the last column of the table. In order

Table 3.1 Number of students in a school

Class	*Number of students*	*Cumulated number*	*Assigned range*
1	100	100	1–100
2	83	183	101–183
3	71	254	184–254
4	65	319	255–319
5	57	376	320–376
6	50	426	377–426
7	43	469	427–469
8	40	509	470–509
9	35	544	510–544
10	30	574	545–574

to select a random sample of 10 students, we consult three-digit random numbers. The random numbers are

240	284	212	007	462
556	350	058	454	499

Thus students with roll number 57 in class 3, 30 in class 4, 29 in class 3, 7 in class 1, 36 in class 7, 12 in class 10, 31 in class 5, 58 in class 1, 28 in class 7, and 30 in class 8 are in the sample.

3.3 ESTIMATION BASED ON SIMPLE RANDOM SAMPLING

We will now illustrate how random samples can be used for estimating population parameters such as means, totals, and proportions. The basic rule is that a population mean or proportion is estimated by the corresponding mean or proportion in the sample. A population total is estimated by multiplying the sample mean by the number of units in the population.

EXAMPLE 3.4

Average employment of small establishments In the 1958 Greek census of non-agricultural establishments 493 small establishments were listed in a section of Athens. A simple random sample of 30 establishments gives the following data on employment y.

2	2	2	2	3	5	2	3	4	5
2	4	2	6	6	7	5	4	5	3
2	2	2	2	5	2	2	2	2	2

We have

$$N = 493 \qquad Sy_i = 97$$

$$n = 30 \qquad Sy_i^2 = 385$$

$$\frac{n}{N} = 0.0609 \qquad \frac{1}{n}(Sy_i)^2 = 313.63$$

$$1 - \frac{n}{N} = 0.9391 \qquad S(y_i - \bar{y})^2 = 71.37$$

$$s_y^2 = \frac{1}{n-1} S(y_i - \bar{y})^2 = 2.461$$

The sample mean $\bar{y} = 97/30 = 3.23$ is an estimate of the average number of employees per establishment. The sample estimate of the variance of $\bar{y}$ is

$$\tfrac{1}{30} \times 0.9391 \times 2.461 = 0.077029$$

leading to a standard error of 0.277 and a coefficient of variation of $27.7/3.23 = 8.6$ percent. The total employment in all the 493 establishments will be estimated as $493 \times 3.23 = 1{,}592$ persons with a standard error of $493 \times 0.277 = 137$ persons, the coefficient of variation being the same as 8.6 percent. It may also be stated that the system of intervals $3.23 \pm (2 \times 0.277) = (2.68, 3.78)$ contains the true average with a high probability of 0.95 (see Sec. 2.3).

EXAMPLE 3.5

Percentage of establishments which are too small We want to estimate the proportion of establishments which employ five or fewer persons in the population considered in the previous example. We have

$$N = 493 \qquad 1 - \frac{n}{N} = 0.9391$$

$$n = 30 \qquad \frac{p(1-p)}{n-1} = 0.003103$$

$$n_0 = 27$$

$$p = \frac{n_0}{n} = 0.90 \qquad v(p) = \left(1 - \frac{n}{N}\right)\frac{p(1-p)}{n-1} = 0.002914$$

The sample estimate of the proportion of small establishments is 0.90 and its variance is estimated at 0.002914, which gives a standard error of 0.054 and a coefficient of variation of 6 percent. The reason that the estimate is so precise is that there is a preponderance of small establishments in the universe considered. If we make an estimate of the number of the larger establishments (employing six or more persons), the estimate is 10 percent and its coefficient of variation is as large as 54 percent.

Remark When N, the number of units in the population, is large, the coefficient of variation of the sample proportion p is given approximately by $\sqrt{(1-P)/nP}$. This is large when P, the proportion in the population, is small. Thus the relative error of the estimate is large when the trait to be measured is rare in the population.

EXAMPLE 3.6

How many persons in the sample? Suppose it is believed that about 20 percent of persons in a large population suffer from a disease. How many persons should be selected in the random sample in order that the coefficient of variation of the estimate (of the proportion P) be 10 percent or less? In this case we have to find n such that

$$\frac{1}{P}\sqrt{\frac{P(1-P)}{n}} \leqslant \frac{1}{10}$$

which means that

$$n \geqslant \frac{100(1-P)}{P}$$

Substituting $P = 0.20$, it is found that the sample size should exceed 400 persons.

Remark If the coefficient of variation of the sample proportion is 10 percent, this means that the estimate is close to the population proportion within 20 percent (that is, twice the coefficient of variation) with a high probability.

Remark One should not insist on a high degree of precision without regard to the purpose for which the estimate is needed. For example, the sample size needed is as large as 90,000 if it is insisted that the relative error be no more than 1 percent for $P = 0.10$ in the population.

EXAMPLE 3.7

How many stores in the sample? Suppose that previous information suggests that the average sales of retail stores is $\mu_y = \$22{,}348$ with a standard deviation of $\sigma_y = \$153{,}069$

and a consequent coefficient of variation of $CV(y) = 6.85$. A new sample of stores is to be selected for estimating the average sales correct to 10 percent. How many stores should be selected at random? In this case the sample size n is to be found from the formula

$$\left[\frac{1}{n}(1-f)\right]^{1/2} CV(y) = 0.05$$

where f is the sampling fraction. When the number of stores in the population is quite large, $1 - f = 0$. Substituting $CV(y) = 6.85$, the desired number of stores in the sample is found to be 18,769.

Remark It may be noted that it is hard to obtain the sample size n without some knowledge of $CV(y)$. This is the reason that a sampling statistician always makes it a point to record the value of $CV(y)$ for the populations on which he has worked. This leads to better sampling practice in the future.

3.4 SUBPOPULATION ANALYSIS

Quite often we make estimates for subgroups of a population in addition to the entire population. Thus we may wish to know the average earnings of females in addition to average earnings per person. The females form a subgroup of the population of all persons. Another subgroup or subpopulation may consist of males working in agriculture and so on. How to analyze the data for subgroups is the subject matter of Examples 3.8 and 3.9.

EXAMPLE 3.8

Average earnings of a subgroup A demographic sample survey was conducted in the area of Ikoyi in the city of Lagos. The sampling fraction was 2 percent. Table 3.2 gives the information available for the 111 households in the sample.

Table 3.2 Household data from a Nigerian survey

Household size	*Number of households*	*Number of males*	*Number of females*	*Household income (£)*	
				Sy	*Sy*^{2}
1	15	14	1	869	124,785
2	19	28	10	1,392	244,716
3	17	30	21	678	70,535
4	17	32	36	2,384	578,772
5	12	33	27	1,674	284,587
6	15	50	40	1,685	341,903
7	5	18	17	764	188,257
8	6	27	21	646	122,413
9 or more	5	25	26	301	56,501
Total	111	257	199	10,393	2,012,469

The sample mean $10{,}393/111 = £93.6$ is an estimate of the average income per household and the standard error of the estimate is £9.13. Consider now the subgroup of households with two or fewer members. The number of such households, N_g, is not known but a random sample of 34 such households is available. We have for this subgroup

$$n_g = 34 \qquad Sy_i = 2{,}261$$

$$\frac{n_g}{N_g} = 0.02 \qquad Sy_i^2 = 369{,}501$$

$$\tfrac{1}{34}(Sy_i)^2 = 150{,}356.5$$

$$1 - f = 0.98 \qquad \tfrac{1}{33}S(y_i - \bar{y})^2 = 6{,}640.74$$

The average income of this subgroup is estimated as $2{,}261/34 = £66.5$ and the standard error is

$$(0.98 \times \tfrac{1}{34} \times 6{,}640.74)^{1/2} = £13.8$$

Further, the proportion of single-member households in this subgroup is $15/34 = 0.44$ and its standard error is

$$\left(0.98 \times \frac{0.44 \times 0.56}{33}\right)^{1/2} = 0.086$$

giving a coefficient of variation of about 19 percent.

Remark For estimating the mean of a subgroup we calculate the sample mean relating to the subgroup. For estimating the variance use the usual formula in which the sampling fraction applying to the entire population is inserted. The same procedure is used for estimating a proportion in a subgroup.

EXAMPLE 3.9

Total earnings of a subgroup It is more difficult to estimate the total earnings of a subgroup from the data of Example 3.8, since the total number of such households in the population is not known. The best thing to do is to use the total sample from the entire population and assume that all households not belonging to the subgroup have a value of zero for y. Consider the subgroup of households with two or three members. We have (4, p. 199)

$$N = 5{,}550 \qquad Sy_i = 2{,}070$$

$$n = 111 \qquad Sy_i^2 = 315{,}251$$

$$\frac{n}{N} = 0.02 \qquad \tfrac{1}{111}(Sy_i)^2 = 38{,}602.702$$

$$S(y_i - \bar{y})^2 = 276{,}648.298$$

Hence the estimated total income is $50 \times 2{,}070 = £103{,}500$ and its variance is

$$5{,}550 \times 50 \times 0.98 \times \frac{276{,}648.298}{110} = 683{,}950{,}776$$

leading to a standard error of £26,152 and a coefficient of variation of about 25 percent.

3.5 STABILITY OF VARIANCE ESTIMATION

In all the previous examples the variance of the estimator has been estimated from the sample itself to determine the precision of the estimate. Since it is calculated from the sample, the variance estimator is subject to sampling errors. We thus want to know how stable the variance estimator is. This becomes important when, for example, the variance estimated from a sample is to be used for comparing one method with another or when it is to be used for determining the sample size needed to achieve a given level of precision. This question is examined in Examples 3.10 and 3.11.

EXAMPLE 3.10

Is the variance estimator stable? Let the true proportion of unemployed persons in a population be $P = 0.10$ and suppose that a sample of $n = 400$ persons has been taken to estimate P. It can be shown that if the population is quite large, the coefficient of variation of the variance estimator is given by (4, p. 191)

$$\left\{\frac{1}{n}\left[\frac{1}{P(1-P)} - 4\right]\right\}^{1/2}$$

This works out as 13 percent for $P = 0.10$ and $n = 400$. Considering a variance estimator to be satisfactory if the coefficient of variation is 20 percent or less, the present sample size is more than sufficient for the purpose of satisfactory variance estimation.

Remark Table 3.3 gives the sample size n if the coefficient of variation of the variance estimator is to be 20 percent or less when the objective is to estimate a population proportion P.

Table 3.3 Sample size needed for trustworthy variance estimation

P	n	P	n
0.05	426	0.70	19
0.10	178	0.80	56
0.20	56	0.90	178
0.30	19	0.95	426

Thus a sample of about 56 units should suffice, provided the proportion to be estimated is not too small, that is, P lies between 0.20 and 0.80.

EXAMPLE 3.11

Is the variance estimator for the mean stable? In Example 3.4, we estimated the average employment and its standard error from a sample of 30 establishments. We shall now determine whether the estimate of the variance obtained is stable. To determine this we make the following calculations on the population:

Mean = 3.27 Variance $\mu_2 = 2.90$

Third moment $\mu_3 = 7.66$ Fourth moment $\mu_4 = 41.03$

$$\beta_1 = \frac{\mu_3}{\mu_2^{3/2}} = 1.55 \qquad \frac{\mu_4}{\mu_2^2} = \beta_2 = 4.88$$

The coefficient of variation of the variance estimator is given approximately by (4, p. 191)

$$\sqrt{\frac{1}{n}\left[\beta_2(y) - \frac{n-3}{n-1}\right]}$$

This works out as

$$\sqrt{\tfrac{1}{30}(4.88 - \tfrac{27}{28})} = 0.36 \text{ or } 36 \text{ percent}$$

We may consider a coefficient of variation of 20 percent or less as satisfactory for the estimation of variance. In that case the minimum sample necessary to select from the population is around 97.

Remark Table 3.4 gives for different values of $\beta_2(y)$ the sample size n needed for estimating the population mean or total with the variance estimator having a coefficient of variation of 20 percent or less.

Table 3.4 Sample size for different values of $\beta_2(y)$

$\beta_2(y)$	*Sample size* n	$\beta_2(y)$	*Sample size* n
2	25	100	2,475
3	50	200	4,975
4	75	300	7,475
5	100	400	9,975
10	225	500	12,475
20	475	600	14,975
50	1,225	700	17,475

3.6 SYSTEMATIC SAMPLING

A more convenient method of sample selection when the units are numbered from 1 to N is the following. Suppose that a sample of 20 establishments is to be selected from a list of 500 establishments classified by industry. This means that 1 out of 25 establishments in the population is to be chosen for the sample, or that the sampling interval is 25. By using random numbers we select a number at random between 1 and 25. Suppose the number selected is 07. Then the establishment numbered 07 is selected in the sample. The other establishments in the sample are obtained by adding the sampling interval 25 successively to 07. Thus the sample contains the following establishments:

07	32	57	82	107
132	157	182	207	232
257	282	307	332	357
382	407	432	457	482

An advantage of the method is that the sample is evenly distributed over the various industry groups. A simple random sample taken from the population does not possess this property.

EXAMPLE 3.12

Sample of blocks The town of Ibadan is divided up into $N = 576$ blocks which are numbered in a serpentine fashion; that is, neighboring blocks have consecutive numbers. A 10 percent sample of blocks is to be taken, which gives a sampling interval of $k = 10$. If the random number between 1 and 10 is 03, the blocks with numbers

03, 13, 23, 33, 43, 53, . . . , 573

are in the sample. If, however, the random number is 08, the sample consists of the blocks with numbers

08, 18, 28, 38, 48, 58, . . . , 568

The first sample gives 58 blocks while the second sample 57 blocks only. Thus the sample size may differ by 1 from sample to sample when N is not exactly divisible by k.

Remark The probability that a specified block is selected in the sample is 1/10 whatever the size of the sample.

3.7 ESTIMATION IN SYSTEMATIC SAMPLING

The use of the systematic sampling method in survey work will now be illustrated. For estimating the population total the sample total is multiplied by the sampling interval. This when divided by N gives an estimate of the population mean. As before, the population proportion is estimated in the same way as the mean. The question of estimating the variance from the sample is more intricate. If the arrangement of units in the population can be considered to be random, the systematic sample behaves like a simple random sample.

EXAMPLE 3.13

Average employment in industry There are 169 industrial establishments employing 20 or more persons in the town of Piraeus in Greece. The following are the employment figures based on a 1-in-5 systematic sample.

35	88	35	36	156	25	24	237	80
468	22	139	163	37	37	27	25	26
38	24	62	331	28	31	81	121	49
23	34	23	22	53	50	50		

We have $n = 34$, $Sy_i = 2{,}680$. Since the sampling fraction is 1/5, an estimate of the total employment is given by $5 \times 2{,}680 = 13{,}400$ persons. This gives an average employment of 79.3 persons per establishment. A rigorous estimate of the sampling variance cannot be made from a single systematic sample. However, if we assume that the numbering of the establishments is random, we can treat it as a random sample. In that case

$$N = 169 \qquad Sy_i^2 = 521{,}042$$

$$n = 34 \qquad S(y_i - \bar{y})^2 = 309{,}795$$

$$Sy_i = 2{,}680 \qquad \frac{1}{n-1} S(y_i - \bar{y})^2 = 9{,}387.73$$

The variance estimate for the mean is

$$\tfrac{1}{34} \times \tfrac{4}{5} \times 9{,}387.73 = 220.89$$

leading to a standard error of 14.8 and a coefficient of variation of 18.7 percent. The same coefficient of variation applies to the estimate of the population total.

EXAMPLE 3.14

Populations with trend Apart from convenience, a great advantage of systematic sampling is that the sample is spread evenly over the population when the units in the population have been numbered suitably. Thus the sample estimate will be precise if there is some kind of trend in the population. Consider, for example, the rather academic problem of estimating the mean of the population $y_i = i$ ($i = 1, 2, 3, \ldots, 100$) from a sample of 10. We have

$$N = 100 \qquad n = 10 \qquad S_y^2 = \frac{\sum (y_i - \mu)^2}{99} = 841.75$$

The 10 systematic sample means are 46, 47, 48, 49, 50, 51, 52, 53, 54, and 55, giving a variance of 0.825 for the estimate of the average. However, the variance of the sample mean based on a random sample of 10 units is

$$\tfrac{1}{10} \times \tfrac{9}{10} \times 841.75 = 75.76$$

which is quite large. This shows that systematic sampling is better than random sampling for handling populations with linear trend.

EXAMPLE 3.15

The question of periodicity In case there is periodic variation in the population, one has to be very careful in the use of systematic sampling. Consider the following group of 10 households located on a particular street. The members within households are listed with respect to age. The sex of adults is denoted by a capital letter and that of children with a small letter. The size of each household is four and the head of the household is always a male. It is desired to estimate the sex ratio from a sample of 10 persons.

1	2	3	4	5	6	7	8	9	10
M	M	M	M	M	M	M	M	M	M
F	F	f	F	M	M	F	F	F	F
m	f	m	m	f	f	f	m	f	m
f	m	m	m	f	m	f	f	m	m

The proportion of males in the group is 23/40. But the four systematic samples with a sampling interval of 4 give these proportions as 10/10, 2/10, 5/10, 6/10 respectively, leading to a variance of 0.9219 for the estimated proportion. A simple random sample of 10 persons will, however, give the estimated proportion with a variance of

$$\frac{1}{n}\left(1-\frac{n}{N}\right)\frac{NPQ}{N-1}=\frac{N-n}{n}\frac{PQ}{N-1}$$

which is as small as 0.0188. The reason for the poor performance of the systematic sample is that there is periodicity in the data, and the sampling interval has fallen in line with it.

Remark It is not uncommon to come across populations with periodic variation. Temperatures over a 24-hour period, sales of stores over a week, and postal articles received in a post office over the week are some of the examples. One has to be sufficiently acquainted with the data in order to be able to decide upon the sampling interval if systematic sampling is to be used. Consider, for example, the problem of estimating the number of vehicles passing over a bridge during a certain month. We expect that traffic over the bridge exhibits periodicity during the day, there being hours that are very busy and others when there is very little traffic. Suppose we select an hour at random and examine the traffic over this hour and subsequent 24-hour periods. If the hour selected happens to be the peak hour, the sample will contain all peaks and this will produce a very high figure. On the other hand, if the first hour selected shows poor traffic, all observations taken at this time during the subsequent days are expected to be below the average, thereby producing a low figure.

EXAMPLE 3.16

Rigorous estimate of variance If an unbiased estimate of the variance is needed from a systematic sample, this can be done by selecting more than one systematic sample from the population. Consider the following example. Ten independent systematic samples, each being a 1-in-50 sample, are selected from the universe of industrial establishments (employing 20 or more persons) in the town of Piraeus. The total employment t_i in the sample establishments by systematic sample is as follows:

Sample number	1	2	3	4	5	6	7	8	9	10
t_i	78	169	679	141	141	216	154	123	534	141

We have

$$St_i = 2{,}376 \qquad St_i^2 = 925{,}986 \qquad S(t_i - \bar{t})^2 = 361{,}448$$

An estimate of the total employment is

$$50 \times \tfrac{1}{10} \times 2{,}376 = 11{,}880$$

and the variance estimate is

$$(50)^2 \times \tfrac{1}{10} \times \tfrac{1}{9} \times 361{,}448$$

leading to a standard error of 3,170 persons and a coefficient of variation of 26.7 percent.

3.8 SAMPLING WITH UNEQUAL PROBABILITIES

When the sampling units vary considerably in size, a simple random or a systematic sample of units does not produce a good estimate due to the high variability of the units for the characteristic y under study. Examples are areas of farms and populations of cities. In such a situation we can get a better estimate by giving higher probability of selection to the larger units. The sample may be selected with replacement or without replacement. Several methods of selecting the sample with probability proportionate to size (pps) are available. Some of these techniques are explained in the examples that follow.

The method of sampling with probability proportionate to size is usually used for the selection of large units such as cities, villages, farms. Generally the units are divided up into groups or strata (Sec. 2.7) and two units are selected from each stratum (Sec. 9.5). This is the reason that most of the examples given relate to the selection of two units in the sample.

EXAMPLE 3.17

Sampling with replacement Table 3.5 gives the population of 20 blocks with the number of households x in each block known from a previous inquiry. A sample of two blocks is to be taken. The probability that a block is selected in a sample of one is to be made proportional to x. To make the selection the sizes of the blocks are cumulated and

Table 3.5 Number of persons in 20 blocks

Block	x	*Cumulated* x	*Range*	*Block*	x	*Cumulated* x	*Range*
1	18	18	1–18	11	18	198	181–198
2	9	27	19–27	12	40	238	199–238
3	14	41	28–41	13	12	250	239–250
4	12	53	42–53	14	30	280	251–280
5	24	77	54–77	15	27	307	281–307
6	25	102	78–102	16	26	333	308–333
7	23	125	103–125	17	21	354	334–354
8	24	149	126–149	18	9	363	355–363
9	17	166	150–166	19	19	382	364–382
10	14	180	167–180	20	12	394	383–394

ranges are assigned to the blocks on the basis of the sizes. Since the total size is a three-digit number, random numbers consisting of three digits are used. All numbers exceeding 394 are rejected. The random numbers are 008 and 214. These numbers fall within the ranges of blocks 1 and 12 respectively. Hence these are the two blocks in the sample. It is obvious that the probability of selecting block 1 is 18/394 and that for block 12 is 40/394.

Remark In this method the same block may be selected twice in the sample.

Remark When the number of units in the population is large, cumulating the sizes is a laborious process. The following simpler method may be used (2). Take a pair of random numbers, one 2-digit (as there are 20 blocks) and the other 2-digit (as the maximum size is 40). Reject all numbers exceeding 20 for the first random number and all exceeding

40 for the second number. The first random number selects the block provisionally. If the second number is less than or equal to the size of this block, the block is finally selected. If not, it is rejected and the process starts all over again. Using the tables of random numbers (Appendix 2), the pairs of random numbers are

(07, 40) (05, 18)

the second pair selecting block 5. For the second selection the pairs of random numbers are

(01, 19) (17, 12)

the second pair giving rise to block 17 in the sample. The two sample blocks are 5 and 17.

EXAMPLE 3.18

Systematic sampling If it is important that the units selected be all different, a number of methods can be used. Consider the universe of 10 cities in a part of Greece with 1940 population x lying between 12,000 and 19,000. The 1951 populations are denoted by y. All figures are in hundreds (see Table 3.6).

Table 3.6 Populations of 10 cities (in hundreds)

City	y	x	*Cumulated* x	*City*	y	x	*Cumulated* x
1	177	127	127	6	176	150	836
2	166	130	257	7	164	155	991
3	177	139	396	8	200	159	1,150
4	185	141	537	9	180	169	1,319
5	150	149	686	10	170	189	1,508

The purpose is to select two different cities such that the probability that a city enters the sample is in proportion to the 1940 population x. We may use the generalization of systematic sampling to select the sample. The measures of size x are cumulated, the total being $X = 1{,}508$. Since the sample size is $n = 2$, the sampling interval is $1{,}508/2 = 754$. A three-digit number is selected at random between 1 and 754. Let the random number be 602. This number falls in the range of city 5, which is selected in the sample. We add the sampling interval 754 to the random number 602 to get 1,356. This falls in the range of city 10, which is therefore selected in the sample. Thus our sample comprises the cities numbered 5 and 10. It is obvious that the cities are selected with probability proportional to x.

EXAMPLE 3.19

Sampling with pps of remainder There is another method of selecting a sample of two different units. Consider the population of Table 3.6. A random number between 1 and 1,508 is taken. This happens to be 0383, which falls in the range of city 3 and leads to its selection in the sample. This city is removed from the population; that is, the numbers 258 to 396 are ignored at the time of the second selection. Another random number is taken between 1 and 1,508. This is 1,416, which brings the tenth city into the sample. Thus our sample comprises the third and tenth cities.

EXAMPLE 3.20

One unit per randomized group There is yet another method of selecting the sample. Consider the population of Table 3.6. Two different cities are to be selected. We divide the ten cities at random into two groups each containing five cities. The two groups are shown in Table 3.7. The measures of size x of the units within each group are cumulated. One unit is selected from each group with probability proportional to x by following the usual procedure. For the first group a random number is taken between 1 and 704 and for the second between 1 and 804. Using the table of random numbers (Appendix 2), city 2 and city 10 are selected in the sample.

Table 3.7 Selection of one unit per group

Group 1				*Group 2*			
Serial number	y	x	*Cumulated* x	*Serial number*	y	x	*Cumulated* x
2	166	130	130	4	185	141	141
1	177	127	257	6	176	150	291
5	150	149	406	7	164	155	446
8	200	159	565	9	180	169	615
3	177	139	704	10	170	189	804

3.9 ESTIMATION IN UNEQUAL PROBABILITY SAMPLING

The problem of estimating population totals, means, and proportions is more intricate when the sample is selected with unequal probabilities. The analysis is simpler when sampling is with replacement, but this may involve some loss of efficiency. Greater precision may be obtained by selecting different units in the sample (sampling without replacement). In the first case the value of y for the selected unit is divided by the probability of selection in a sample of one. This when aggregated over the sample and divided by the sample size provides an estimate of the population total for y. In the second case, y is divided by the probability with which the unit is selected in the whole sample of size n and the ratio is aggregated over the sample to produce an estimate of the total of y for the population. The examples that follow will make the ideas clear.

EXAMPLE 3.21

Sampling with replacement Consider the population of Table 3.5 from which blocks 1 and 12 were selected in the sample. The object is to estimate the present number of households y in the area. We shall denote the two selections 1 and 2 respectively. The number of households in the two blocks are found on actual count to be 19 and 37 respectively. We have

$$y_1 = 19 \qquad\qquad y_2 = 37$$

$$p_1 = \tfrac{18}{394} \qquad\qquad p_2 = \tfrac{40}{394}$$

$$\frac{y_1}{p_1} = 394 \times 1.055 \qquad\qquad \frac{y_2}{p_2} = 394 \times 0.925$$

$$\frac{1}{2}\left(\frac{y_1}{p_1} + \frac{y_2}{p_2}\right) = 390 \qquad\qquad \frac{1}{2}\left|\frac{y_1}{p_1} - \frac{y_2}{p_2}\right| = 25.61$$

The sample estimate of the number of households in the area is 390 and its standard error is 25.61.

Remark If we take all possible samples of two blocks with pps, the variance of the estimate is found to be 3,247 (4, p. 48). The comparable variance in simple random sampling is 17,122, which is more than five times as large. This makes it clear how sampling with unequal probabilities may achieve dramatic reduction in the variance as compared with equal probability sampling.

Remark If n units are selected with pps, the estimate of the population total is $\hat{Y} = (1/n)S(y_i/p_i)$ and an estimate of the variance is (4, p. 49)

$$\frac{1}{n(n-1)} S\left(\frac{y_i}{p_i} - \hat{Y}\right)^2$$

EXAMPLE 3.22

Systematic sampling Consider the universe of Table 3.6 from which two cities were selected systematically with pps. The cities selected are 5 and 10. The probabilities of inclusion are 149/754 and 189/754 respectively. The joint probability that both cities are selected is 121/754. We have (4, p. 55)

$$y_1 = 150 \qquad\qquad y_2 = 170$$

$$p_1' = \tfrac{149}{754} \qquad\qquad p_2' = \tfrac{189}{754}$$

$$\frac{y_1}{p_1'} = 759.06 \qquad\qquad \frac{y_2}{p_2'} = 678.20$$

The estimate of the population total for y is

$$759.06 + 678.20 = 1{,}437$$

and an unbiased estimate of its variance is

$$\left[\frac{(149/754)(189/754)}{121/754} - 1\right] \qquad (759.06 - 678.20)^2$$

The estimate of the variance turns out to be negative. This is a disadvantage with this procedure. This situation can be avoided if the cities are arranged at random before making the systematic selection (1, 3).

Remark A quick estimator of the variance, which tends to overstate the true variance, can be obtained by assuming sampling with replacement. Thus

$$p_1 = \frac{149}{1{,}508} \qquad\qquad p_2 = \frac{189}{1{,}508}$$

$$\frac{1}{2}\left(\frac{y_1}{p_1} + \frac{y_2}{p_2}\right) = 1{,}437 \qquad\qquad \frac{1}{2}\left|\frac{y_1}{p_1} - \frac{y_2}{p_2}\right| = 81$$

The standard error is 81 and the relative error is 5.6 percent. In a survey of unemployment in Greece, the author (3) observed that the quick estimator overstated the variance by about 16 percent.

EXAMPLE 3.23

Selection with pps of remainder The analysis with this method is considerably simpler. Consider again the selection of two cities from the population of Table 3.6 (see Example 3.19). We have (4, p. 60)

$$y_1 = 177 \qquad y_2 = 170$$

$$p_1 = \frac{139}{1{,}508} \qquad p_2 = \frac{189}{1{,}369}$$

$$\frac{y_1}{p_1} = 1{,}920 \qquad \frac{y_2}{p_2} = 1{,}231$$

$$t_1 = \frac{y_1}{p_1} = 1{,}920 \qquad t_2 = y_1 + \frac{y_2}{p_2} = 1{,}408$$

An estimate of the population total for y is

$$\tfrac{1}{2}(1{,}920 + 1{,}408) = 1{,}664$$

and its standard error is

$$\tfrac{1}{2}(1{,}920 - 1{,}408) = 256$$

leading to a coefficient of variation of 15.4 percent.

Remark The estimate of the variance given by this method is always nonnegative.

Remark There is another method of selecting the sample which is equivalent to the method of Example 3.23. It is called inverse sampling with unequal probabilities. In this method sampling is done with replacement with probability proportionate to size. The selection continues until $n + 1$ different units are taken. The last selection is rejected and the n different units selected are said to form the sample. It can be shown that this is equivalent to selecting the first unit with pps, the second with pps of the remaining units, and so on (4, p. 230).

EXAMPLE 3.24

One unit per randomized group In this method one unit is selected with pps from each group after the units are divided up at random into groups (5). We shall take the population of Table 3.6 from which the sample was selected in Example 3.20. We shall make a distinction between p_1 (where the divisor is 1,508) and p_1' (where the divisor is 704, the total size of the group). The cities in the sample are the second and the tenth (selections 1 and 2 respectively). We have (4, p. 230)

$$N_1 = 5 \qquad N_2 = 5$$

$$y_1 = 166 \qquad y_2 = 170$$

$$p_1 = \frac{130}{1{,}508} \qquad p_2 = \frac{189}{1{,}508}$$

$$p_1' = \frac{130}{704} \qquad p_2' = \frac{189}{804}$$

$$\frac{y_1}{p_1} = 1{,}925.6 \qquad \frac{y_2}{p_2} = 1{,}356.4$$

$$\frac{y_1}{p_1'} = 898.9 \qquad \frac{y_2}{p_2'} = 723.2$$

$$\frac{y_1^2}{p_1 p_1'} = 1{,}730{,}922 \qquad \frac{y_2^2}{p_2 p_2'} = 980{,}948$$

$$N_1^2 + N_2^2 - N = 40 \qquad N^2 - (N_1^2 + N_2^2) = 50$$

The population total for y is estimated as

$$898.9 + 723.2 = 1{,}622$$

and the variance of the estimate is

$$\tfrac{40}{50}[1{,}730{,}922 + 980{,}948 - (898.9 + 723.2)^2] = 64{,}530$$

This gives a standard error of 254 persons and a coefficient of variation of 15.77 percent.

Remark We have outlined the procedure when two units are selected in the sample. The method is the same when more than two units are to be taken. For example, four groups, each containing about the same number of units, will be formed if the intended sample size is four. The estimation procedure is similar to the one given above.

PROBLEMS

3.1. A simple random sample of $n = 2{,}070$ farms is taken from a population of $N = 20{,}700$ farms and information is collected on the number of cattle y on each farm. The following data are obtained.

$$Sy_i = 25{,}881 \qquad Sy_i^2 = 599{,}486$$

Estimate the average number of cattle per farm and the total number of cattle in the population. Find the coefficients of variation of the estimates. How many farms should be selected in a future survey in order that the relative error of the estimates be no more than 5 percent?

3.2. A simple random sample of $n = 100$ households selected from a village containing $N = 850$ households gave 30 households possessing transistor radios. A record of the number of members y in the 30 households produced $Sy_i = 140$, $Sy_i^2 = 976$. Estimate the average size of a household possessing a transistor radio and the total number of persons in such households. Find the standard errors of your estimates.

3.3. In a certain town containing 16,000 households a simple random sample of $n = 1{,}000$ households gave 766 immigrant and 234 native households. The total number of persons in the sample was 3,323, of which 881 were natives. In all, 407 households owned the dwelling they were living in; of these 350 were immigrants with 1,177 persons in them. There were 1,265 persons in the 416 immigrant households renting the dwelling they were living in and 636 persons in the native households renting their accommodation. Estimate (*a*) the proportion of native households, (*b*) the proportion of dwelling units occupied by natives out of all tenant-occupied dwelling units, (*c*) the average size of an immigrant household, and (*d*) the number of persons in native households owning their accommodation. Find the standard errors of the estimates in (*a*) and (*b*).

3.4. A questionnaire is to be sent to a random sample of employees in a factory employing 10,000 persons. The purpose is to ascertain attitudes in terms of proportions which may lie anywhere between 10 percent and 60 percent. (*a*) Find the standard deviation of the estimate based on 400 questionnaires. (*b*) What sample size should be used in order that the standard deviation of the estimate be no greater than 2 percent? (*c*) Find the sample size if the coefficient of variation of the estimate is desired to be 10 percent or less. (*d*) How close is the estimate in (*c*) to the population proportion?

3.5. The purpose is to estimate the total number of households in an area containing 20 blocks. The eye-estimated x number of households in each block is obtained by making a rapid survey of the area. The actual y number of households is later obtained through careful fieldwork.

i	1	2	3	4	5	6	7	8	9	10
y_i	19	9	17	14	21	22	27	35	20	15
x_i	18	9	14	12	24	25	23	24	17	14
i	11	12	13	14	15	16	17	18	19	20
y_i	18	37	12	47	27	25	25	13	19	12
x_i	18	40	12	30	27	26	21	9	19	12

(*a*) Find the variance of the estimate when a sample of two blocks is selected with replacement with probability proportional to x. (*b*) Compare it with equal probability sampling with replacement. (*c*) If the blocks selected in (*a*) are 4 and 17, find the estimate and its standard error.

3.6. A small population is built up of four farms. The areas x of the farms are 10, 20, 30, and 40 acres and the cultivated areas y are 5, 12, 21, and 32 acres respectively. A sample of two farms is selected in the following manner. The first selection is made with pp to x and the second with pp to the remaining x. If the estimator of Example 3.23 is used, find the estimate of variance for each sample of two farms and hence obtain the variance of the estimate.

3.7. A village is divided up into 20 areas and the number of dwellings x in each area is counted. The areas are listed in some order and a systematic sample of two areas is taken with pp to x. The number of dwellings occupied by renters y is determined for the sample areas. Estimate the total number of rented dwellings in the village and find the standard error of the estimate. You may use the following data. Areas numbered 3 and 11 are in the sample.

i	1	2	3	4	5	6	7	8	9	10
x_i	23	18	33	89	114	66	61	25	46	58
y_i	19	17	25	84	91	48	48	20	34	42
i	11	12	13	14	15	16	17	18	19	20
x_i	149	10	30	90	56	42	113	45	16	66
y_i	131	6	23	79	47	34	97	30	11	45

3.8. There is a long list of holdings giving the number of parcels in each holding. Five independent 1 percent systematic samples are selected with equal probability from the list. The following results are obtained.

Sample	1	2	3	4	5
Number of holdings	104	104	104	103	103
Number of parcels	1,212	1,347	1,238	1,249	1,310

Estimate the total number of parcels in all the holdings and find the standard error of the estimate.

3.9. The purpose is to estimate the total area under wheat in a district. Information on cultivated area x in each village in the district is available from a previous census. A sample of 33 villages is selected with replacement with pp to x and the area under wheat y in the village determined. The data obtained are given below. Estimate the area under wheat in the district and give its sampling error. The total cultivated area, based on the census, is known to be 78,019 acres.

y	52	149	289	381	278	111	634	278	112	355	99
x	401	634	1,194	1,770	1,060	827	1,737	1,060	360	946	470
y	498	111	6	399	79	105	27	515	249	85	221
x	1,625	827	96	1,304	377	259	186	1,767	604	701	524
y	133	144	103	179	330	219	62	79	60	100	141
x	571	962	407	715	845	1,016	184	282	194	439	854

3.10. Information on the yield of a crop is known for each of the 80 plots into which an area has been divided. The 10 systematic samples each containing 8 plots give the following totals.

970	943	955	973	935
968	980	1,009	1,042	1,022

The variability of yield per plot is found to be $S_y^2 = 107.57$. Compare the precision of a systematic sample with a simple random sample of the same size for estimating the average yield per plot. Find the intrasample correlation coefficient in the case of systematic sampling.

REFERENCES

1. Hartley, H. O., and J. N. K. Rao (1962), Sampling with Unequal Probabilities without Replacement, *Ann. Math. Statist.*, **33.**
2. Lahiri, D. B. (1951), A Method of Sample Selection Providing Unbiased Ratio Estimates, *Bull. Intern. Statist. Inst.*, **33.**
3. Raj, D. (1964), The Use of Systematic Sampling with Probability Proportionate to Size in a Large Scale Survey, *J. Am. Statist. Assoc.*, **59.**
4. Raj, D. (1968), "Sampling Theory," McGraw-Hill Book Company, New York.
5. Rao, J. N. K., et al. (1962), A Simple Procedure of Unequal Probability Sampling without Replacement, *J. Roy. Statist. Soc.*, (B)**24.**

4
Use of Supplementary Information

4.1 INTRODUCTION

In sampling work it is not unusual to find that some information is already available for the various units comprising the population. Such information is generally based on previous censuses or large-scale surveys. For example, the number of inhabitants in different villages may be known from a previous census of population. The geographical areas based on cadastral maps may be available at the time of sample selection and so on. Under these circumstances it is important to make use of this information for improving the precision of the estimates. The supplementary information x may be used in a variety of ways. In one method the sample is selected with probability proportionate to x. Another use is to stratify the population on the basis of x. Or the auxiliary information may be used to form a ratio estimate, difference estimate, or regression estimate. An elementary discussion of some of these techniques was presented in Chap. 2. A more detailed description of these methods is given in this chapter.

4.2 STRATIFICATION

In this method the units in the population are allocated to groups or strata on the basis of information on x. An attempt is made to make the strata internally homogeneous by placing in the same stratum units which appear to be similar. By selecting a sample of a suitable size from each stratum it is possible to produce an estimate for the population characteristic y which is considerably better than that given by a simple random sample from the entire population. The following example makes the ideas clear.

EXAMPLE 4.1

Stratification of cities We shall use information on the 1941 population of 69 cities in Greece to divide the cities into four strata. The strata boundaries based on x are shown in column 2 of Table 4.1. The number of cities in each stratum, the average population in 1951, and the standard deviation of population are included in this table. The smallest cities are in the first stratum and the largest ones in the fourth stratum. There are few very large cities; many of the cities are small. The average population of a city varies from 13,500 to 56,600 among the strata made. The smaller cities are less variable; the larger ones are much more variable. But the coefficients of variation are not drastically different. These are the characteristics of many stratified populations we encounter in practice.

Table 4.1 Stratification of 69 cities in Greece

Stratum h	*Boundaries based on x (hundreds)*	*Number* N_h	*Mean* $\bar{Y}_h$	*Variance* S_{yh}^2	*Standard deviation* S_{yh}
(1)	(2)	(3)	(4)	(5)	(6)
1	66–142	30	135	698	26
2	142–219	20	218	3,825	62
3	219–364	12	318	2,818	53
4	364–623	7	566	23,927	155

Suppose the purpose is to estimate the total population of these cities by taking a sample of 20 cities. The first question to answer is how the total sample size should be distributed to the strata. There are a number of possibilities. The sample size in a stratum may be taken in proportion to the number of units, N_h, in the stratum (proportional allocation). The same number of units may be taken from each stratum irrespective of the size of the stratum (equal allocation). The number of units in the sample from a stratum may be made proportional to the product of N_h and the standard deviation S_{yh} (Neyman's allocation). Or, any arbitrary allocation may be used.

Let us try these allocations and compare the results. It is clear that an estimate of the total population of the cities in stratum h is $N_h\bar{y}_h$ where $\bar{y}_h$ is the sample mean based on n_h cities selected at random from stratum h. The variance of this estimate can be shown to be given by (2, p. 35)

$$V(N_h\bar{y}_h) = N_h^2\frac{1}{n_h}\left(1-\frac{n_h}{N_h}\right)S_{yh}^2$$

which depends on the sample size n_h. When the estimate $N_h\bar{y}_h$ is added over the strata, the estimate of the total population of all the 69 cities is obtained. The variance of this estimate is calculated by adding over the strata the variance of $N_h\bar{y}_h$. It appears that it is wiser to take a larger sample from a stratum which is more variable, that is, for which S_{yh}^2 is higher. The results obtained for the three allocations are given in Table 4.2.

Table 4.2 Allocation of sample to strata

Stratum	*Neyman's allocation*		*Equal allocation*		*Proportional allocation*	
h	n_h	$V(N_h\bar{y}_h)$	n_h	$V(N_h\bar{y}_h)$	n_h	$V(N_h\bar{y}_h)$
1	4	136.110	5	104,721	9	48,874
2	7	142,071	5	229,500	6	178,398
3	3	101,448	5	47,356	3	101,448
4	6	27,915	5	67,063	2	418,789
Total	20	407,544	20	448,640	20	747,509
CV (%)	3.94		4.13		5.34	

The coefficients of variation of the estimate of the total population y are given in the last row of the table. It is apparent that in this case proportional allocation is inferior to the other two allocations. Actually it can be proved that Neyman's allocation is the best when a sample of a specified size is to be allocated to the strata (2, p. 66). It is of interest to calculate the variance of the estimate from a simple random sample of 20 cities taken from the population without stratifying it. Calculations show that the coefficient of variation in this case is as high as 11.6 percent. This makes it clear how useful it is to stratify the population before sample selection.

4.3 ESTIMATION FROM STRATIFIED SAMPLES

The question of making estimates of population totals, means, and proportions will now be considered. The basic principle is to form an estimate from each stratum making use of the rules discussed in Chap. 3. By adding over the strata the population total can be estimated; from this estimation the mean is estimated by dividing by the number of units in the population. The examples that follow illustrate these principles.

EXAMPLE 4.2

Average number of cattle per farm A district contains 2,072 farms which have been divided up into five strata on the basis of area in acres. From the N_h farms in a stratum a random sample of n_h farms is taken (see Table 4.3) and the number of cattle on each farm in the sample is determined. The sample mean $\bar{y}_h$ and the sample variance s_{yh}^2 are calculated from the sample in each stratum. The object is to estimate the average number of cattle per farm.

Table 4.3 Stratification of farms by area

Stratum	N_h	n_h	$\bar{y}_h$	s_{yh}^2	$\frac{N_h(N_h - n_h)}{n_h}$
Under 16	635	84	4.24	27.54	4,165.30
16–30	570	125	11.63	55.84	2,029.20
31–50	475	138	15.95	71.70	1,159.96
51–75	303	112	23.59	192.32	516.72
Over 75	89	41	29.61	334.93	104.20
Total	2,072	500			

We have (2, p. 63)

$$\sum N_h \bar{y}_h = 26{,}680.81 \qquad \frac{1}{N}\sum N_h \bar{y}_h = 12.87$$

$$\sum N_h^2 \frac{1}{n_h}\left(1 - \frac{n_h}{N_h}\right) s_{yh}^2 = 445{,}467$$

Thus the average number of cattle per farm is 12.87 and the variance of this estimate is 445,467/4,293,184 = 0.103761. The standard error is 0.322 and the coefficient of variation of the estimate is 2.5 percent. Furthermore, an estimate of the total number of cattle in the district is 26,681 and its standard error is 668.

EXAMPLE 4.3

Proportion of renting households The 2,026 households in a city are divided up into four strata on the basis of declared income. Simple random samples of households are selected from within strata and the proportion of households renting the house they live in is found by interview. The data obtained are given in Table 4.4. We shall estimate the proportion of households living in rented houses.

Table 4.4 Stratification of households by declared income

Stratum based on income	*Stratum number* N_h	*Sample size* n_h	*Number renting*	*Proportion* p_h	*Weight* W_h	$\frac{p_h(1 - p_h)}{n_h - 1}$	$1 - f_h$	W_h^2
(1)	(2)	(3)	(4)	(5)	(6)	(7)	(8)	(9)
Under 50	1,190	40	30	0.7500	0.5874	0.004808	0.9664	0.3450
50–100	523	35	18	0.5143	0.2581	0.007347	0.9331	0.0666
100–200	215	35	7	0.2000	0.1061	0.004706	0.8372	0.01126
Over 200	98	40	5	0.1250	0.0484	0.002804	0.5918	0.002343
Total	2,026	150	60		1.0000			

The sample estimate of the proportion of households renting the house is given in column 5 for each stratum. The weight $W_h = N_h/N$ of a stratum is given in column 6 while column 7 shows the estimate of the variance of p in each stratum. By multiplying the corresponding entries in columns 5 and 6, an estimate of the proportion P of households renting the house is found to be $\sum W_h p_h = 0.60$ or 60 percent. To obtain an estimate of its variance, the corresponding entries in the last three columns are multiplied and added over the strata. We have (2, p. 64)

$$\sum W_h^2(1 - f_h)\frac{p_h(1 - p_h)}{n_h - 1} = 0.002107$$

Thus the sample estimate of the variance is 0.002107, the standard error being 0.046.

Remark It may be noted that the sample allocation used in this example is very poor. The best allocation makes n_h proportional to $N_h\sqrt{P_h(1 - P_h)}$. Using the sample estimates of P_h, the best allocation works out as 86, 44, 14, and 6 households in the four strata respectively. Thus the sample size should be doubled in the first stratum and only a quarter of the sample size taken is needed in the last two strata.

Remark If the proportions P_h are not used for sample allocation and proportional allocation is employed, the sample numbers in the four strata are 88, 39, 16, and 7 respectively. This allocation does not differ much from the optimum allocation. This shows that proportional allocation is ordinarily adequate when the purpose is to estimate a population proportion from a stratified sample.

EXAMPLE 4.4

Gain from stratification When a survey has been conducted using a particular stratification, it is possible to determine from the stratified sample the gain due to stratification. This is done by estimating from the sample the variance of the estimate when no stratification is used. This variance can be compared with the variance obtained from stratified sampling. Consider the data of Example 4.2 regarding the number of cattle per farm. The stratified sample gives a variance of 0.104. To estimate the variance from an unstratified sample of 500 farms, the following calculations are needed (1).

$$\sum W_h s_{yh}^2 = 82.71 \qquad \sum W_h \bar{y}_h^2 = 219.91$$

$$\sum \frac{W_h^2 s_{yh}^2}{n_h} = 0.15 \qquad (\sum W_h \bar{y}_h)^2 = 165.64$$

$$\sum \frac{W_h s_{yh}^2}{n_h} = 0.94 \qquad \frac{1}{n}(1 - f) = 0.0015$$

The within-strata variance is estimated as 82.71 and a biased estimate of the between-strata variance is 219.91 − 165.64. When corrected for bias, the estimate of the variance based on unstratified simple random sampling is

$$0.0015(82.71 + 219.91 + 0.15 - 165.64 - 0.94)$$

which works out as 0.204. Thus stratification has reduced the variance to about 50 percent.

We shall now take the data of Example 4.3 and estimate the gain due to stratification. The sample proportion p_h plays the role of $\bar{y}_h$ and $n_h p_h(1 - p_h)/(n_h - 1)$ plays the role of s_{yh}^2. We have

$$\sum W_h n_h \frac{p_h(1-p_h)}{n_h-1} = 0.202242 \qquad \sum W_h p_h^2 = 0.403681$$

$$\sum W_h^2 \frac{p_h(1-p_h)}{n_h-1} = 0.002208 \qquad (\sum W_h p_h)^2 = 0.360720$$

$$\sum W_h \frac{p_h(1-p_h)}{n_h-1} = 0.005356 \qquad \frac{1}{n}(1-f) = 0.006173$$

The estimate of the variance based on unstratified random sampling is

$$0.006173(0.202242 + 0.403681 + 0.002208 - 0.360720 - 0.005356) = 0.001494$$

Comparing it with the variance of 0.002107 obtained from stratified sampling, it is clear that stratification has increased the variance by 50 percent. The explanation lies in the poor allocation of the total sample size to the strata.

EXAMPLE 4.5

Sampling skew populations The question of the proper allocation of the sample size to the strata becomes very important when the population is very skew with respect to the characteristic under study. Consider the population of stores, stratified by employment, given in Table 4.5. More than 97 percent of the stores are very small and only 0.1 percent of the stores are large; that is, they employ 100 or more persons. The standard deviation of retail sales (measured in some units) is low for the small stores but very high for the large stores. The purpose is to estimate the average sales by taking a sample of 1,000

Table 4.5 A skew population

Stratum based on number of employees	*Total employment* X_h	*Number of stores* N_h	*Standard deviation of retail sales* S_{yh}	*Allocation*		
				Optimum	*X-proportional*	*N-proportional*
Under 10	160,358	105,420	25	887	642	977
10–99	64,125	2,380	107	86	258	22
100 or more	25,517	100	803	27	100	1
Total	250,000	107,900		1,000	1,000	1,000

stores. If the allocation based on the number of stores N_h is used, the standard deviation of the estimate is found to be 1.222. Neyman's allocation using the product of N_h and the standard deviation S_{yh} produces a standard deviation of 0.864. In case the standard deviation S_{yh} is not known, it is possible to employ another allocation called X-proportional allocation. Here the sample size in a stratum is proportional to its measure of size X_h (employment in this case). With this allocation the standard deviation of the estimate is calculated as 0.971.

Remark Note that the sample size in the last stratum is too small with N-proportional allocation. This is undesirable since this stratum is very variable. This is the reason that its performance is poor.

Remark The X-proportional allocation differs considerably from the optimum (Neyman's) allocation; yet the standard deviation of the estimate has increased by 12 percent only. This fact is expressed by saying that the optimum is flat. Small departures from the optimum do not bring about large increases in the variance.

Remark In case the cost of collecting information differs from stratum to stratum, the sample size to be taken from a stratum is small if the cost per unit is large. It can be shown that the best sample size n_h is proportional to $N_h S_{yh}/\sqrt{c_h}$, where c_h is the cost per unit in stratum h (2, p. 65).

EXAMPLE 4.6

Sample size Suppose it is desired to find the minimum sample size n needed to estimate the proportion of households renting the house they live in (Table 4.4) with a standard error of 0.03. We shall use proportional allocation as we have found that it is quite adequate for estimating a proportion. Assuming n to be small as compared with N, the initial sample size n' is calculated from the formula (2, p. 68)

$$n' = \frac{\sum W_h p_h(1 - p_h)}{(0.03)^2} = \frac{0.19688}{0.0009} = 219$$

The final sample size n is obtained as

$$n = \frac{n'}{1 + n'/N} = \frac{219}{1.108} = 198$$

Thus a total sample of 198 households is needed to produce an estimate of P with a standard error of 0.03.

EXAMPLE 4.7

Stratification after selection The proportions of persons in a community belonging to four educational levels are known but there is no list of persons for each educational level from which to select the sample. Thus a random sample of 2,000 persons is taken from the entire population of 50,980 and data on income and educational level are collected. The following are the means and variances (in some units) as obtained from the sample

Table 4.6 Sample data on income

Educational level	*Weight* W_h	*Sample mean* $\bar{y}_h$	*Sample variance* s_{yh}^2	*Sample size* n_h
1	0.8033	4.1	34.8	1,585
2	0.1315	13.0	92.2	250
3	0.0407	25.0	174.2	105
4	0.0245	38.2	320.4	60
Total	1.0000			2,000

(Table 4.6). The purpose is to estimate the average income per individual. The sample estimate of the population mean is (2, p. 232)

$$\sum W_h \bar{y}_h = 6.96$$

since a random sample taken from the population provides random samples from the strata, although the sample sizes within strata are not known in advance. If a proportionate sample of units could be selected from strata, the variance of the estimate is estimated as

$$\frac{1}{n}\sum W_h s_{yh}^2 = \frac{55.0189}{2{,}000} = 0.02751$$

To this must be added the quantity

$$\frac{1}{n^2}\sum (1 - W_h)\, s_{yh}^2 = \frac{566.5811}{4{,}000{,}000} = 0.00014$$

since units could not be selected from within strata. Thus the variance of the estimate is 0.02765.

Remark It may be noted that the contribution from the second part is small. Thus the precision of this procedure is about the same as stratified sampling with proportionate allocation.

4.4 RATIO ESTIMATION

We shall now discuss the problem of estimating from a sample the population ratio R of Y to X such as the ratio of males to females. An allied problem is the use of the prior information on x for improving the precision of an estimate for y such as the mean of y or the total for y in the population. This is usually done by first estimating the ratio of y to x in the population and then multiplying the estimate obtained by the known population total for x. The estimate obtained in this manner is called a *ratio estimate*. The rationale of the procedure is that because of the close relationship between y and x, the ratio of y to x may be less variable than y. For example, let y be the rent paid by a family and x the income of the family. If the random sample of families selected gives a low average rent, the average income is likely to be low too and hence the ratio of the two may be close to the true ratio in the population. The sample estimate based on y alone is likely to be more variable than the ratio estimate. Several examples of this technique follow.

EXAMPLE 4.8

Rate of growth of cities The populations x of the 69 urban areas of Greece (excluding the three largest cities) are known from a previous census in 1941, the total population being $X = 13{,}559$ in hundreds. Twenty cities are selected at random and their present populations y are obtained through fieldwork. The sample gives the following information on y and x. We want to use this information for estimating the rate of growth of urban areas. The ratio R of y to x in the population is estimated by forming a similar ratio from the sample, namely, $\bar{y}/\bar{x}$. This when multiplied by the past population total X gives an estimate of the total present population Y. The estimate of the mean is obtained by dividing the estimate for Y by N, the number of units in the population. For the calculation of sampling errors we need the sum and sum of squares of the x values and the y values

Table 4.7 Past *x* and present *y* populations of 20 towns

x	y	x	y	x	y	x	y
122	126	130	148	83	121	139	177
290	257	87	119	153	151	623	790
70	108	198	338	156	224	150	176
141	185	347	368	327	410	217	236
497	421	151	221	304	295	91	129

separately as well as the sum of products of x and y values. The calculations are given below (2, p. 93).

We have

$$N = 69 \qquad Sy^2 = 1{,}741{,}114 \qquad \frac{n}{N} = 0.2899$$

$$n = 20 \qquad Sx^2 = 1{,}316{,}956 \qquad 1 - f = 0.7101$$

$$Sy = 5{,}000 \qquad Sxy = 1{,}488{,}332 \qquad \bar{x} = 213.8$$

$$Sx = 4{,}276 \qquad \hat{R} = \frac{Sy}{Sx} = 1.16932 \qquad \bar{X} = 196.51$$

$$Sy^2 + \hat{R}^2 Sx^2 - 2\hat{R}Sxy = 61{,}127$$

The rate of growth is estimated as $\hat{R} = 1.169$ with a standard error of

$$\frac{1}{196.51}\left(\frac{0.7101}{20 \times 19} \times 61{,}127\right)^{1/2} = 0.0543$$

and a coefficient of variation of 4.64 percent. If the object is to estimate the total population of the cities, the sample estimate is $X\hat{R} = 15{,}855$. The standard error of this estimate is $13{,}559 \times 0.0543 = 736$.

Furthermore, if information on x is not used, the estimate of Y based on the simple average is $N\bar{y} = 17{,}250$ with a standard error of 2,090 or a relative standard error of 12.1 percent. This shows how beneficial information on x has been in making a more precise estimate of the total number of persons in the universe considered.

Remark We have used information on the past populations x of the cities for making an improved estimate of the present total population of the urban areas. It is sometimes asked whether the ratio estimate will always be an improvement on the simple average of the y values. The answer is that the improvement depends on the strength of association between y and x. A good rule to use is the following. If the correlation coefficient $\rho(x,y)$ between x and y exceeds $(1/2)\text{CV}(x)/\text{CV}(y)$, the ratio estimate is more precise than the simple average of the y values. Here $\text{CV}(x)$ and $\text{CV}(y)$ stand for the coefficient of variation of x and y respectively (2, p. 92). In the present case x and y are expected to have about the same coefficient of variation. Hence the ratio estimate is better if the correlation coefficient exceeds 0.50. This analysis applies when the regression line passes through a point not far from the origin.

Remark It is not essential that information on the auxiliary variate x be perfect. There may be biases present in the data, such as systematic undercount or overcount of persons.

These errors do not vitiate the use of the ratio estimate. In some countries local officials feel hesitant about using population figures for forming the ratio estimate on the ground that these figures are serious undercounts or overestimates. Such fears are baseless.

EXAMPLE 4.9

Sex ratio A simple random sample of 30 households is selected from an area containing 1,500 households. The number of males y and the number of females x are determined for each sample household. The purpose is to estimate the sex ratio, that is, ratio of males to females in the area considered. The following calculations are made from the sample.

$$Sy = 53 \qquad Sy^2 = 117 \qquad n = 30$$

$$Sx = 51 \qquad Sxy = 89 \qquad N = 1{,}500$$

$$Sx^2 = 109 \qquad 1 - \frac{n}{N} = 0.98 \qquad \hat{R} = 1.039$$

$$Sy^2 + \hat{R}^2\,Sx^2 - 2\hat{R}Sxy = 49.72 \qquad \bar{x} = 1.7$$

The sex ratio is $\hat{R} = 1.039$ or 104 males per 100 females. The standard error of $\hat{R}$ is

$$\frac{1}{1.7}\left(0.98 \times \frac{49.72}{30 \times 29}\right)^{1/2} = 0.139$$

and the relative standard error is 13.4 percent.

EXAMPLE 4.10

Determination of sample size It is known that the ratio estimate, say that of the rate of growth or of sex ratio, is biased in the sense that the expected value of the ratio estimate $\hat{R}$ is not equal to the population ratio R. Thus we want to determine the size of the sample to use which makes the bias unimportant. A working rule to use is to make the bias less than 10 percent of the standard deviation of the estimate. Now it can be proved that the ratio of the bias of the ratio estimate $\hat{R}$ and its standard deviation $\sigma(\hat{R})$ is less than the coefficient of variation of $\bar{x}$, the sample mean for x. Hence the sample size n is to be so chosen that (2, p. 87)

$$\frac{1}{n}\left(1 - \frac{n}{N}\right)\frac{S_x^2}{\bar{X}^2} \leqslant \frac{1}{100}$$

For the data of Example 4.8, calculations show that

$$\bar{X} = 196.51 \qquad S_x = 124 \qquad \mathrm{CV}(x) = 0.631$$

Hence

$$\left(\frac{1}{n} - \frac{1}{69}\right)(0.631)^2 \leqslant \tfrac{1}{100}$$

or

$$n \geqslant 25$$

If 25 or more cities are selected in the sample, the bias of the ratio estimate is not important. Confidence statements can then be made as if no bias were present.

Remark Assuming that the square of the coefficient of variation of the number of females per household for the data of Example 4.9 is 0.27, about 27 households should be selected as the sample to ensure that the bias of the ratio estimate is negligible.

Remark The bias of the ratio estimate is negligible in large samples. These examples show that a sample of 30 units is large enough for the type of populations studied here.

EXAMPLE 4.11

Subpopulation analysis A random sample of 150 land areas is taken from a population containing 3,250 such areas. In each sample area the number of women x and those working for a living y are determined. It is later discovered that it is of great interest to estimate the proportion of women working for a living in a part of the population from which only 40 land areas have been taken in the original sample. The following calculations are made on the sample from this subpopulation.

$$n = 40 \qquad Sy^2 = 22{,}048 \qquad 1 - \frac{n}{N} = 0.9538$$

$$Sy = 801 \qquad Sx^2 = 328{,}356 \qquad \bar{x} = 85.625$$

$$Sx = 3{,}425 \qquad Sxy = 76{,}987 \qquad \hat{R} = 0.2339$$

$$Sy^2 + \hat{R}^2 Sx^2 - 2\hat{R}Sxy = 3{,}998$$

The sample estimate of the proportion of women working for a living is $\hat{R} = 0.2339$ or about 23 percent. The standard error of this estimate is

$$\frac{1}{85.625}\left(0.9538 \times \frac{3{,}998}{40 \times 39}\right)^{1/2} = 0.0183$$

Remark If we could have taken a simple random sample of 3,425 women directly from the population (without using land areas), the standard error of the proportion would have been approximately

$$\left(\frac{0.9538 \times 0.2339 \times 0.7661}{3{,}424}\right)^{1/2} = 0.007$$

which is much smaller than the error obtained when the sampling unit is a land area and not an individual.

4.5 RATIO ESTIMATION IN STRATIFIED SAMPLING

When the population has been stratified and a simple random sample is taken from each stratum, two types of ratio estimates may be formed. We may estimate the ratio in each stratum and form a weighted average of the strata ratios. In the second case, the numerator Y is estimated from the stratified design and so is the denominator X. The ratio of the two provides an estimate of the population ratio (2, p. 105).

Since the ratio estimate is biased, the first method may lead to an estimate whose bias is not negligible as compared with the standard error. This will be so when the sample sizes within strata are small. And quite often we can only afford to take a sample of two units per stratum. In that case the second method of forming the ratio estimate should be used. An example illustrating this method is given.

EXAMPLE 4.12

Total enrollment in schools A province contains 10 districts, the number of schools in a district varying from 15 to 42. Two schools are selected at random from each district for estimating the total enrollment for 1969 (the variate y). Information on total enrollment in 1965 (the variate x) is known for each school, X being 290,000 for all the 294 schools in the province. Table 4.8 gives the data from the sample. In this table, N_h is the number

Table 4.8 Enrollment data from 20 schools

Stratum	N_h	y_{h1}	x_{h1}	y_{h2}	x_{h2}	$N_h \frac{y_{h1} - y_{h2}}{2\hat{Y}}$	$N_h \frac{x_{h1} - x_{h2}}{2\hat{X}}$	$1 - f_h$
1	35	1,600	1,052	350	810	0.0820	0.0166	0.9429
2	26	651	685	792	807	−0.0069	−0.0062	0.9231
3	15	500	1,481	2,534	1,406	−0.0572	0.0022	0.8667
4	39	715	672	823	785	−0.0079	−0.0086	0.9487
5	42	608	580	750	730	−0.0112	−0.0123	0.9524
6	17	1,498	1,406	1,263	1,209	0.0075	0.0066	0.8824
7	26	1,150	1,097	1,230	1,160	−0.0039	−0.0032	0.9231
8	38	902	852	816	803	0.0061	0.0036	0.9474
9	40	605	575	702	655	−0.0073	−0.0063	0.9500
10	16	1,252	1,201	1,168	1,106	0.0021	0.0030	0.8750

of schools in stratum h; y_{h1} and y_{h2} are the enrollment figures for 1969 for the two schools in the sample, the corresponding figures for 1965 being x_{h1} and x_{h2}. The sample estimate of the total enrollment for 1969 is

$$\hat{Y} = \tfrac{1}{2} \sum N_h(y_{h1} + y_{h2}) = 266{,}698$$

and that for 1965 is

$$\hat{X} = \tfrac{1}{2} \sum N_h(x_{h1} + x_{h2}) = 255{,}625$$

Thus the estimate of the ratio of Y to X is 1.0433 and the ratio estimate of the enrollment in 1969 is

$$1.0433 \times 290{,}000 = 302{,}557$$

In order to estimate the coefficient of variation of the ratio estimate, we calculate for each stratum the quantity (2, p. 238)

$$(1 - f_h)\left(N_h \frac{y_{h1} - y_{h2}}{2\hat{Y}} - N_h \frac{x_{h1} - x_{h2}}{2\hat{X}}\right)^2$$

When added up over strata, this gives the square of the coefficient of variation. Calculations from Table 4.8 show that the coefficient of variation of the ratio estimate obtained is 0.084 or 8.4 percent.

4.6 DIFFERENCE ESTIMATION

The ratio estimate works well when there is reasonable proportionality between the main variate y and the auxiliary variate x; that is, y/x is k where the range of k is small. If this is not so and the regression line passes through a point far from the origin, we have

$$E(y) = a + kx$$

In this situation it is better to estimate $\bar{Y} - k\bar{X}$ from the random sample and to this add the known value $k\bar{X}$ in order to estimate the population mean for y. This estimator

$$\bar{y} - k(\bar{x} - \bar{X})$$

is called the *difference estimator*. Any value of k can be used. But the best value of k is the regression coefficient of y on x. The following examples illustrate the use of this method.

EXAMPLE 4.13

Total population of urban areas We shall consider the universe formed by the 69 urban areas in Greece, which was discussed in Example 4.8. A simple random sample of 20 cities is selected. We have

$$n = 20 \qquad \bar{X} = 196.5 \qquad S(y_i - \bar{y})^2 = 491{,}114$$

$$1 - \frac{n}{N} = 0.7101 \qquad \bar{x} = 213.8 \qquad S(x_i - \bar{x})^2 = 402{,}747$$

$$X = 13{,}559 \qquad \bar{y} = 250.0 \qquad S(y_i - \bar{y})(x_i - \bar{x}) = 419{,}332$$

Let $k = 1$. For estimating the population mean $\bar{Y}$, the difference estimate is

$$250 - (213.8 - 196.5) = 232.7$$

An estimate of the total population of the 69 cities is

$$69 \times 232.7 = 16{,}056$$

Further

$$S(y_i - \bar{y})^2 + k^2 S(x_i - \bar{x})^2 - 2kS(x_i - \bar{x})(y_i - \bar{y}) = 55{,}197$$

Thus the variance estimator for the mean is (2, p. 100)

$$\frac{1}{20 \times 19} \times 0.7101 \times 55{,}197 = 103.1447$$

and the standard error is 10.15. The standard error of the estimate of the total population is $69 \times 10.15 = 700$.

Remark When the ratio estimate is used for estimating the total population Y, the standard error of the estimate is 736 (see Example 4.8).

Remark Several values of k can be tried for making the difference estimate. The results obtained for selected values of k are presented in Table 4.9.

Table 4.9 Standard error of the difference estimate for different values of *k*

k	*Standard error*
0.8	832
0.9	745
1.0	700
1.1	703
1.2	757

EXAMPLE 4.14

Average number of cattle per farm There is a population of 75,000 farms in a region. Information on the geographical area x of each farm is available, the average area per farm being 31 acres. A random sample of 2,000 farms is taken and information collected on the number of cattle y on each farm. The problem is to estimate the average number of cattle per farm. The following calculations are made from the sample.

$$Sx = 62{,}756 \qquad Sxy = 1{,}146{,}301 \qquad N = 75{,}000$$

$$Sy = 25{,}650 \qquad S(y_i - \bar{y})^2 = 267{,}274 \qquad n = 2{,}000$$

$$Sx^2 = 2{,}937{,}801 \qquad S(x_i - \bar{x})^2 = 968{,}644 \qquad \bar{y} = 12{,}825$$

$$Sy^2 = 596{,}235 \qquad S(x_i - \bar{x})(y_i - \bar{y}) = 341{,}456 \qquad \bar{x} = 31.378$$

$$f = 0.0267 \qquad 1 - f = 0.9733 \qquad \bar{X} = 31.000$$

Let $k = 0.5$. The difference estimate of the average number of cattle per farm is

$$12.825 - 0.5(31.378 - 31.00) = 12.64$$

Furthermore

$$S(y_i - \bar{y})^2 + k^2 S(x_i - \bar{x})^2 - 2kS(x_i - \bar{x})(y_i - \bar{y}) = 167{,}979$$

Hence an estimate of the variance is

$$\frac{1}{2{,}000 \times 1{,}999} \times 0.9733 \times 167{,}979 = 0.040893$$

and the standard error is 0.202, the coefficient of variation being 1.6 percent.

4.7 REGRESSION ESTIMATION

Instead of making a guess of the value of k for using the difference estimate, we may as well calculate the regression coefficient b from the sample and use this in place of k. The estimator thus obtained is called the *regression estimator*. For estimating the population mean of y from a random sample, the regression estimator is (2, p. 100)

$$\bar{y} - b(\bar{x} - \bar{X})$$

where $\bar{X}$ and $\bar{x}$ are the population mean and the sample mean of x respectively. This estimator is expected to be more precise than the difference or the ratio estimators. But it is more cumbersome to calculate. Only approximate formulas for its bias and variance are known.

Regression estimation is a classical method of estimation. Suppose it is possible to collect information on each unit of a population rather inexpensively. More objective and possibly more expensive observations on the variate of interest, y, are taken on a sample selected from the population. The use of the regression coefficient b can then produce a fairly precise estimate of the population mean for y. Examples of this technique follow.

EXAMPLE 4.15

Average number of persons per city We shall use the data of Example 4.8 to estimate the average number of persons per city by taking a random sample of 20 cities. The regression coefficient b is found to be

$$b = \frac{S(x_i - \bar{x})(y_i - \bar{y})}{S(x_i - \bar{x})^2} = 1.04118$$

The regression estimate of the mean is

$$250.0 - 1.04118(213.8 - 196.5) = 232$$

Furthermore

$$S(y_i - \bar{y})^2 - bS(x_i - \bar{x})(y_i - \bar{y}) = 54{,}514$$

The estimate of the variance is approximately given by

$$\frac{1}{20} \times 0.7101 \times \frac{54{,}514}{18} = 107.52$$

and the standard error is 10.3 persons. For all the 69 cities the estimate of the total population is 16,008 with a standard error of 711 persons.

Remark The ratio, difference, and regression estimates all give standard errors of about the same magnitude for the data of this example.

EXAMPLE 4.16

Average number of cattle per farm The data of Example 4.14 will now be considered to estimate the average number of cattle per farm when auxiliary information on the area of the farm is used. The regression coefficient b is found to be 0.3525. The estimate of the mean is

$$12.825 - 0.3525(31.378 - 31.000) = 12.69$$

Further calculations show that

$$S(y_i - \bar{y})^2 - bS(x_i - \bar{x})(y_i - \bar{y}) = 146{,}911$$

Thus an approximate estimate of the variance is

$$\frac{1}{2{,}000} \times 0.9733 \times \frac{146{,}911}{1{,}998} = 0.035783$$

the standard error being 0.19, leading to a relative standard error of 1.5 percent.

Remark Note that the divisor used is $n - 2$, that is, 1,998 in the present example.

PROBLEMS

4.1. The farms in a province are divided into six strata on the basis of their areas reported in the census. The means and variances of the yields of cash crops are given below.

Stratum h	N_h	$\bar{Y}_h$	S_{yh}^2
1	50,000	0.13	0.25
2	23,000	0.72	2.89
3	20,000	3.34	72.25
4	5,300	18.03	1,225
5	1,500	68.85	9,025
6	200	786.00	40,000

(*a*) Design a sample to yield minimum variance for about 3,000 farms in the sample for estimating the average yield per farm. (*b*) If all 200 farms in the last stratum must be included in the sample, what will the sample allocation be? (*c*) Compare the variance with that obtained from a proportionate sample of the same size.

4.2. All the tire dealers in an area are assigned to strata according to the number of new tires held at a previous census. A random sample of 4,170 dealers is selected from within strata and the number of new tires ascertained. The results obtained are given below.

Stratum boundaries	N_h	n_h	$\bar{y}_h$	s_{yh}^2
1–9	19,850	3,000	4.1	34.8
10–19	3,250	600	13.0	92.2
20–29	1,007	340	25.0	174.2
30–39	606	230	38.2	320.4

(*a*) Estimate the average number of new tires per dealer and find the standard error of the estimate. (*b*) Estimate the gain in precision due to stratification.

4.3. The following information on literacy is available for an area.

Age-group	*Number of persons*	*Proportion literate*
15–24	25,200	0.50
25–34	19,100	0.30
35–49	36,300	0.10
50 or over	19,400	0.01

(*a*) If a proportionate stratified sample is to be used in the near future for estimating the proportion of literate persons with a coefficient of variation of 10 percent, find the sample size needed. (*b*) Compare its performance first with an unstratified simple random sample and then with a stratified random sample with optimum allocation.

4.4. The 242,800 parcels in a district are allocated to five strata on the basis of topographical features and size of locality. The strata sizes in thousands are 82.0, 52.6, 41.6, 37.8, and 28.8. The standard deviations of agricultural area in the strata are 27.6, 18.6, 21.2, 28.5, and 16.8 thousands respectively. The cost of collecting information per parcel in dollars is found to be 2.25, 2.75, 3.00, 3.50, and 2.50 in the respective strata. The total budget for the survey is $9,500. Find the number of parcels to be selected in the sample from the different strata for the purpose of estimating the total agricultural area in the district.

4.5. The 340 villages in a tract are allocated to four strata on the basis of agricultural area. The number of villages N_h in each stratum, the average area under wheat $\bar{Y}_h$, and the standard deviation of area under wheat S_h are given.

Stratum	N_h	$\bar{Y}_h$	S_h
1	63	112	56
2	199	277	116
3	53	558	186
4	25	960	361

The purpose is to estimate, from a sample of 34 villages, the average area w under wheat per village and the difference z in the averages of the larger zones formed by combining strata 1, 2 and 3, 4 respectively. (*a*) What is the allocation which minimizes the sum of the variances of $\hat{w}$ and $\hat{z}$? Find the coefficients of variation of $\hat{w}$ and $\hat{z}$. If the variance of $\hat{w}$ is to be minimized, find the relative sampling errors of $\hat{w}$ and $\hat{z}$. (*b*) Find the sample size if it is desired that the relative errors of $\hat{w}$ and $\hat{z}$ be 5 percent and 7.5 percent respectively. (*c*) It is stipulated that the coefficient of variation of $\hat{w}$ be approximately two-thirds the coefficient of variation of $\hat{z}$ for a sample of 34 villages. Find the sample allocation in this case.

4.6. It is proposed to divide each of the two populations

$$(a)\ \exp(-y), y \geqslant 0 \qquad (b)\ 2(1-y), 0 \leqslant y \leqslant 1$$

into two strata by cutting the frequency functions at suitable points y_1 and y_2 respectively. Two rules are to be used for making the strata. With the first rule the strata are to be made of the same size (weight) and with the second stratification is to be done such that the variance of the estimate of the mean is a minimum. The sample allocation is equal in either case. Find the cutoff points and the variance of the estimate based on a sample of n units. Compare your results with unstratified random sampling.

4.7. Information based on a previous census is available on the number of cattle x on each of the 75,308 farms comprising a population. The average number of cattle per farm is found to be $\bar{X} = 11.72$. A simple random sample of 2,055 farms is taken from the population and current information collected on the number of cattle y on each farm. The following information is available from the sample.

$$Sy_i = 25{,}751 \qquad Sx_i = 23{,}642$$

$$Sy_i^2 = 596{,}737 \qquad Sx_i^2 = 504{,}150$$

$$Sx_i y_i = 499{,}172$$

Use the methods of ratio, difference, and regression estimation to determine the average number of cattle per farm. Compare the relative performance of the three methods. Find the standard error of the estimate when the information in the census is not used. What sample size is needed for achieving a coefficient of variation of 5 percent?

4.8. A simple random sample of 48 cities is selected from a population of 200 cities and the present population y of each city is found through fieldwork. Information on the population of these cities at the time of the previous census x is available, the average population per city being 116.94×10^3. The sample gives

$$Sy_i = 6{,}187 \times 10^3 \qquad Sx_i = 5{,}006 \times 10^3$$

$$Sy_i^2 = 1{,}522{,}257 \times 10^6 \qquad Sx_i^2 = 1{,}042{,}200 \times 10^6$$

$$Sx_i y_i = 1{,}248{,}030 \times 10^6$$

Estimate the rate of growth of the cities and find its standard error.

4.9. A trained investigator makes an eye estimate of the area of each parcel in a commune containing 200 parcels. This exercise produces an area of 1,160 stremmas. The areas are actually measured on a random sample of 10 parcels with the following results:

Parcel	1	2	3	4	5	6	7	8	9	10
Estimated area	4.8	5.8	6.0	5.9	7.6	6.7	4.7	5.8	4.4	5.2
Actual area	4.5	5.3	5.8	6.1	7.1	6.7	4.2	5.7	3.9	5.0

Estimate the total area in the commune and find the standard error of the estimate.

4.10. The number of persons x in each household and the number of males y is determined from a random sample of 30 households selected from a population of 15,000. Calculations show that

$$Sx_i = 108 \qquad Sy_i = 54$$

$$Sx_i^2 = 436 \qquad Sy_i^2 = 122$$

$$Sx_i y_i = 221$$

Estimate the proportion male and find the standard error of the estimate. If a frame listing all persons in the population is available, find the number of persons that should be selected in the sample to achieve comparable precision. Comment on your results.

REFERENCES

1. Cochran, W. G. (1963), "Sampling Techniques," John Wiley & Sons, Inc., New York.
2. Raj, D. (1968), "Sampling Theory," McGraw-Hill Book Company, New York.

5
Sampling in Clusters

5.1 INTRODUCTION

As pointed out in Sec. 2.8, the sample is usually selected in clusters or groups of elementary units since frames listing elementary units are rarely available. For example, an agricultural survey may be based on a sample of villages when the list of fields is not available. From the viewpoint of efficiency it is better to select a sample of fields; such a sample is likely to be scattered over the entire population and thereby provide a better cross section than a sample of villages. On the other hand it is more expensive to collect information from fields which are scattered all over the population than from fields located in a few villages. Thus an important problem is to determine the best size of the cluster for a specified cost of the survey. The problem can be solved if the cost of the survey and the variance of the estimate can be expressed as functions of the size of the cluster.

5.2 CLUSTER SAMPLING

A number of examples of the use of cluster sampling have been given in the previous chapters. A few more examples are included in this section.

EXAMPLE 5.1

Number of cattle per farm The 3,510 farms in a village are allocated to 90 clusters; the number of farms in different clusters is not necessarily the same. A simple random sample of 15 clusters is selected and the number of cattle (the variate y) determined. The purpose is to estimate the average number of cattle per farm. The sample data are given in Table 5.1.

Table 5.1 A sample of clusters of farms

Cluster	*Number of farms* x	*Number of cattle* y	*Average cattle per farm* z
1	35	418	11.94
2	25	402	16.08
3	48	360	7.50
4	30	394	13.13
5	70	515	7.36
6	55	910	16.55
7	66	600	9.09
8	18	316	17.56
9	30	288	9.60
10	32	350	10.94
11	64	784	12.25
12	24	290	12.08
13	48	795	16.56
14	40	478	11.95
15	82	906	11.05
Total	667	7,806	183.64

A simple estimate of the average number of cattle per farm can be obtained by calculating the mean of the cluster averages. This works out as 12.24 cattle per farm. The estimate is biased since the number of farms per cluster is not the same. In order to estimate its variance, we have

$$Sz_i = 183.64 \qquad Sz_i^2 = 2{,}395.80 \qquad S(z_i - \bar{z})^2 = 147.56$$

Thus an estimate of the variance is

$$\frac{1}{15 \times 14} \times 0.8333 \times 147.56 = 0.5855$$

Another estimate of the average can be made by calculating the ratio of y to x from the sample. This ratio estimate is 7,806/667 or $\hat{R} = 11.70$ cattle per farm. The following calculations are needed to estimate its variance.

$$Sy^2 = 4{,}759{,}890 \qquad Sx^2 = 34{,}883 \qquad Sxy = 393{,}716$$

$$Sy^2 + \hat{R}^2 Sx^2 - 2\hat{R}Sxy = 322{,}156$$

The standard error of the ratio estimate is

$$\frac{1}{44.47}\left(\frac{0.8333}{15 \times 14} \times 322{,}156\right)^{1/2} = 0.8039$$

Further, an unbiased estimate of the population mean is

$$\frac{90}{3{,}510}\frac{7{,}806}{15} = 13.34$$

with an estimated variance of

$$\left(\frac{90}{3{,}510}\right)^2 \frac{1}{15 \times 14} 697{,}648 = 2.1840$$

Remark Of the three estimators considered, the unbiased estimate has the largest estimated variance and the mean of the cluster means the smallest.

EXAMPLE 5.2

Number of cattle per farm A sample of 15 clusters is selected from the population of Example 5.1. The clusters are selected with replacement with probability proportionate

Table 5.2 Number of cattle on a sample of areas

Cluster	*Number of farms* x	*Number of cattle* y	*Average cattle per farm* z
1	83	906	10.92
2	61	619	10.15
3	31	331	10.68
4	28	326	11.64
5	46	697	15.15
6	28	392	14.00
7	64	784	12.25
8	56	930	16.61
9	83	906	10.92
10	51	586	11.49
11	19	168	8.84
12	30	360	12.00
13	31	331	10.68
14	55	914	16.62
15	53	739	13.94
Total	719	8,989	185.89

to the number of farms in the cluster. The number of cattle on the farms in the sample are found by fieldwork. The data obtained are given in Table 5.2. We have

$$Sz_i = 185.89 \qquad Sz_i^2 = 2{,}381.3205 \qquad S(z_i - \bar{z})^2 = 77.6477$$

The sample estimate of the average number of cattle per farm is $185.89/15 = 12.39$. This is an unbiased estimate of the population average. An estimate of its variance is (1, p. 49)

$$\frac{1}{15 \times 14} \times 77.6477 = 0.3698$$

which is smaller than the variances of the estimators considered in Example 5.1.

EXAMPLE 5.3

Efficiency of cluster sampling When the data in the sample clusters are recorded by elements, it is possible to estimate from the sample itself the efficiency of the cluster as a sampling unit. Suppose a human population is divided up into $N = 160$ area segments each containing four households. A simple random sample of $n = 20$ segments is selected

Table 5.3 Number of persons by household in selected segments

Segment	*Number of persons by household* y_{ij}				*Total* y_i
1	6	5	2	4	17
2	3	5	5	6	19
3	4	2	13	5	24
4	6	4	2	7	19
5	2	10	3	4	19
6	10	5	2	5	22
7	3	3	2	5	13
8	5	6	1	8	20
9	3	6	4	4	17
10	5	4	8	3	20
11	4	7	4	9	24
12	6	7	3	6	22
13	4	7	7	6	24
14	5	4	1	6	16
15	3	4	4	8	19
16	3	3	3	5	14
17	7	5	5	5	22
18	4	6	6	4	20
19	2	4	4	3	13
20	5	4	5	9	23

and information collected on the number of persons in the sample households. The data obtained are given in Table 5.3. We have

$$Sy_i = 387 \qquad SSy_{ij}^2 = 2{,}239$$

$$Sy_i^2 = 7{,}721 \qquad SS(y_{ij} - \bar{y})^2 = 366.89$$

$$S(y_i - \bar{y})^2 = 232.55 \qquad SS(y_{ij} - \bar{y}_i)^2 = 308.75$$

$$SSy_{ij} = 387 \qquad SS(\bar{y}_i - \bar{y})^2 = 58.14$$

$$A = \frac{N}{n} SS(y_{ij} - \bar{y}_i)^2 + \frac{N-1}{n-1} SS(\bar{y}_i - \bar{y})^2 = 2{,}956.54$$

The sample estimate of the total number of persons is $8 \times 387 = 3{,}096$. An estimate of its variance is

$$\frac{160 \times 160}{20}\left(1 - \frac{1}{8}\right)\frac{232.55}{19} = 13{,}708.21$$

which gives a standard error of 117. Since the data are available by household, it is possible to estimate the variance if a direct sample of 80 households had been selected from the 640 in the population. The mean square between clusters is estimated by 58.14/19 while an estimate of the mean square within clusters is $308.75/(20 \times 3)$. This gives an estimate of the total corrected sum of squares in the population as (1, p. 112)

$$159 \times \frac{58.14}{19} + 160 \times 3 \times \frac{308.75}{20 \times 3} = 2{,}958.54$$

from which an estimate of the variance per household is obtained as 2958.54/639. Thus an estimate of the variance obtained from a direct sample of 80 households is

$$\frac{640 \times 640}{80}\left(1 - \frac{1}{8}\right)\frac{2{,}958.54}{639} = 20{,}742.40$$

giving a standard error of 143.9. This shows that sampling in clusters has given a more precise estimate than sampling by individual households. In fact the relative gain in precision achieved through cluster sampling as compared with sampling individual households is

$$\frac{1/13{,}708 - 1/20{,}742}{1/20{,}742} = 51 \text{ percent}$$

Remark The reason that cluster sampling gives a lower variance is that there is negative intraclass correlation between the number of persons in households within the same cluster. To demonstrate this we calculate the intraclass correlation coefficient. Since the cluster size is 4, we have

$$4SS(\bar{y}_i - \bar{y})^2 - SS(y_{ij} - \bar{y})^2 = -134.33$$

$$SS(y_{ij} - \bar{y})^2 = 366.89$$

The intraclass correlation coefficient is estimated as (1, p. 110)

$$\hat{\rho} = \frac{-134.33}{3 \times 366.89} = -0.122$$

Remark It can be shown that for the same number of elements in the sample, the variance for clusters of size M is $[1 + (M - 1)\rho]$ times the variance for a simple random of elements taken from the population (1, p. 109).

5.3 MULTISTAGE SAMPLING

When the clusters are large it is difficult to enumerate them completely. At the same time it is unnecessary to collect information on every element in the sample clusters. We may further select a sample from each of the clusters selected. This procedure is called subsampling or two-stage sampling. For example, we may select a sample of villages for an agricultural survey. Then a sample of fields may be selected from the villages in the sample. The villages are the first-stage or primary sampling units (psu's) and the fields are the second-stage units. Furthermore, a plot of a suitable size may be selected at random from each field in the sample. The plot is then the third-stage unit. The entire procedure will be called three-stage sampling. The main advantage is that a frame has to be prepared for only those units which are in the sample. The fieldwork becomes less expensive since the survey is to be carried out in the selected psu's only. Supervision becomes relatively easy when the work is restricted to just a few clusters.

In multistage sampling the process of estimation can be carried out stage by stage, using the appropriate methods of estimation at each stage. Suppose, for example, that a 10 percent simple random sample of blocks is selected from a city and a fraction of households taken from each block, giving the following information on the number of persons.

Sample block	1	2	3	4	5
Sampling fraction for households	1/10	1/5	1/5	1/10	1/10
Number of persons in sample	30	40	50	25	35
Estimated population of block	300	200	250	250	350

We can then estimate the total number of persons in each block (see last row) from which the total population of the sample blocks is put at 1,350. Hence the sample estimate of the total population of the city is $10 \times 1{,}350 = 13{,}500$.

The variance of the estimate in multistage sampling is built up of two components. One is the variation between primary sampling units and the other is the variation within primary sampling units. The sum of the two components gives the variance of the estimate. The examples below illustrate the estimation procedure when the psu's are selected with equal probability.

EXAMPLE 5.4

Number of persons in an area From a population of $N = 159$ enumeration districts (EDs) a random sample of $n = 20$ EDs was drawn. From the sample ED containing M_i households, m_i were selected in the sample and information collected on the number of persons in the households. The data obtained are given in Table 5.4.

Table 5.4 Data from a two-stage population sample

ED	M_i	m_i	y_{ij}									
1	63	6	5	2	4	3	1	5				
2	25	3	5	5	6							
3	43	4	2	13	5	5						
4	57	6	4	2	7	2	7	2				
5	23	2	10	3								
6	96	10	5	2	5	5	3	1	1	4	2	4
7	30	3	3	2	5							
8	45	5	6	1	8	7	16					
9	27	3	6	4	4							
10	46	5	4	8	3	2	2					
11	39	4	7	4	9	1						
12	62	6	7	3	6	3	4	4				
13	44	4	7	7	6	6						
14	49	5	4	1	6	3	5					
15	31	3	4	4	8							
16	25	3	3	3	5							
17	67	7	5	5	5	9	4	6	2			
18	44	4	6	6	4	3						
19	23	2	4	4								
20	48	5	4	5	9	6	4					

We shall estimate the total number of persons in the area. As a first step we calculate the estimated number of persons $M_i\bar{y}_i$ in each ED and the estimated variance s_{wi}^2. This is done in Table 5.5. An estimate of the total number of persons in the area is

$$\hat{Y} = \tfrac{159}{20} SM_i\bar{y}_i = \tfrac{159}{20} \times 4{,}166.90 = 33{,}127$$

Table 5.5 Calculation of sampling errors in two-stage sampling

ED	$M_i\bar{y}_i$	s_{wi}^2	*ED*	$M_i\bar{y}_i$	s_{wi}^2
1	210.00	2.67	11	204.75	12.25
2	133.33	0.34	12	279.00	2.70
3	268.75	22.25	13	286.00	0.33
4	228.00	6.00	14	186.20	3.70
5	149.50	24.50	15	165.33	5.34
6	307.20	2.62	16	91.67	1.34
7	100.00	2.34	17	344.57	4.48
8	342.00	29.30	18	209.00	2.25
9	126.00	1.34	19	92.00	0.00
10	174.80	6.20	20	268.80	4.30

In order to estimate its variance, we calculate (1, p. 116)

$$S\left(M_i\bar{y}_i - \frac{1}{n}SM_i\bar{y}_i\right)^2 = 123{,}146$$

from which the between-psu contribution to the variance is obtained as

$$\frac{N^2}{n}\left(1 - \frac{n}{N}\right)\frac{1}{n-1} \times 123{,}146 = 7{,}162{,}236$$

The within-psu contribution is

$$\frac{N}{n}S\frac{M_i^2}{m_i}\left(1 - \frac{m_i}{M_i}\right)s_{wi}^2 = \tfrac{159}{20} \times 50{,}395 = 400{,}640$$

Adding the two components the variance estimate is obtained as 7,562,876 which gives a standard error of 2,750 persons and a coefficient of variation of 8.3 percent.

Remark A less rigorous but simpler estimate of the variance can be obtained by assuming that each ED provides an independent estimate of the total population of the area. Under this model an estimate of the total number of persons in the area is obtained as the mean of the ED estimates, namely,

$$\tfrac{1}{20}S(159M_i\bar{y}_i) = 33{,}127$$

and an estimate of the variance is

$$\frac{159 \times 159}{20 \times 19}S\left(M_i\bar{y}_i - \frac{1}{n}SM_i\bar{y}_i\right)^2 = 8{,}192{,}773$$

which gives a standard error of 2,862 persons and a coefficient of variation of 8.6 percent. This procedure is appropriate when the first-stage sampling fraction is small. If the first-stage sampling fraction is not small, the method overstates the variance. To this extent it is a conservative method of estimating the variance.

EXAMPLE 5.5

Average size of a household We shall now estimate the average number of persons per household from the data of Example 5.4. An estimate of the total number of persons in the area has been found as $\hat{Y} = 33{,}127$. Similarly an estimate of the total number of households is

$$\hat{X} = \tfrac{159}{20}SM_i = \tfrac{159}{20} \times 887 = 7{,}052$$

Thus the average size of a household is estimated as $\hat{R} = \hat{Y}/\hat{X} = 4.698$ persons. This is a ratio estimate. In order to estimate the variance (1, p. 127) we note that the within-psu contribution is simply $1/\hat{X}^2$ times the corresponding contribution calculated in Example, 5.4. Thus this contribution is 0.008056. In order to calculate the between-psu contribution we first obtain $M_i(\bar{y}_i - \hat{R})$ for each sample psu (see Table 5.6).

Table 5.6 Calculation of between-psu component

ED	1	2	3	4	5	6	7
$M_i(\bar{y}_i - \hat{R})$	−85.97	15.88	66.74	−39.79	41.45	−143.81	−40.94
ED	8	9	10	11	12	13	14
$M_i(\bar{y}_i - \hat{R})$	130.59	−0.85	−41.31	21.53	−12.28	79.29	−44.00
ED	15	16	17	18	19	20	
$M_i(\bar{y}_i - \hat{R})$	19.69	−25.78	29.80	2.29	−16.05	43.30	

This gives

$$\frac{N^2}{n}\left(1-\frac{n}{N}\right)\frac{1}{n-1}SM_i^2(\bar{y}_i-\hat{R})^2=4{,}038{,}217$$

and the between-psu contribution is $1/\hat{X}^2$ times this, which works out as 0.081201. Adding the two, we get an estimate of the mean-square error of $\hat{R}$ as 0.089257 which gives a standard error of 0.299 and a coefficient of variation of 6.4 percent.

Remark An approximate estimate of the variance can be obtained by assuming that the primary sampling units have been selected with replacement. This estimator is (1, p. 128)

$$\frac{N^2}{\hat{X}^2}\frac{1}{n(n-1)}SM_i^2(\bar{y}_i-\hat{R})^2$$

In the present example, it works out as

$$\frac{159\times 159}{(7{,}052)^2}\times\frac{1}{20\times 19}\times 69{,}432.300=0.092888$$

which is about 4 percent larger than the variance obtained by adding the between-psu and the within-psu components.

5.4 PRIMARY SAMPLING UNITS SELECTED WITH REPLACEMENT

Quite often the primary sampling units are large and vary considerably in size. In a population survey of rural areas, for example, the village may be used as a psu but there are big villages and there are small villages. One method of controlling the variation due to the size of the village is to select the villages with probability proportionate to some measure of size as obtained from a previous census. The selection may be done with replacement in order to make the analysis simple. The primary sampling units are subsampled by making use of the maps or lists available. The following examples illustrate the estimation procedure.

EXAMPLE 5.6

Total number of cattle A district consists of 53 villages from which 14 villages are selected with replacement with probability p_i proportional to area. Within each selected village a list is made of the farms M_i from which a random sample of m_i farms is taken, the sampling fraction being 1/4. The total number of cattle on the farms in the sample is obtained by interview. The purpose is to estimate the total number of cattle in the district. The following information is obtained (see Table 5.7).

Each of the entries in column 8 of Table 5.7 is an estimate of the total number of cattle in the district. Thus their average is taken to find a better estimate of the total number of cattle, Y. Hence $\hat{Y}=397{,}631/14=28{,}402$. In order to estimate the variance, we first calculate the sum of the squares of the deviations of the entries in column 8 from their mean. This is given by (1, p. 120)

$$S\left(\frac{M_i y_i}{p_i m_i}\right)^2-\frac{1}{14}\left(S\frac{M_i y_i}{p_i m_i}\right)^2=1{,}523{,}512{,}313$$

Table 5.7 Data from a two-stage agricultural sample

Village	p_i	*Number of farms* M_i	m_i	*Number of cattle* y_i	M_i/p_i	$M_i/p_i m_i$	$M_i y_i/p_i m_i$	$(M_i/p_i)\hat{R}$
(1)	(2)	(3)	(4)	(5)	(6)	(7)	(8)	(9)
1	0.0026	19	5	14	7,308	1,461.6	20,462	83,019
2	0.0098	23	5	82	2,347	469.4	38,491	26,662
3	0.0146	31	8	207	2,123	265.4	54,938	24,117
4	0.0167	40	10	124	2,395	239.5	29,698	27,207
5	0.0187	54	13	113	2,888	222.2	25,109	32,808
6	0.0187	54	13	113	2,888	222.2	25,109	32,808
7	0.0220	39	10	114	1,773	177.3	20,212	20,141
8	0.0249	55	14	242	2,209	157.8	38,188	25,094
9	0.0258	46	12	203	1,783	148.6	30,166	20,255
10	0.0298	83	20	256	2,785	139.2	35,635	31,638
11	0.0362	74	19	272	2,044	107.6	29,267	23,220
12	0.0370	70	17	131	1,892	111.3	14,580	21,493
13	0.0465	60	15	208	1,290	86.0	17,888	14,654
14	0.0465	60	15	208	1,290	86.0	17,888	14,654
Total					35,015		397,631	

from which the variance estimate is calculated as

$$\frac{1{,}523{,}512{,}313}{14 \times 13} = 8{,}370{,}946$$

The standard error is 2,893 and the coefficient of variation is 10.2 percent. A sample of 14 villages and 176 farms has provided an estimate of the total number of cattle correct to about 20 percent.

If the purpose is to estimate the total number of farms in the district, the entries in column 6 will be used. Their average is

$$\bar{X} = \frac{1}{n} S \frac{M_i}{p_i} = \frac{35{,}015}{14} = 2{,}501$$

and the sum of squares of the deviations from the mean is 28,308,829. This gives the variance estimator as

$$\frac{28{,}308{,}829}{14 \times 13} = 155{,}543$$

leading to a standard error of about 394 for the estimated number of farms, the coefficient of variation being 15.8 percent.

Furthermore, the following calculations are needed for estimating the average number of cattle per farm.

$$\frac{\hat{Y}}{\hat{X}} = \frac{28{,}402}{2{,}501} = 11.356 \qquad S\left[\frac{M_i}{p_i}(\bar{y}_i - \hat{R})\right]^2 = 5{,}515{,}408{,}609$$

This gives the estimate as 11.36 cattle per farm, the variance estimate being

$$\frac{5,515,408,609}{14 \times 13(2,501)^2} = 4.8448$$

leading to a standard error of 2.2 cattle and a coefficient of variation of 19.4 percent.

Remark The estimate of the total number of farms in the district is poor. This is the reason that the estimate of the average number of cattle per farm is not precise.

Remark The calculation of standard errors is simple when this method is used. This method is a version of the well-known method of interpenetrating subsamples.

EXAMPLE 5.7

How many towns in the sample? Consider the problem of estimating the average expenditure per household per week on durable goods for towns with census population lying between 4,000 and 10,000 persons in a certain country. The range of expenditure is 20 to 40 drachmas† from town to town while within towns this range is much wider and may be taken as 0 to 100 drachmas. The data available for the 10 towns in the population are given in Table 5.8. We have

$$M_0 = \sum M_i = 16,000 \qquad \sum M_i R_i = 506,800$$

$$R = \frac{506,800}{16,000} = 31.675 \qquad \sum M_i(R_i - R)^2 = 412,972$$

Table 5.8 Size of psu and average expenditure per household

Town	*Number of households* M_i	*Average expenditure* R_i	*Town*	*Number of households* M_i	*Average expenditure* R_i
1	1,000	20	6	1,600	30
2	1,150	25	7	1,750	33
3	1,300	28	8	2,000	35
4	1,450	30	9	2,500	40
5	1,650	32	10	1,600	31

We shall assume that the variability of household expenditure within towns is the same, that is, $S_w^2 = 300$. If a sample of n towns is selected with replacement with probability proportionate to M_i and m households are investigated from a selected town, it can be shown that the variance of the estimate (of expenditure per household) is (1, p. 128)

$$\frac{1}{nM_0}\left[412,972 + 300\left(\frac{M_0}{m} - 10\right)\right]$$

of which the first part is the between-town contribution and the second the within-town contribution. Table 5.9 gives comparative figures on variances for selected values of n and m, using a sample of 100 households in each case. We note that the precision of the

† 30 drachmas = US\$1.

Table 5.9 Variance for different number of towns in the sample

Number of towns in sample n	*Number of households per town* m	*Variance within towns*	*Variance between towns*	*Total variance*
1	100	2.812	25.811	28.623
2	50	2.906	12.905	15.811
3	33	2.938	8.604	11.542
4	25	2.953	6.453	9.406
5	20	2.963	5.162	8.125

estimate is considerably increased by taking more towns in the sample and fewer households from each town. This is so although the towns differ moderately from each other and households within towns differ markedly. This is the reason that, within the allowed budget, an attempt is always made to spread the sample over as many primary sampling units as possible.

5.5 THE SELF-WEIGHTING SYSTEM

Considerable simplification in the analysis of the data can be achieved when the sample selected is made self-weighting. This is done by adjusting the size of the sample at the last stage of selection in such a manner that the estimate can be made by multiplying the total value of y in the sample by a known constant such as 100 or 200. To give an illustration, consider the population of Example 5.6. The estimator of the total number of cattle in the district is

$$\hat{Y} = \frac{1}{n} S \frac{M_i}{p_i} \frac{y_i}{m_i}$$

when m_i farms are selected from the M_i in the village which is selected with probability p_i proportional to area. If the overall sampling fraction is 1/100, the numbers m_i can be chosen such that

$$\frac{1}{14} S \frac{M_i}{p_i} \frac{y_i}{m_i} = 100\, S y_i$$

This gives

$$m_i = \frac{M_i}{1{,}400\, p_i}$$

Thus the number of farms to be selected from the first village (see Table 5.7) is $19/(1{,}400 \times 0.0026) = 5$ and that from village 6 is $54/(1{,}400 \times 0.0187) = 2$. If the last-stage units are chosen in this manner, no weights are needed for making the estimates. It can also be shown that, under certain conditions, this is an efficient procedure (1, p. 130).

We shall consider a very common situation in which the population is stratified and two primary sampling units are selected from each stratum. The selection is made without replacement with probability proportionate to some measure of size. Depending upon the material available, the primary units are further subsampled. It is possible to choose the number of last-stage units such that the final estimate is self-weighting. In a two-stage design, for example, let p_i' be the probability that the primary unit U_i is selected in the sample from a stratum and let m_i be the number of households to be selected from the M_i in the psu. If the overall sampling fraction is 1/200, the m_i are determined from the identity

$$S \frac{M_i}{p_i'} \frac{Sy_{ij}}{m_i} = 200 \; SSy_{ij}$$

which gives

$$m_i = \frac{M_i}{200 \, p_i'}$$

For example, the number of households to be selected is 8 when there are 160 households in the psu and the probability of selection p_i' is 0.1. As the following examples show, the analysis of the data becomes simple when the self-weighting procedure is used.

EXAMPLE 5.8

Total number of persons There are 1,000 villages in a country. The villages are allocated to 20 strata on the basis of population. From each stratum two villages are selected with probabilities proportionate to size (based on census population) and without replacement. Each village is divided up into blocks from which a random sample of five blocks is taken. Lists of households are made in the sample blocks. From each block a certain number of households m_i is selected at random from which information is collected on the number of persons and the number unemployed. The number of households m_i to be selected from a block is so determined that the final estimate of the total for the characteristic y is given by $\hat{Y} = 100$ (total of y in the sample). The following information is available from the survey by primary sampling unit (see Table 5.10).

We have

$$\sum (y_1 + y_2) = 4{,}190 \qquad \sum (y_1 - y_2)^2 = 76{,}380$$

$$\sum (z_1 + z_2) = 19{,}131 \qquad \sum (z_1 - z_2)^2 = 1{,}889{,}989$$

$$\sum (w_1 + w_2) = 981 \qquad \sum (w_1 - w_2)^2 = 13{,}459$$

Since the overall sampling fraction in a stratum is 1/100, that for a single primary sampling unit is 1/200. Thus the estimate of the total number of households in the area is

$$200 \times \tfrac{1}{2} \sum (y_1 + y_2) = 419{,}000$$

Its standard error is approximately given as

$$200 \times \tfrac{1}{2}[\sum (y_1 - y_2)^2]^{1/2} = 27{,}600$$

Table 5.10 Population data from a self-weighting sample

	Number of households		*Number of persons*		*Number unemployed*	
Stratum	y_1	y_2	z_1	z_2	w_1	w_2
1	105	60	517	302	26	06
2	70	120	366	583	17	33
3	123	95	499	383	42	26
4	187	115	350	480	19	20
5	63	130	303	638	08	57
6	115	60	542	285	33	04
7	180	88	902	357	31	19
8	78	171	301	843	05	38
9	110	90	486	473	17	32
10	95	101	457	430	38	07
11	63	93	317	440	17	29
12	102	54	476	303	24	23
13	130	115	617	402	58	04
14	121	143	434	704	19	25
15	57	180	300	875	25	56
16	173	45	813	208	47	12
17	26	65	145	326	08	19
18	79	38	419	201	25	17
19	148	112	706	548	37	13
20	130	160	613	787	37	08
Total	2,155	2,035	9,563	9,568	533	448

which leads to a relative error of 6.59 percent. Similarly, the total number of persons is estimated as

$$200 \times \tfrac{1}{2} \sum (z_1 + z_2) = 1{,}913{,}100$$

with a standard error of

$$200 \times \tfrac{1}{2}[\sum (z_1 - z_2)^2]^{1/2} = 137{,}500$$

and a coefficient of variation of 7.19 percent. Furthermore, the number unemployed is

$$100 \sum (w_1 + w_2) = 98{,}100$$

and its standard error is

$$100[\sum (w_1 - w_2)^2]^{1/2} = 11{,}600$$

the relative error being 11.82 percent.

Remark The analysis of data from a self-weighting sample is exceedingly simple when two primary sampling units are selected from a stratum.

EXAMPLE 5.9

Average size of a household We shall now estimate the average size of a household from the self-weighting sample of Example 5.8. Table 5.11 is made for this purpose.

Table 5.11 Calculation of sampling errors

Stratum	$y_1 - y_2$	$z_1 - z_2$	$w_1 - w_2$	$z_1 - z_2 - r_1(y_1 - y_2)$	$w_1 - w_2 - r_2(z_1 - z_2)$
(1)	(2)	(3)	(4)	(5)	(6)
1	45	215	20	9.530	8.970
2	−50	−217	−16	11.300	−4.868
3	28	116	16	−11.848	10.049
4	72	−130	−1	−458.752	5.669
5	−67	−335	−49	−29.078	−31.814
6	55	257	29	5.870	15.816
7	92	545	12	124.928	−15.959
8	−93	−542	−33	−117.362	−5.195
9	20	13	−15	−78.320	−15.667
10	−6	27	31	54.396	29.615
11	−30	−123	−12	13.980	−5.690
12	48	173	1	−46.168	−7.875
13	15	215	54	146.510	42.970
14	−22	−270	−6	−169.548	7.851
15	−123	−575	−31	−13.382	−1.502
16	128	605	35	20.552	3.963
17	−39	−181	−11	−2.926	−1.735
18	41	218	8	30.794	−3.183
19	36	158	24	−6.376	15.895
20	−30	−174	29	−37.020	37.926

The average size of a household is obtained by finding the ratio of the total number of persons and the number of households. This works out as $r_1 = 19{,}131/4{,}190 = 4.566$. In order to estimate its variance, we first calculate

$$\sum [(z_1 - z_2) - r_1(y_1 - y_2)]^2 = 305{,}674.12$$

its square root being 552.8. The standard error then is

$$\frac{200 \times \frac{1}{2}\,(552.8)}{200 \times \frac{1}{2}\,(4{,}190)} = 0.1319$$

the relative error being about 2.9 percent. This shows that the household size has been estimated quite precisely.

EXAMPLE 5.10

Proportion unemployed The estimation of a proportion from a self-weighting multistage sample follows the same lines as the estimation of a mean. Both are ratio estimates. From the data of Table 5.10 we can estimate the proportion unemployed as

$$r_2 = \frac{981}{19{,}131} = 0.0513$$

giving an unemployment rate of about 5 percent. The additional calculations needed for estimating its variance are given in column 6 of Table 5.11. We have

$$\sum [(w_1 - w_2) - r_2(z_1 - z_2)]^2 = 6{,}628.32$$

its square root being 81.4. Thus the standard error of the proportion unemployed is

$$\frac{81.4}{19{,}131} = 0.00425$$

the relative error being about 8.3 percent.

5.6 THE METHOD OF COLLAPSED STRATA

Suppose the cost structure of a survey is such that only 50 primary sampling units can be taken into the sample. There is considerable auxiliary information available on the basis of which the population is divided up into 50 homogeneous strata. Thus only one psu is taken from each stratum. When the data have been collected, the estimate of the population total for the characteristic y is made by adding the strata estimates. An estimate of the variance can be made by pairing the strata and adding the squares of the differences within pairs. The pairing is to be done before the data are available. The estimate of the variance obtained in this manner will be an overestimate since its expected value contains the true differences between the strata forming the pair (1, p. 74). Thus pairing should be done by putting together strata which are believed to be about equal with respect to y. Example 5.11 illustrates the point.

EXAMPLE 5.11

Collapsed-stratum technique A population is divided up into 40 strata and one psu is taken in the sample from each stratum. The strata are allocated to 20 pairs on the basis of population as obtained from a previous census. Each primary sampling unit is subsampled to produce a 1 percent self-weighting sample of households from each stratum. The number of persons in the sample from each stratum is given in Table 5.12. The purpose is to estimate the total number of persons in the population.

Since the data are based on a 1 percent sample from each stratum, the estimate of the total number of persons is

$$100(4{,}027 + 3{,}763) = 779{,}000$$

Using the collapsed-stratum method, an estimate of the standard deviation is

$$100(46{,}268)^{1/2} = 21{,}510$$

giving a relative error of 2.76 percent.

Remark The sixth pair is unfortunate. But this type of pair can occur in any survey.

Table 5.12 Number of persons in the sample from each stratum

Pair number	Number of persons y_1	y_2	$(y_1 - y_2)^2$	Pair number	Number of persons y_1	y_2	$(y_1 - y_2)^2$
1	210	268	3,364	11	170	148	484
2	133	100	1,089	12	329	288	1,681
3	342	270	5,184	13	163	201	1,444
4	126	149	529	14	56	85	841
5	174	205	961	15	223	232	81
6	275	165	12,100	16	118	137	361
7	286	270	256	17	160	149	121
8	184	97	7,569	18	209	183	676
9	344	281	3,969	19	352	297	3,025
10	92	139	2,209	20	81	99	324
Total					4,027	3,763	46,268

PROBLEMS

5.1. All farms in an area are allocated to 53 clusters, from which a simple random sample of 14 clusters is taken. Information collected on the number of cattle in each cluster in the sample is given below.

Number of farms	*Number of cattle*	*Number of farms*	*Number of cattle*
64	784	32	351
55	914	18	316
83	906	30	287
48	793	24	284
66	598	48	359
40	489	30	393
25	401	70	516

The average number of farms per cluster in the population is 39.09. Find the average number of cattle per farm and its standard error. Compare the following estimators: (*a*) unweighted mean of means, (*b*) weighted mean of means, and (*c*) ratio estimate.

5.2. All farms in an area are allocated to 53 clusters, from which a sample of 14 clusters is selected with replacement with pp to area. Information is collected on the number of cattle on each cluster in the sample. The following data are obtained.

Chance of selection	*Number of cattle*	*Chance of selection*	*Number of cattle*
0.0249	914	0.0098	330
0.0465	1,124	0.0146	537
0.0167	489	0.0220	495
0.0362	975	0.0258	697
0.0187	574	0.0370	516
0.0298	906	0.0465	1,124
0.0026	66	0.0187	574

Find the average number of cattle per farm and its standard error.

5.3. The following method was used for estimating the area under a crop in a certain district containing 190 villages. The entire area was divided into 32,000 clusters of 8 contiguous plots each, and 13 villages were selected with replacement with pp to the number of clusters in the village. From each village four clusters were selected at random and were completely enumerated for the area under the crop. The data obtained are given below.

	Area under the crop (acres)			
Village	1	2	3	4
1	549	1,154	1,002	86
2	136	329	0	337
3	98	522	938	0
4	610	298	357	188
5	109	296	915	105
6	652	0	79	190
7	510	94	605	682
8	0	497	0	0
9	195	410	107	161
10	213	126	575	561
11	1,382	177	574	0
12	0	0	673	368
13	0	682	161	173

Estimate the total area under the crop and find its standard error. Find the number of villages required to be sampled in a future survey in order to estimate the area under the crop with a relative error of 5 percent with four clusters per village.

5.4. A volume containing 2,825 pages lists 50,000 persons whose biographies are given. A random sample of 20 pages is taken and 2 persons are selected at random on each page. The ages of the selected persons are noted, from which the total age of all persons listed on the selected page is estimated. The following estimates are obtained.

577.5	845.5	836.0	768.0	832.0
731.5	774.0	900.0	720.0	945.0
1,350.0	729.0	1,092.5	1,254.0	621.0
824.0	880.0	655.5	950.0	836.0

Estimate the average age of listed persons and find the standard error of the estimate.

5.5. It is proposed to estimate the average household expenditure per week on durable goods in an area containing 500 villages. A sample of n villages is to be selected with replacement with pp to the number of households in the village. From a selected village m households selected at random are to be asked to report expenditure on durable goods. A preliminary estimate of the mean is known to be 30 drachmas. The variance of the sample average is given by

$$V(\hat{\mu}) = \frac{128}{n} + \frac{1{,}500}{nm}$$

The cost function is known to be $350 = 2n + nm$. Find the values of n and m which minimize the variance for the specified cost. Calculate the minimum variance.

5.6. An area is divided up into blocks each containing about 30 dwelling units. A simple random sample of 200 blocks is to be selected for estimating the proportion of families with income exceeding $15,000 per annum. A preliminary estimate of this proportion is known to be 0.10, the intrablock correlation being 0.20. Find the expected relative standard error of the estimate for these two situations: (*a*) All dwelling units in selected blocks are enumerated; (*b*) a subsample of five dwelling units per block is enumerated.

5.7. A sample of 14 blocks is selected with pp to census population from an area. Households M_i within selected blocks are listed, from which a simple random sample of one in eight households m_i is selected to collect information on monthly expenditure y_i on vegetables. The data obtained are given below.

p_i	M_i	m_i	y_i	p_i	M_i	m_i	y_i
0.0440	78	10	114	0.0197	45	5	82
0.0291	63	8	207	0.0334	81	10	124
0.0596	166	20	256	0.0930	119	15	208
0.0930	119	15	208	0.0499	110	14	242
0.0052	38	5	14	0.0374	108	13	113
0.0741	141	17	131	0.0515	91	12	203
0.0374	108	13	113	0.0724	148	19	272

Obtain the per household monthly expenditure on vegetables and compute the standard error of the estimate.

5.8. The villages in a population are allocated to 10 strata and a field-to-field complete enumeration is made to find the total area under a particular crop. From each stratum two villages are selected with replacement with pp to the number of fields growing the crop. From a selected village a random sample of two fields growing the crop is taken and a plot of a given size and shape is put up at random in the field. The plot yields are given below. Using the unweighted mean in the stratum, find the yield per plot and its standard error.

Stratum	Area under the crop	Village 1		Village 2	
		Plot 1	Plot 2	Plot 1	Plot 2
1	2,850	32	247	125	41
2	375	60	72	54	68
3	1,458	31	145	167	111
4	6,643	205	171	59	110
5	7,214	43	237	206	180
6	2,920	167	113	178	193
7	2,767	92	100	349	169
8	1,583	81	69	87	99
9	3,504	365	379	330	320
10	1,451	86	182	141	185

5.9. In order to estimate the number of inhabitants y in a rural area consisting of 19 villages, the villages are allocated at random to 3 groups containing 6, 6, and 7 villages respectively. From each group a village is selected with pp to census population x. The probabilities of selection of the selected villages are 7,521/42,379, 7,393/42,174, and 6,131/46,372 in the three respective groups. The selected villages consist of 16, 18, and 16 enumeration areas (EAs). From each village two enumeration areas are selected at random and their present populations y are found to be

Village	1		2		3	
Population of selected pair of EAs	330	500	430	292	304	272

Estimate the total number of persons in the area and find the relative standard error of the estimate.

5.10. A city is divided up into 100 blocks from which 10 blocks are selected with replacement with pp to the number of households enumerated in a previous census. An expected sampling fraction of 5 percent is used for selecting a random sample of households from each block. The sample gives the following data on the number of rooms and the number of persons in them.

Block	1	2	3	4	5	6	7	8	9	10
Rooms	45	63	55	56	71	53	56	60	55	76
Persons	92	107	95	75	108	110	90	116	80	130

Estimate the number of persons per room and find the standard error of the estimate.

REFERENCES

1. Raj, D. (1968), "Sampling Theory," McGraw-Hill Book Company, New York.

6
Double Sampling and Errors in Surveys

6.1 DOUBLE SAMPLING

We have considered a number of methods of using auxiliary information by which the precision of the estimate can be increased. If auxiliary information is not available, it may be advantageous to conduct the inquiry in two phases. In the first phase auxiliary information is collected on the variate x on a fairly large sample assuming that the variate x is cheaper to enumerate. Then a subsample is taken and information collected on the variate of interest, y. The two samples are used in the best possible manner to produce an estimate which may be more precise per unit cost than the one based on a sample for y alone. How this can be achieved will be clear from the examples that follow.

EXAMPLE 6.1

Double sampling for difference estimation In order to estimate the total cultivated area in a commune containing $N = 850$ parcels of land, a random sample of $n' = 100$ parcels is selected and eye estimates x of cultivated area obtained by going from parcel to parcel. A subsample of $n = 30$ parcels is selected and each parcel measured for cultivated area y.

The data obtained are shown in Table 6.1. The sample mean $\bar{x}'$ of eye estimates on the 100 parcels was found to be 4.31 stremmas.†

Table 6.1 Eye estimates x and measurements y of area

Parcel	x	y	*Parcel*	x	y
1	2	2.0	16	4	4.9
2	1	1.4	17	4	4.1
3	2	1.3	18	2	2.0
4	10	9.5	19	1	1.0
5	7	6.7	20	5	8.5
6	3	3.0	21	3	3.0
7	4	4.6	22	2	1.7
8	4	3.6	23	4	5.9
9	2	4.0	24	7	6.6
10	7	7.4	25	1	1.3
11	1	0.8	26	1	1.4
12	2	2.1	27	13	12.0
13	2	2.3	28	4	4.0
14	5	6.8	29	3	3.0
15	2	2.5	30	10	10.2

Let u stand for $y - x$. We have

$$\bar{y} = 4.25 \qquad s_y^2 = \frac{S(y_i - \bar{y})^2}{n-1} = 9.027$$

$$\bar{x} = 3.93 \qquad s_u^2 = \frac{S(u_i - \bar{u})^2}{n-1} = 0.843$$

$$S(y_i - \bar{y})^2 = 261.79 \qquad \frac{n'}{N} = 0.118 \qquad \frac{n}{n'} = 0.30$$

The average cultivated area per parcel based on eye estimation is to be corrected for its bias. The bias involved is 3.93 − 4.25. Hence our estimate of the average cultivated area is (7, p. 140)

$$4.25 - 3.93 + 4.31 = 4.63 \text{ stremmas}$$

An unbiased estimate of its variance calculated from the sample is

$$\tfrac{1}{100}(1 - 0.118)\,9.027 + \tfrac{1}{30}(1 - 0.3)\,0.843$$

which amounts to 0.09929, giving a standard error of 0.315 and a coefficient of variation of 6.8 percent.

† 10 stremmas = 1 hectare.

Remark Suppose that it costs 10 cents to make an eye estimate of area of a parcel and \$1.00 to measure the area. The total cost of the survey by the double sampling method is \$40.00. If a single sample is used for area measurement, as many as 40 parcels can be measured for the same cost. In the latter case the sample estimate of the variance of $\bar{y}$ is

$$\tfrac{1}{40}(1 - \tfrac{4}{85})9.027 = 0.215055$$

which gives a standard error of 0.464 and a coefficient of variation of 10 percent. This explains the superiority of the double sampling technique over the single sampling method.

EXAMPLE 6.2

Double sampling for ratio estimation We shall now use the second-phase sample to estimate the ratio of y to x. This when multiplied by $\bar{x}'$ can be used as an estimate of the mean of y in the population. For the data of Example 6.1, this estimate works out as (7, p. 148)

$$\frac{4.25}{3.93} \times 4.31 = 4.66 \text{ stremmas}$$

Being a ratio estimate it is subject to bias. Only an approximate estimate of its mean-square error can be obtained from the sample. The following calculations are needed.

$$\hat{R} = \frac{4.25}{3.93} = 1.0814 \qquad \frac{n}{N} = 0.035 \qquad \frac{n'}{N} = 0.118$$

$$Sy_i^2 = 804.52 \qquad Sx_i^2 = 726 \qquad Sx_i y_i = 751.5$$

$$\frac{Sy_i^2 + \hat{R}^2 Sx_i^2 - 2\hat{R}Sx_i y_i}{n-1} = 0.9710 \qquad s_y^2 = 9.027$$

An estimate of the mean-square error is (7, p. 149)

$$\tfrac{1}{30}(1 - 0.035)\,0.9710 + \tfrac{1}{100}(1 - 0.118)(9.027 - 0.971)$$

which works out as 0.102291. This is about the same as the estimated variance of the difference estimator (Example 6.1).

EXAMPLE 6.3

Double sampling for regression estimation Another method of using the second-phase sample is to estimate the regression coefficient b of y on x. This regression coefficient is then used to form the estimator

$$\bar{y} - b(\bar{x} - \bar{x}')$$

Using the data of Example 6.1, we have

$$b = \frac{S(y_i - \bar{y})(x_i - \bar{x})}{S(x_i - \bar{x})^2} = \frac{250.425}{261.87} = 0.9563$$

Thus the sample estimate of the population mean is

$$4.25 - 0.9563(3.93 - 4.31) = 4.61$$

In order to estimate its variance, the following calculations are made.

$$r = \frac{S(y_i - \bar{y})(x_i - \bar{x})}{[S(x_i - \bar{x})^2\, S(y_i - \bar{y})^2]^{1/2}} = \frac{250.425}{(261.87 \times 261.79)^{1/2}} = 0.9564$$

$$s_x^2 = \frac{261.87}{29} = 9.03 \qquad b^2 s_x^2 = 8.2579$$

$$s_y^2(1 - r^2) = 0.77$$

An approximate value of the variance estimate then is (7, p. 150)

$$\frac{0.77}{30} + \frac{8.2579}{100} = 0.10825$$

It may be noted that the three methods of estimation—difference, ratio, and regression—have produced estimates of comparable precision in this case.

EXAMPLE 6.4

Double sampling for stratification The first-phase sample can also be used for stratification to obtain higher precision. Consider, for example, the $N = 210{,}000$ income tax returns filed by the totality of taxpayers in a country. In order to present advance estimates of income and tax to the government a first-phase sample of $n' = 42{,}000$ returns was selected at random. These returns were stratified on the basis of declared family income x. For each stratum a sample of returns was taken and the correct family income y was determined by the income tax authorities. The weights w_h, the means $\bar{y}_h$, and the sample variances s_h^2 for each stratum are given in Table 6.2. Family income is expressed in units of thousand drachmas.

Table 6.2 Data on family income from a two-phase sample

Family income (thousand drachmas)	w_h	s_h^2	$\bar{y}_h$	n_h
Under 20	0.167	4.2312	16.829	353
20–50	0.556	67.0597	32.923	2,357
50–100	0.200	183.3316	68.040	1,693
100–400	0.072	4,005.6241	158.330	3,041
400 or more	0.005	316,226.0000	746.547	217

An estimate of the average family income is given by (7, p. 151)

$$\bar{y} = \sum w_h \bar{y}_h = 50.637 \text{ thousand drachmas}$$

In order to estimate its variance, we calculate

$$\sum \frac{w_h^2 s_h^2}{n_h} = 0.05738$$

$$\frac{1}{n'}\left(1 - \frac{n'}{N}\right) \sum w_h(\bar{y}_h - \bar{y})^2 = \frac{1}{42{,}000} \times \frac{4}{5} \times 3{,}570.8286 = 0.06802$$

$$\frac{1}{n'}\left(1 - \frac{n'}{N}\right) \sum \frac{w_h s_h^2}{n_h} = \frac{1}{42{,}000} \times \frac{4}{5} \times 7.42063 = 0.00014$$

The estimate of the variance equals (7, p. 243)

$$\frac{n'}{n' - 1}(0.05738 + 0.06802 - 0.00014) = 0.12526$$

The standard error is 0.354 thousand drachmas and the coefficient of variation is 0.70 percent.

Remark Since the sample has not been selected directly from within strata, the variability between stratum means is a component of the variance of the estimate.

Remark Soon after presenting the advance estimates to government, all the 210,000 returns were corrected and stratified on the basis of declared income. The true weights w_h' were found to be 0.140, 0.577, 0.203, 0.076, and 0.004. If random samples of sizes n_h (Table 6.2) are taken from within strata, the estimate of the variance is

$$\sum w_h'^2 s_h^2 \frac{1}{n_h}\left(1 - \frac{n_h}{N_h}\right) = 0.037485$$

Thus the standard error based on a single-phase sample is 0.193, which is about one-half the standard error given by the two-phase sample used for preparing advance estimates.

6.2 SAMPLING OVER SEVERAL OCCASIONS

Sample surveys nowadays are not limited to one-time inquiries. The same population is enumerated at different points of time to estimate changes that are taking place or to estimate the average over the period. For example, the United States Current Population Survey (CPS) is carried out every month for collecting information on employment, unemployment, and allied topics. The Indian National Sample Survey (NSS) has been a continuing operation since 1950 and a great deal of information has been collected in the different rounds of the survey. The advantages of such continuing surveys are that they are less expensive than one-time inquiries and provide up-to-date information on trends. These surveys afford an opportunity to introduce improvements from time to time by making use of experience from earlier surveys.

A new feature of continuing surveys is the structure of the sample on each occasion. A number of alternatives can be considered.

Fixed sample The same sample is used on each occasion.
Independent sample A new sample is taken each time.
Partial replacement A part of the sample is retained, the remainder being replaced for the next occasion.

Generally there will be high positive correlation between observations on the same unit at two successive occasions. Thus it will be best to take an independent sample at each occasion if the purpose is to obtain the average over the period. If the main object is to estimate the change, it is best to use a fixed sample on both occasions.† For estimating the mean relating to the most recent occasion it is better to replace the sample partially.

† This is called the *panel* method. There are several types of panels, e.g., the consumer purchasing panel (Sec. 15.8), the consumer product testing panel (Sec. 20.4), the retail audit panel (Sec. 20.4), and the radio and television audience panel (Sec. 20.4).

Operationally, it may not be feasible to use the same sample every time. The units in the sample will resist being questioned every time their neighbors are always out of the sample. Second, the sample units are likely to change their character if they are required to participate in the survey indefinitely. Thus it would be better to spread the burden of response evenly over the population. In the Indian National Sample Survey the sample villages are retained for a few rounds but new households are selected every time. In the United States CPS the segments in the sample are rotated following a definite pattern (Sec. 14.12). The same procedure is used in the Canadian labor force survey.

What fraction of the sample should be replaced on each occasion will depend on the strength of association between observations on the same unit at two successive occasions. If the correlation is large, a smaller fraction should be retained. By and large, more than 50 percent of the units should be replaced on the first occasion for use at the second. This fraction should settle down to 1/2 on later occasions (7, p. 162).

6.3 SOME USEFUL RESULTS

Let a sample of size n be selected on each occasion from a large population with variance σ^2. We shall assume that the variance remains the same on each occasion. Let $n\lambda$ units be retained for the second occasion and $n\mu$ be selected anew. The means of the two samples may be denoted by $\bar{y}'$ and $\bar{y}''$ respectively. The corresponding sample means on the first occasion are $\bar{x}'$ and $\bar{x}''$. It can be proved (7, p. 156) that the best estimate of the mean M_2 on the second occasion is

$$\hat{M}_2 = \frac{1}{1 - \rho^2 \mu^2} [\lambda\mu\rho(\bar{x}'' - \bar{x}') + \lambda\bar{y}' + \mu(1 - \rho^2 \mu)\,\bar{y}'']$$

with a variance of

$$\frac{1 - \rho^2 \mu}{1 - \rho^2 \mu^2} \frac{\sigma^2}{n}$$

The estimate of the sum $\sum = M_1 + M_2$ is

$$\hat{\sum} = \frac{1}{1 + \rho\mu} [\mu(1 + \rho)\,(\bar{y}'' + \bar{x}'') + \lambda(\bar{y}' + \bar{x}')]$$

with a variance of

$$\frac{2(1 + \rho)}{(1 + \mu\rho)} \frac{\sigma^2}{n}$$

The estimate of the difference $\Delta = M_2 - M_1$ is

$$\hat{\Delta} = \frac{1}{1 - \mu\rho} [\mu(1 - \rho)\,(\bar{y}'' - \bar{x}'') + \lambda(\bar{y}' - \bar{x}')]$$

with a variance of

$$\frac{2(1-\rho)}{1-\mu\rho}\frac{\sigma^2}{n}$$

A numerical example follows.

EXAMPLE 6.5

Sampling on two occasions Fifty blocks are selected at random from a town containing a large number of blocks. Information is collected on the number of persons x in each block during a particular week. A subsample of 25 blocks is taken and information on the number of persons a year later, y, is collected. At the same time a fresh sample of 25 blocks is selected and the number of person in each block is determined. The data are given in Table 6.3. We shall use these data to estimate the average number of persons

Table 6.3 Number of persons per block on two occasions

x	y	x	y	x	y	x	y	x	y	x	y	x	y	x	y
12	...	60	...	40	...	18	20	36	39	...	25	...	14	...	31
24	...	15	...	31	...	40	45	8	11	...	16	...	19	...	15
25	...	11	...	26	...	30	32	31	34	...	25	...	15	...	29
17	...	27	...	42	...	26	29	16	21	...	42	...	47	...	30
14	...	46	...	6	...	9	10	38	40	...	33	...	33	...	16
12	...	17	...	9	11	12	16	40	43	...	18	...	17		
27	...	21	...	14	17	8	9	12	15	...	24	...	21		
21	...	18	...	18	18	23	20	18	20	...	31	...	36		
19	...	22	...	23	25	42	48	15	18	...	12	...	18		
30	...	13	...	24	23	27	30	27	31	...	22	...	41		

per block on each occasion, the average over the period, and the change in the average from the second to the first occasion. We have

$$\bar{x}'' = 23.84 \qquad \bar{x}' = 22.56$$

$$\bar{y}'' = 25.20 \qquad \bar{y}' = 25.04$$

$$S(x_i - \bar{x})^2 = 6{,}588 \qquad S(y_i - \bar{y})^2 = 5{,}429$$

$$s_x^2 = 134.449 \qquad r = 0.987$$

On the first occasion the estimate of the mean is

$$\hat{M}_1 = \frac{1{,}160}{50} = 23.2$$

and the estimate of its variance is

$$v(\hat{M}_1) = \frac{134.449}{50} = 2.689$$

when the finite population correction is ignored. We shall assume that the value of the correlation coefficient ρ has been found to be 0.95 on the basis of past experience. This gives

$$1-\rho^2\mu = 0.5488 \qquad 1-\rho^2\mu^2 = 0.7744 \qquad \lambda = \tfrac{1}{2} = \mu$$

Thus the best estimate of M_2, the mean at the second occasion, is

$$\hat{M}_2 = \frac{1}{0.7744}\left(\frac{0.95 \times 1.28}{4} + \frac{25.04}{2} + 0.2744 \times 25.20\right) = 25.489$$

and

$$v(\hat{M}_2) = \frac{0.5488}{0.7744} \frac{122.6224}{50} = 1.738$$

In case the first sample is not used for estimating the mean at the second occasion, we have

$$\hat{M}_2 = 25.12 \qquad v(\hat{M}_2) = \frac{1}{50} \frac{5{,}429}{49} = 2.216$$

Thus the nonuse of the first sample increases the variance estimate by about 28 percent.

The estimate for the first occasion can now be revised in the light of information on the second occasion. Making use of this information, the best estimate of M_1 is

$$\hat{M}_1 = \frac{1}{0.7744}\left(\frac{0.95}{4} \times 0.16 + \frac{22.56}{2} + 0.2744 \times 23.84\right) = 23.063$$

with a variance of 1.738. As an estimate of the sum $\sum = M_1 + M_2$ on the two occasions, we have

$$\hat{\sum} = \hat{M}_1 + \hat{M}_2 = 23.063 + 25.489 = 48.552$$

and the variance estimate is

$$\frac{2(1+\rho)}{1+\mu\rho} \frac{s^2}{n} = \frac{2 \times 1.95}{1.475} \frac{122.6224}{50} = 6.4844$$

Thus the average number of persons per block over the period is 24.276 with a variance of 1.6211. The best estimate of the change $\Delta = M_2 - M_1$ in the average number of persons per block is

$$\hat{\Delta} = \hat{M}_2 - \hat{M}_1 = 2.426$$

and an estimate of its variance is

$$\frac{2(1-\rho)}{1-\mu\rho} \frac{s^2}{n} = \frac{0.10}{0.525} \frac{122.6224}{50} = 0.4671$$

In case simple averages are used on each occasion, the estimate of the change Δ is

$$25.12 - 23.20 = 1.92$$

and an estimate of its variance is

$$2(1-\lambda\rho)\frac{s^2}{n} = 1.05 \times \frac{122.6224}{50} = 2.575$$

which is more than five times the variance when the proper estimator is used. This shows how important it is to use the more appropriate method.

6.4 NONSAMPLING ERRORS

Apart from sampling errors there are many other types of errors in the data collected from surveys. Such errors are called *nonsampling errors* and are present in censuses as well as in sample surveys. Take for example the information collected on the age of a person in a household survey. The person may not know his age exactly, especially when the birth was not registered. Even if the exact age is known, it may be reported carelessly or deliberately understated (as is the case with some young women) or overstated (as is the case with some old men).

This brings about response errors in the data. Second, the person may report his age correctly but the interviewer may record it wrongly or it may be coded incorrectly in the office or wrongly punched in the machine room. Furthermore, the person in the sample may not cooperate and may refuse to give information on his age or he may not be available at home at the time of the survey. This brings about nonresponse errors in the survey data. It should be remembered that no survey is perfect; there are errors in all surveys. A good survey differs from a bad one in two ways. First, the errors are under control. Second, the relative magnitudes of the nonsampling and sampling errors can be ascertained.

6.5 RESPONSE ERRORS

Suppose a person is asked about the number of cigarettes he smoked yesterday. Let the true number be y_j. We shall assume that the response x_{jt} obtained is a random variable, as if the person gives different answers with known probabilities (2). The average of the responses is $\bar{X}_j$ and the variance of the responses is σ_j^2. If at trial t the response is x_{jt}, the difference $x_{jt} - \bar{X}_j$ is the individual response deviation. Now suppose that a population contains N individuals. Associated with these N individuals we have the values $\bar{X}_1$, $\bar{X}_2, \ldots, \bar{X}_N$, which are the average individual responses to the question, "How many cigarettes did you smoke yesterday?" The mean of these responses, namely, $\sum \bar{X}_j/N = \bar{X}$, is called the *expected survey value* obtained under the essential conditions of the survey. (Different conditions will produce different values of $\bar{X}$.) The average of the true values y_j is $\sum Y_j/N = \bar{Y}$. If a random sample of n individuals is taken the sample mean Sx_{jt}/n will have an expected value of $\bar{X}$. The difference of the two, namely, $\bar{X} - \bar{Y}$, is called the response bias of the survey when the purpose is to estimate $\bar{Y}$, the average number of cigarettes smoked by the N individuals in the population. If we believe that the conditions are such that the response bias will be substantial, there is no point in going ahead with the survey.

The mean-square error of $\bar{x} = Sx_{jt}/n$ around the true mean $\bar{Y}$ will be given by $E(\bar{x} - \bar{Y})^2$. This can be expressed as (7, p. 180)

$$E(\bar{x} - \bar{Y})^2 = V\left[\frac{1}{n}S(x_{jt} - \bar{X}_j)\right] + V\left[\frac{1}{n}S(\bar{X}_j - \bar{X})\right]$$

$$+ 2\,\text{Cov}\left[\frac{1}{n}S(x_{jt} - \bar{X}_j), \frac{1}{n}S(\bar{X}_j - \bar{X})\right] + (\bar{X} - \bar{Y})^2$$

In the first term of this equation the deviations are taken from individual average responses while in the second term the deviations of individual mean responses are taken from the population average $\bar{X}$. The first term is called the response variance and the second the sampling variance. The third term reflects the correlations between response and sampling deviations. The last term is the square of the response bias. We thus note that the mean-square error of the sample mean is built up of the response variance, the sampling variance, the response bias, and correlations between sampling and response deviations. It is the total error which is to be made small and not simply the sampling variance.

6.6 THE ROLE OF INTERVIEWERS

Modern sample surveys are usually conducted through interviewers or investigators. The interviewers are supposed to put the question to the respondent and record the answer. The answer obtained depends on the manner the question is asked, the mood of the respondent, and so on. The answer recorded is the result of the interaction between the investigator and the respondent. Thus the average response $\bar{X}_j$ is determined both by the respondent and the interviewer and so is the individual response bias $\bar{X}_j - y_j$. The chief difficulty with interviewers is that they size up the respondent in the light of responses obtained during the early part of the interview and use this information for interpreting the answers given during the latter part of the conversation. Sometimes no questions are asked at all when the interviewer has made sure what type of answers to expect. The interviewer answers them by herself or himself later on. Second, the interviewer may not stick to the exact wording of the question and this may introduce an error. The answer obtained may be recorded wrongly or ambiguously or it may be deliberately distorted to suit the tastes of the interviewer. All this points to the fact that interviewers are a potential source of error in surveys.

6.7 GROSS AND NET ERRORS

It is true that many of the individual response deviations may be very large. But this does not mean that the response bias of the survey will necessarily be large. Some of these errors may be positive, others may be negative, and the response bias may be small. This is expected to be the case when the over- or underreporting is not deliberate. In case the errors are mostly systematic in the same direction, there may be appreciable response bias. The present author undertook a study of response errors in the field of agriculture (6). Five communes were selected and all land in the communes was measured along with the name of the agricultural holder. Then the farmers in the communes were asked how many parcels of land they operated within these communes and the area operated by them. These reports were compared with figures obtained from the ground survey. Table 6.4 gives information on gross response errors. It will be found that about 65 percent of the response errors on the number of parcels were in the negative direction and only 22 percent were in the positive direction. It was different in the case of areas where 48 percent of the errors were negative and 45 percent were positive. These are gross errors, some of which were very large. The net effect of these errors was the following. The average number of parcels reported per farmer was found to be 6.6 while the true figure was 8.7, giving a response bias of –2.1 parcels. With respect to areas, the average area operated per farmer was reported to be 37.8 stremmas and the true figure was 42.2, giving a response bias of –4.4 stremmas.

Table 6.4 Gross response errors in an agricultural survey

Number of parcels			*Area (stremmas)*		
Response deviation	*Percent of reports*	*Cumulated percentage*	*Response deviation*	*Percent of reports*	*Cumulated percentage*
–11 or less	3	3	–21 or less	8	8
–10 to –6	9	12	–20 to –11	10	18
–5 to –3	20	32	–10 to –6	11	29
–2 to –1	33	65	–5 to –1	19	48
0	13	78	0	7	55
1 to 2	12	90	1 to 5	19	74
3 to 5	16	96	6 to 10	7	81
6 to 10	3	99	11 to 20	10	91
11 or more	1	100	21 or more	9	100

6.8 CONTROL OF RESPONSE ERRORS

Confronted with these response errors the survey statistician must take positive steps to see that they are kept down to a minimum. Some of these steps are outlined below.

TRAINING OF INTERVIEWERS

Since the interviewers are a potential source of error, great care is taken in recruiting the right type of persons: those who do not have strong opinions on the purposes of the survey and those who are prepared to follow instructions. When recruited, they are trained thoroughly in the purposes of the survey and the methods of measurement. The definitions of the terms used are explained and the procedures of investigation are standardized. Their work is supervised to make sure that they adhere to the instructions given.

CONSISTENCY CHECKS

Certain details are introduced in the questionnaire which furnish information on the consistency of the data collected. For example, the respondent may be asked how old he is and later on his date of birth may be asked. In an expenditure survey, additional information on income may be collected to determine whether the data on expenditure are reasonably consistent. At the same time household assets may be recorded to get further confirmation of his status.

REINTERVIEWS

The supervisory staff of the survey may be required to take a sample of the respondents already interviewed by the investigator and collect new information using perhaps superior techniques. A comparison of the two sets of figures will throw light on the response errors involved. This method is being used in the United States Current Population Survey. If the number of reinterviews is fairly large, parallel estimates may be made of some of the key items in the survey. This will give information on the response bias of the survey.

RECORD CHECKS

In addition to collecting information by interview, checks may be made against records when they are available. For example, a person's age may be compared with his birth record available at the municipal office; his declared income may be compared with the entry shown in the employer's payroll, and so on. Although such checks are difficult to make, it is worth trying them from time to time to make sure that the errors are under control.

6.9 THE METHOD OF INTERPENETRATING SUBSAMPLES

It has been stated that the data collected by interviewers cannot be treated as an independent random sample of responses. The reason is that there are correlations within interviewer assignments. In such a situation the method of interpenetrating subsamples (4) can be used with advantage. In this method the total sample is divided up at random into m groups and an interviewer is assigned at random to each group. Since the m subsamples are random samples from the same population, the interviewer means should agree apart from fluctuations of sampling. If they do not, the survey is not under statistical control. When they do, the average of the m interviewer means can be used as an estimate of the population mean and an unbiased estimate of the variance can be obtained from the sample itself. The following example illustrates the procedure.

EXAMPLE 6.6

Interpenetrating subsamples A random sample of $n = 120$ households was selected from a large rural area. This sample was divided up into 12 equal subsamples and an interviewer was assigned at random to a subsample, each interviewer collecting information

Table 6.5 Value of livestock per household by subsample

Subsample i	*Value of livestock owned* x_{ij}										*Mean* $\bar{x}_i$
1	721	64	664	134	546	610	29	278	432	510	398.8
2	850	393	856	415	372	965	175	263	378	595	526.2
3	480	370	284	294	119	980	652	581	836	224	482.0
4	627	976	169	764	697	109	557	275	69	385	462.8
5	569	704	265	748	498	308	917	612	205	318	514.4
6	812	305	271	100	76	385	617	302	105	107	308.0
7	393	426	841	665	724	308	132	405	552	427	487.3
8	1,205	785	428	227	385	186	590	486	772	390	545.4
9	112	560	395	654	290	512	860	365	431	385	456.4
10	369	785	655	947	332	842	249	371	361	207	511.8
11	458	1,126	577	1,062	663	814	908	197	542	203	655.0
12	273	428	310	295	178	1,080	178	291	543	382	395.8

on the value of livestock owned by 10 households under his charge. The data obtained are shown in Table 6.5. We have

$$m = 12 \qquad n = 120 \qquad \bar{n} = 10 \qquad SSx_{ij} = 57{,}439 \qquad SSx_{ij}^2 = 36{,}035{,}173$$

Total sum of squares $\quad SS(x_{ij} - \bar{x})^2 = 8{,}541{,}517$

Sum of squares between subsamples $\quad S\bar{n}(\bar{x}_i - \bar{x})^2 = 833{,}868$

Sum of squares within subsamples $\quad SS(x_{ij} - \bar{x}_i)^2 = 7{,}707{,}649$

The analysis of variance is shown in Table 6.6.

Table 6.6 Analysis of variance

Source of variation	*Degrees of freedom*	*Sum of squares*	*Mean square*	F	$F_{0.05}$
Between interviewers	11	833,868	$s_b^2 = 75{,}806$	1.06	1.87
Within interviewers	108	7,707,649	$s_w^2 = 71{,}367$		
Total	119	8,541,517			

The analysis of variance shows that there is no significant difference between the means of the interviewers. Apart from systematic errors, the survey is under statistical control. As an estimate of the value of livestock per household, we have

$$\hat{\mu} = \bar{x} = \tfrac{1}{12}S\bar{x}_i = \frac{57{,}439}{120} = 478.7$$

The variance estimate is

$$v(\bar{x}) = \frac{S(\bar{x}_i - \bar{x})^2}{m(m-1)} = \frac{833{,}868}{120 \times 11} = 631.7182$$

and the standard error is 25.13.

Remark It is possible to find from the sample the correlation within the work of the interviewers. The sample estimate of the covariance is (7, p. 172)

$$\frac{s_b^2 - s_w^2}{\bar{n}} = 443.9$$

and the intraclass correlation coefficient is

$$r = \frac{443.9}{443.9 + 71{,}367} = 0.0062$$

Remark If the work of some interviewers is found to be very discrepant and the survey director is satisfied that they have not adhered to instructions, the data supplied by them may be disregarded.

Remark There are several other uses of this method. If several questionnaires are to be tried to find the best one, each questionnaire may be directed to a subsample and tabulated by subsample. If the object is to find out whether highly educated persons make better interviewers than less educated ones, the two sets of interviewers may be allocated to randomly selected subsamples and the differences studied.

Remark The method of interpenetrating samples has its limitations. If there is a constant bias common to all interviewers, it cannot be detected by this technique. Second, interviewer differences cannot be spotted unless they are really very large.

6.10 BIAS REDUCTION BY DOUBLE SAMPLING

The method of double sampling can be employed for reducing the response bias in survey results. The main survey is conducted by using measurement techniques which are inexpensive and not very intensive. A subsample is taken on which more intensive and accurate measurements are made. The subsample is then used for diminishing the bias of the results obtained from the main sample. As an example, consider a survey of establishments in which the purpose is to estimate the total number of paid employees. Questionnaires can be mailed to a sample of establishments requesting information on employment during a certain period. True figures on employment are then collected from a subsample of the establishments by asking interviewers to examine the records of the establishments in the subsample. If we assume that the establishments cooperate and the interviewers can get at the truth, it is possible to use this method to reduce the response bias to zero and to estimate the response variance at the same time. The following example illustrates the method.

EXAMPLE 6.7

Estimation of response variance From a directory listing 3,500 large manufacturing establishments a random sample of $n' = 158$ establishments was taken and questionnaires mailed to obtain information on the number of paid employees x. This gave an average of $\bar{x}' = 46.99$ employees per establishment. A random subsample of $n = 30$ establishments was taken and interviewers were sent to examine the payrolls of these establishments and

Table 6.7 Reported x and true y figures on employment

x	y	x	y	x	y	x	y	x	y	x	y
35	41	121	125	28	33	69	65	27	31	28	33
61	69	26	31	56	65	22	26	47	52	52	51
139	135	23	32	50	50	57	63	36	34	71	76
30	28	35	34	20	18	21	20	60	60	23	25
25	30	44	47	95	109	46	57	65	65	24	24

obtain more accurate data on employment y. Table 6.7 presents the figures for x and y on the subsample. We have

$$Sy = 1{,}529 \qquad Sx = 1{,}436 \qquad d_i = y_i - x_i$$

$$\bar{y} = 50.97 \qquad \bar{x} = 47.87 \qquad Sx^2 = 93{,}388$$

$$s_y^2 = 870.6551 \qquad s_d^2 = 20.1621 \qquad Sxy = 97{,}846$$

The sample estimate of μ, the average employment per establishment, is (8)

$$\hat{\mu} = 50.97 - 47.87 + 46.99 = 50.09$$

This estimate is free of response bias in as much as the records are accurate. The estimate of the variance of $\hat{\mu}$ is built up of two components. One is the pure sampling variance given by

$$\left(\frac{1}{n'} - \frac{1}{N}\right) s_y^2 = \left(\frac{1}{158} - \frac{1}{3{,}500}\right) 870.6551 = 5.2617$$

The other is the response variance represented by

$$\tfrac{1}{30}(1 - \tfrac{30}{158})\, 20.1621 = 0.5445$$

The total variance is 5.806, of which 9.4 percent is the response variance. We can now choose n' and n suitably for a future survey to minimize the total variance for a given cost function. The estimate of the total variance is given by

$$\left(\frac{1}{n'} - \frac{1}{N}\right) s_y^2 + \left(\frac{1}{n} - \frac{1}{n'}\right) s_d^2$$

6.11 THE PROBLEM OF NONRESPONSE

Another source of error in surveys is noncontact or refusals. The selected family may not be available at home when the interviewer calls. The selected person may refuse to cooperate, saying that he has no time to answer questions or that he considers the purposes of the survey to be senseless. Persuasion and further recalls are therefore necessary for achieving complete coverage of the sample. But it is expensive to call and call again. At the same time we cannot afford to neglect the nonrespondents. Any results based on the respondents alone will not apply to the entire population from which the sample was selected. Experience from different surveys shows that the non-respondents generally differ from the respondents in several respects. Neglecting them will introduce a bias in the results. Under these circumstances one solution is to take a small subsample of the nonrespondents and use all the persuasion, ingenuity, and other resources at our command to get a response from them. The two samples can then be combined suitably to get a better estimate of the population parameter (7, p. 79).

EXAMPLE 6.8

Subsampling of nonrespondents The purpose is to estimate the average wages of skilled workers in manufacturing establishments of all types. A random sample of $n = 300$ establishments is selected from the directory listing $N = 1{,}000$ establishments. A questionnaire is mailed to get information on the number of skilled workers on the payroll last month and the total wages paid to them. After three reminders it is found that $n_1 = 185$ establishments have responded, giving an average wage of 850 drachmas. A random subsample of $u = 30$ is selected from the $n_2 = 115$ nonrespondents and visits are made to collect the information. The following calculations are made on the two samples.

$$\bar{y}_1 = 850 \qquad \frac{S(y_i - \bar{y}_1)^2}{184} = 1{,}403{,}600 \qquad w_1 = 0.62$$

$$\bar{y}_2 = 610 \qquad \frac{S(y_i - \bar{y}_2)^2}{29} = 1{,}568{,}200 \qquad w_2 = 0.38$$

The quantities w_1 and w_2 are the estimated relative weights of the strata of respondents and nonrespondents respectively. As an estimate of the population mean μ we take

$$\hat{\mu} = \frac{185 \times 850 + 115 \times 610}{300} = 758$$

Note the multiplier 115 in place of $u = 30$. To proceed further, we shall first estimate $S_y{}^2$, the variance of y in the population. A good approximation is provided by

$$\sum w_h(\bar{y}_h - \bar{y})^2 + \sum w_h s_h{}^2 = 1{,}479{,}719$$

If there had been no nonresponse, the estimate of the variance is

$$\frac{1}{300}\left(1 - \frac{300}{1{,}000}\right) 1{,}479{,}719 = 3{,}452.68$$

To this should be added the quantity

$$\frac{(115/30) - 1}{300} 0.38 \times 1{,}568{,}200 = 5{,}621.47$$

which is the price paid for collecting information from a fraction only of the nonrespondents. The sum of the two components is an estimate of the variance of $\hat{\mu}$. We have

$$v(\hat{\mu}) = 9{,}074.15$$

which gives a coefficient of variation of 12.56 percent.

6.12 WEIGHTING BY AT-HOME PROBABILITIES

In many household surveys the major cause of nonresponse is nonavailability of the head of the household at home when the interviewer has called. If recalling is expensive, it is possible to bring into the sample the not-at-homes without call-backs by using a technique due to Politz and Simmons (5). With this method the interviews are generally held in the evenings when most people are likely to be at home. The interviewer calls only once on the persons in the sample. If the person is available at home, information on y is collected. At the same time a further question is asked whether the person was at home at the same time yesterday, the day before, and so on. This question generally relates to the previous five days. If the person answers "yes" for four days, an estimate p of his availability at home is 5/6. The value obtained for y is then divided by p to adjust for those not available. No information is collected from those who are away when the interviewer has called. They are assigned a value of zero for the character y. The following example illustrates the working of the method.

EXAMPLE 6.9

Weighting by at-home probabilities A random sample of $n = 225$ males is selected from a very large list of persons residing in a city. These persons are visited at their addresses to find out what proportion of the males work for a living. Only one call is made on each person. If a person is available, the at-home probability is estimated by asking whether the person was at home during the previous five days. The following data are collected (Table 6.8). It may be noted that 22 persons were not found at home, from whom no data

Table 6.8 Number working with at-home probabilities

Estimated at-home probability p_i	*Number of males*	*Number working*
1/6	12	1
2/6	20	3
3/6	27	5
4/6	35	10
5/6	48	20
6/6	61	28
Total	203	67

were collected. Denoting y by 1 if the person is working and zero otherwise, the estimate of the population proportion is given by (7, p. 184)

$$\hat{P} = \frac{1}{n} S z_i \qquad z_i = \frac{y_i}{p_i}$$

We have

$$Sz_i = 92 \qquad Sz_i^2 = 162.3$$

$$S(z_i - \bar{z})^2 = 124.682$$

The sample estimate of the proportion working is

$$\hat{P} = \tfrac{92}{225} = 0.41$$

and its variance estimate is

$$v(\hat{P}) = \frac{124.682}{225 \times 224} = 0.002474$$

giving a standard error of 0.05.

Remark In case the at-home probabilities are not used and the estimate is based on respondents alone, the proportion works out as 67/203 = 0.33, which is a serious underestimate of P.

PROBLEMS

6.1. In an investigation in five communes of Greece, all the 617 farmers were asked to report areas y of land operated by them in the commune of residence. These figures were checked against measurements x made from a survey on the ground. Defining $e = y - x$ the following results were obtained.

$$\mu_y = 37.8 \qquad \sigma_x^2 = 680.0$$

$$\mu_x = 42.2 \qquad \sigma_e^2 = 507.0$$

$$\sigma_y^2 = 920.0 \qquad \rho(e, x) = -0.23$$

Simple random samples of sizes 30, 60, and 120 are drawn from this population. Find (*a*) the variance of $\bar{y}$, (*b*) the mean-square error of $\bar{y}$, and (*c*) the probability that the interval $\bar{y} \pm 2\sigma_{\bar{y}}$ will fail to include the true mean μ_x. Comment on your results.

6.2. In an establishment survey, questionnaires are to be mailed to a random sample of n plants. A subsample of the nonrespondents is to be interviewed for the collection of information. The response rate is expected to be 50 percent. The cost of mailing, processing, and personal interview per unit are \$0.10, \$0.40, and \$4.10 respectively. The desired precision is that which a simple random sample of 1,000 plants will give if there is no nonresponse. Determine the best initial sample size and the subsampling rate for nonrespondents. Find the expected cost of the survey.

6.3. A simple random sample of 100 plants is allocated at random to five investigators A, B, C, D, and E, each collecting information on employment from 20 plants. The following results were obtained.

A	3	4	8	11	2	9	21	16	28	14
	11	12	5	9	3	4	16	19	7	8
B	1	3	6	10	1	12	22	14	16	6
	6	5	16	14	3	2	7	3	10	8
C	17	30	18	25	12	4	13	10	6	7
	10	9	20	18	6	5	12	7	14	15
D	2	6	10	7	3	4	12	25	18	16
	7	9	22	14	6	5	12	3	15	8
E	12	15	3	8	5	5	12	18	9	6
	3	5	6	7	1	10	25	17	30	12

Estimate the average size of an establishment and obtain the standard error of the estimate. Do the interviewers differ from one another? Find the intra-interviewer correlation coefficient and the part of the variance attributable to interviewer biases.

6.4. A 1 percent simple random sample of persons is taken from an area to find the proportion of those who caught a cold last month. The number found at home was 400 out of an initial sample of 450. Those found at home were asked the number of nights they were at home during the previous five nights. The numbers who stated that they were at home on 0, 1, . . . , 5 of the five previous nights and the number catching a cold were as follows.

	0/5	1/5	2/5	3/5	4/5	5/5
Total number	16	35	60	72	97	120
Those who caught cold	5	14	25	34	50	62

Estimate the proportion of persons who caught a cold last month and find the standard error of the estimate. Compare it with the estimate based on respondents alone.

6.5. A random sample of 100 households is taken to estimate the proportion of households needing financial assistance. The following procedure is used for collecting information on this personal question. The head of the household is given a spinner with a face marked so that the spinner points to the letter A (assistance needed) with probability 0.8 and to $\bar{A}$ (assistance not needed) with probability 0.2. The respondent is required to spin the spinner unobserved by the interviewer and report only whether or not the spinner points to the letter representing the group to which he belongs. In all, 30 households said that the spinner placed them in the correct group. Estimate the proportion of households needing financial assistance and compute the standard error of the estimate.

6.6. It is proposed to determine the volume of timber in a forest by making eye estimates x of a randomly selected sample of plots of 0.5 acre. Actual measurements y are made on a subsample of plots. The average volume per plot is estimated by making use of the difference estimate. The correlation coefficient of eye estimates and measurements is expected to be 0.9. The cost of collecting information from a plot is $5 for eye estimation and $40 for measurement. The total cost sanctioned for the survey is $9,000. Determine the number of plots to be used for eye estimation and the number to be measured. Compare it with the procedure in which a direct sample of plots is measured and no eye estimates are made.

6.7. A simple random sample of 300 stores taken from a long list gives $18,134 as the average sales for January, the sample variance being $s_y^2 = 1{,}242{,}367 \times 10^3$. An independent sample of 300 stores taken the next month gives $19,568 as the average sales for February and $1{,}250{,}453 \times 10^3$ as the sample variance. The same sample provides information on January sales, the sample mean being $17,874 and the sample variance $1{,}241{,}942 \times 10^3$. The month-to-month correlation of sales is known to be 0.92. Find the best estimate of the average sales for the month of February and compute the standard error of the estimate.

6.8. In order to estimate the proportion of rented dwelling units in an area, a random sample of 368 dwelling units was stratified according to the migration status of the occupants. In all, 287 dwellings were found to be occupied by natives and 81 by immigrants. A random subsample of 74 native and 18 immigrant occupants was taken. In the subsample, 43 natives and 14 immigrants were found to be renting the dwelling in which they lived. Estimate the proportion of rented dwellings in the area and find the standard error of the estimate.

6.9. There is a population of 11 towns divided up into 2 strata. Stratum 1 contains six towns of which B, C, and F are coastal and A, D, and E inland. In the second stratum there are four inland towns, namely, a, b, c, and e, and the coastal town d. One town is to be selected from stratum 1 with pp to 15, 10, 20, 10, 20, and 25 respectively and another from stratum 2 with pp to 15, 30, 10, 25, and 20 respectively. It is desired to maximize the probability of getting both types of towns in the sample. Find a scheme of selection that achieves this.

6.10. A simple random sample of 75,308 farms is selected from a population containing 1,200,000 farms and the area x of each farm copied down from records. This gives an average of 31.25 acres per farm. From this sample a random subsample of 2,055 farms is taken and information collected on the number of cattle y on each farm. The subsample gives

$$Sx = 62{,}989 \qquad Sy = 25{,}751$$

$$s_y^2 = 133.4470 \qquad s_x^2 = 490.4300$$

$$r = 0.679758 \qquad b = 0.354585$$

Find the regression estimate of the average number of cattle per farm and compute the standard error of the estimate.

REFERENCES

1. Cochran, W. G. (1963), "Sampling Techniques," John Wiley & Sons, Inc. New York.
2. Hansen, M. H., et al. (1951), Response Errors in Surveys, *J. Amer. Statist. Assoc.*, **46**.
3. Hansen, M. H., et al. (1953), "Sample Survey Methods and Theory," vol. 1, John Wiley & Sons, Inc., New York.
4. Mahalanobis, P. C. (1946), Recent Experiments in Statistical Sampling in the Indian Statistical Institute, *J. Roy. Statist. Soc.*, **109**.
5. Politz, A. N., and W. R. Simmons (1949, 1950), An Attempt to Get the "Not at Homes" into the Sample without Call-backs, *J. Am. Statist. Assoc.*, **44** and **45**.
6. Raj, D. (1965), Response Bias in Farmers' Reports, *Sankhya*, (B) **27**.
7. Raj, D. (1968), "Sampling Theory," McGraw-Hill Book Company, New York.
8. Raj, D. (1970), A Method of Reducing Response Bias and Estimating Response Variance, *Sankhya* (in press).

part three

Planning and Execution of Surveys

7
Planning of Surveys: Initial Steps

7.1 INTRODUCTION

We now turn to another phase of sample survey design, namely, the planning and preparations that precede the actual operation of the survey. This is an extremely important task since the quality of the survey results depends considerably on the preparations made before the survey is conducted. The amount of planning needed varies greatly with the type of material available and the nature of the information sought. The discussion presented is most relevant to surveys of human populations, economic institutions, and agriculture.

The designing of a survey is in many ways similar to designing a house. The architect starts with a general idea of what the client wants. He then selects his materials and prepares the layout of the rooms and utilities in a manner that will suit the needs of a prospective occupant. In doing this he has an eye on the size and shape of the plot and the amount of money that the client is willing to spend. The architect will try to envisage the various hazards to which the occupant might by subjected and take positive steps to mitigate them. The designer of a survey has to do the same.

7.2 OBJECTIVES OF THE SURVEY

The first step in planning a survey is to lay down its objectives in the clearest possible terms. There must be a need for carrying out the inquiry and it is important to know just what is wanted; after all, the success of the survey is going to depend on whether or not the survey has met this need. The objectives of the survey must be spelled out clearly along with the manner in which the results are going to be used.

Experience shows that the sponsoring agency itself often does not know precisely what it wants or how it is going to use the results. For example, it is not uncommon to be told that the purpose of the survey is to find how agricultural laborers live without spelling out exactly who is an agricultural laborer, what is meant by living, whether income and expenditure are involved, or whether the number of rooms and the facilities provided are to be considered. In such a case it is the job of the statistician to hold a dialogue with the client and produce the detailed specifications of the survey.

Sometimes the objectives of the survey can be put down in simple terms. Consider, for example, the criteria proposed for a sample survey of jute in Bengal (1).

> The reliability of the sample survey must be such that the margin of error of the final estimate of the area under jute should not exceed 5%; secondly, the results must be available sufficiently early in the jute season and preferably by the first or second week of September; and finally, the cost of the sample survey should not be excessive.

The statement of goals and objectives becomes easier if the survey is meant to answer one single question. Modern sample surveys are, however, multipurpose inquiries; there are a number of potential uses for which information is wanted. In that case one may make a good list of the various purposes and assign relative weights to them. This is necessary since some purposes may come in conflict with others. The survey is then designed keeping in view the relative weights assigned. We must know what important estimates are to be made from the survey, how much accuracy is needed, and when the results are to be delivered. These are the specifications of the survey to which the survey designer must try to conform within the budget earmarked.

7.3 THE POPULATION TO BE STUDIED

The statement of the purposes of the survey should ordinarily indicate the population to be sampled. For example, a survey of couple fertility may be restricted to only those couples who have been married for at least 10 years and where both husband and wife are alive at the time of the survey, if the objectives of the survey can be met by considering these couples only. Similarly a survey of opinions on a public issue may be restricted to persons who have passed their twenty-first birthday. For a survey of the jute crop, the population may be the totality of plots, big or small, growing jute during a particular season. In a survey of manufacturing establishments the population may be the totality of establishments operating in the country during a certain period.

These may be called the target populations, the populations intended to be covered by the survey. We may, however, have to restrict these populations somewhat for a variety of reasons. In a population survey, for example, members of the armed forces may have to be disregarded. This is a special class about which information is kept confidential in most countries. Again, this survey may have to be restricted to only those households living at a settled address. The reason is that it is very expensive to list people living in temporary huts, riverboats, etc., and other people who frequently move about from one place to another. If the purposes of the survey can be adequately met even when such people have been excluded, it is best to disregard them. Then there are people in institutions, such as hotels and prisons, who present a special sampling problem. Either they are studied separately in a different manner or they are to be ignored for purposes of the survey. In this manner the population actually sampled becomes a little different from the target population.

Similarly an agricultural survey of farmers may be limited to those whose main profession is farming or those whose agricultural holding is of a certain minimum size. A survey of all parcels growing potatoes may exclude kitchen gardens since these are insignificant, numerous, and difficult to list. An industrial survey may have to be limited to those establishments employing 10 or more workers. The smaller establishments are hard to locate and such establishments may not keep records. In all these cases the grounds for exclusion of certain marginal categories of the population are feasibility and convenience, provided the objectives of the survey can still be met by their exclusion.

When the sampled population differs from the target population, the results of the survey will apply to the sampled population only. Sometimes, information is collected in a different manner from the omitted sector. This is done through inexpensive procedures which are not entirely rigorous but which can throw some light on the subject matter of the survey (3). The users of the data are provided with figures relating to both parts of the universe, along with a description of the limitations under which they were collected.

There is another point to be considered. Statistics collected by countries for their own use are also needed by the international organizations such as the United Nations for presenting a world picture. The recommendations of the United Nations may include certain categories of the population which are of minor importance to the country in question. Instead of redefining the coverage of the population to suit the needs of the United Nations, the country may use its own definition of the population for the survey and provide some kind of information for the part not covered.

It should be pointed out that the definition of the population to be enumerated cannot be considered in isolation from the type of frame available for the conduct of the survey. The two should be considered together (see Sec. 8.1).

If it is found that the population so defined is going to contain units which are out of scope, some means must be found by which such units can be identified and disregarded at the time of analysis. For example, it may be desired to limit a labor force survey to persons 14 or more years old. But it is feared that a number of persons will be missed if the household is asked to name only those members who are 14 years or older. In that case a complete list of all members of the selected households may be made along with the age of the person at the time of the survey. Based on the figures for age, the out-of-scope persons can be identified and thrown out at the analysis stage.

7.4 THE DATA TO BE COLLECTED

It should be possible from the purposes of the survey to derive a fairly broad list of items that could provide information bearing on the problems under investigation. This list should be supplemented by other items that are correlated with the main items and can throw additional light on related questions. For example, in a survey of general attitudes, one might collect information on related items such as marital status, number of children, religion, occupation, political affiliations, and so on. When all the items have been assembled, the practicability of obtaining the information on them should be considered. A number of items can be discarded at this stage or replaced by others which can be measured. Only items relevant to the purposes of the survey should be retained. In this process care should be exercised to make sure that no important item is missing. This can be best done by preparing the blank tables that the survey should fill; this will eliminate irrelevant information and ensure that all essential items find a place.

Consider, for example, a demographic survey conducted in an African country in which one of the aims was to calculate the crude birth and death rates in the country. Every household was asked whether a child had been born alive during the past 12 months and whether any death had occurred during the same period. Analysis of the data showed that too few deaths had been reported and the same applied to births. It was not found possible to resolve this question even partially by tabulating the births by the month of occurrence and seeing if there was any downward trend in the figures which might explain the low birth rate obtained. Nor could the rates be examined by sex. The reason was that certain vital items had not been included in the list, namely, the month of birth and the sex of the child born. These items would not have been omitted if the organization had had experience in handling this type of data and had visualized all the tabulations to be made from the survey.

It should be pointed out that the list of items should not be too long. A long list overburdens the respondent, who may be unwilling to provide the information due to the fatigue involved. In case the list cannot be cut short there is a method of overcoming the difficulty with but a slight loss of efficiency. The questionnaire can be split into two parts. The basic items of information appear on each part but not the other items. Thus information on one group of items is collected from one set of respondents and on another group of items from another set. The loss of efficiency arises from the fact that certain items cannot be correlated such as those that do not appear in the same set (5).

Another point to be remembered is that in some surveys such as surveys of establishments it is important to know the record-keeping practices of the respondents. As far as possible those items of information should be asked which are available or which the establishment is accustomed to using for its own purposes. Frequently it is found that, although this is not exactly what is wanted, a minor modification may make it serve the purpose and data of a better quality can be obtained than if the respondent is asked the information which is difficult for him to supply without considerable work. Consider, for example, the problem of estimating total employment in a class of establishments. The establishments maintain its payroll records in a particular way. These records show the number of employees who were paid for the month, whether newly hired or separated. Although the aim may be to find out the number of employees at a particular date, the kind of count available is sufficiently precise for most purposes and it is much easier to obtain. We may then agree to collect information on those who were paid for the month, whether or not employed at a particular date.

7.5 METHODS OF COLLECTING INFORMATION

There are a number of methods of collecting information: mail questionnaire, personal interview, direct observation, or laboratory test. The decision as to which method is the best will be governed by the nature of information sought and the population under survey. A brief discussion of these methods is given below.

7.6 DIRECT OBSERVATION

Where feasible, this should ordinarily be the best method of collecting information as it is free from memory errors of respondents, exaggeration, and prestige effects. For example, the best way of knowing the number of wells in a village is to go through the village and count the number of wells. But direct observation has a limited role to play in the kind of information we usually collect. We cannot afford to observe a person for a month to find out how often he goes to the movies. Even if we decide to use this method, the person may change his habits for a while to create a certain impression.

The method of observation is useful in studying small communities to find out how people live, their attitudes, and their relationships with each other. Opinions differ as to whether the observer should be an active participant in the life of the community or not. In any case his success will depend on his skill and his personality. This is the one method acceptable to the sociologist for the kind of problems in which he is interested. But there are definite limitations here. The observer may not be objective in reporting what he sees; in spite of best efforts he may report what he infers and not what he perceives (2).

Market research is a field in which the method of direct observation has proved useful. In order to establish sales trends for certain brands, a random sample of retailers is selected and their stocks audited from time to time. The stock count necessitates a thorough personal examination of all parts of the shop premises, including windows, stockrooms, and all other places where stock is likely to be kept for retail sales. Each product is examined carefully to check brand, size, and price. Another application is pantry checks with samples of housewives to discover the brands bought over a period of time and changes in the brands purchased. Dustbins may be checked at the same time to record tins and wrappers in them.

7.7 MAIL QUESTIONNAIRES

This is usually the quickest and the least expensive of the methods of collecting information. A questionnaire is prepared on the subject matter of the survey and mailed to the respondents. The major cost is postage stamps, which should be quite small. If the respondents have the information and are interested in the purposes of the survey, the response can be quick. This method is particularly useful when you want to reach the members of an association who are usually spread over an entire country. And since the respondent provides information unaided by the interviewer, the well-known biases of the interviewers and their whimsical interpretation of the questions do not enter into the results obtained. Furthermore, there are situations in which a considered answer is needed, for example, when an establishment is asked the value of inventory or the cost of raw materials used. In this case the respondent needs time to consult documents to give an acceptable answer. A mailed questionnaire provides the opportunity for such consultations. And then there are questions which one does not like to answer in a face-to-face conversation but which one is prepared to answer through the mail. In all these cases the mailed questionnaire appears to be a convenient vehicle for collecting the information.

On the debit side, you cannot ask all sorts of questions through the mail. Unless the question is simple and straightforward, the respondent is likely to misunderstand the question and that brings about errors. You may try to explain the technical terms used at the back of the questionnaire, but this may not be noticed or may be found to be confusing. In an interview you have a chance to explain yourself again and again, thereby arriving at the final response of the respondent. In a mail survey the response has to be taken as final since ordinarily the respondent will not want to enter into regular correspondence with you on the subject. Sometimes, as in opinion surveys, you do not want the selected individual to discuss the subject with members of his family or friends and report the concensus of opinion. It is the person's own view you are interested in. There is no means of ensuring this in a mailed questionnaire. And then you may not be able to cross check the response by putting the same question differently in different parts of the questionnaire. The respondent can read through the completed questionnaire and correct any errors without divulging the truth.

The greatest drawback in the mailed questionnaire technique is the amount of nonresponse present in such surveys. When a questionnaire is mailed, many of the respondents may not care to complete and return it. The initial response rate may be barely 40 percent if you are lucky. Repeated reminders may push this rate up to 60 percent. What do you do with the remaining 40 percent? You cannot simply ignore them and base the results on the respondents alone. Experience of data collection in several fields shows that the nonrespondents often differ from the respondents in many respects. Their exclusion will introduce systematic errors in the results.

The devices used for persuading the nonrespondents take several forms. A gift may be offered, a postpaid envelope is enclosed, confidentiality of returns is assured, or an appeal is made explaining the purposes of the survey and the importance of the inquiry. This may add another 10 percent to the achieved sample.

One method of handling the remaining 30 percent nonrespondents is to take a small sample of them and send interviewers to collect the information. The interviewers use all the persuasion at their command to obtain a response. The responses obtained through the interview method are treated as a random sample from the 30 percent that had not cooperated. The two samples are combined suitably (see Example 6.8) to make the best estimates.

It may be mentioned that it is not essential to send the entire original questionnaire to the nonrespondents again. A brief version of it, incorporating the essential elements only, may be used. This can be an added inducement for better response. Actually the questionnaire may be shortened at each stage of nonresponse. The data can then be analyzed from the multiphase sample thus obtained.

7.8 INTERVIEWING

This is by far the most common method of measurement in household surveys, particularly in the developing countries. In these countries a sizeable proportion of the population is illiterate, very few records are kept of any kind, and many people cannot think in numerical terms. Any data collected without the help of the interviewer is probably worthless. The success of the survey depends heavily on the skill of the interviewer in eliciting worthwhile response.

The job of an interviewer or enumerator usually consists in traveling to the area selected, making a list of units (households in this case) by canvassing through the area, selecting a sample of households from the list following the instructions laid down, and knocking at the doors of the sample addresses to collect the information sought.

There are two ways of collecting the information from the units in the sample by the method of interview. In one method, a standard set of questions has been prepared and the interviewer is required to follow the same wording, put the questions to the respondent in a uniform manner, and record the answers. This may be called the *questionnaire* method. In the other method a list of items is recorded on the schedule and information is to be collected on these items. But the exact questions to be asked are not standardized. The manner of asking the questions and the probes to be used for eliciting acceptable response are left to the investigator to decide. This may be called the *recording schedule* method.

The kind of information usually collected these days is quite sophisticated; such information is needed for making plans for national development. Thus, items such as unemployment, gainful activity, and operator of an agricultural holding find a place on the list. The social and economic conditions prevailing in the developing countries are such that the respondent may have no clear-cut idea of what is meant by such terms. His understanding may be completely at variance with what is desired to be measured. A calendar of events may have to be presented to him to place him in the proper age-group. In such a situation the questionnaire method is not expected to work; it is the recording schedule which will have to be used to obtain something worthwhile. The intention is not to give the impression that the data collected through the recording schedule will be free of interviewer bias. Nothing could be further from the truth. When the questions are familiar to ordinary people and convey some standard meaning, it is admirable to use standard questions; when not, it is worthwhile to resort to the record schedule system.

At this point a distinction should be made between the reliability and validity of the measurement technique. If all interviewers get the same response on a certain question from a specified individual, the technique used is reliable in the sense that repeated measurements on the same unit tend to agree. But the response obtained will not be valid if it differs from the true value aimed at. When the questions convey some standard meaning to the respondent, the questionnaire technique will be reliable and may be valid. When not, it may be neither of the two. In the latter case the record schedule method should be able to dig deeper and produce more valid results. But then we require interviewers of a higher caliber, more skillful in the art of digging deeper, more tactful in handling difficult situations, and more knowledgeable in the subject matter of the survey. It is not just a matter of following written instructions; rather it is a matter of producing the right tools at the spur of the moment.

There is, however, a strong case for standardizing questions wherever it is practicable to do so. If the interviewers are allowed to ask questions as they like, the answers obtained will not be comparable. In the majority of the situations we need to find the sum of the responses for estimating a parameter of the population. When the responses are not comparable, it is difficult to interpret the sum obtained. Perhaps the assumption will have to be made that the questions have been asked in a uniform manner.

In surveys of the census type where information is collected on familiar items such as the name of the person, age, sex, and marital status, it should be possible to ask standard questions and use standard probes for eliciting the response. It is only in specialized surveys that one may have to decide to depart from the standard pattern. But once this type of inquiry has been conducted over several rounds on the same population, it should be possible to evolve some standard system of asking the questions, making use of the experience that has been gained from round to round. Thus it can be said that, except for one-time specialized surveys, the trend is always in the direction of standardized questionnaires.

It should be mentioned that it is not imperative to use only one of the methods of data collection in a given survey. A combination of the methods may be the best solution in some situations. Thus some items may be observed while others are enumerated by asking a set of questions. The questionnaire may be left with the respondent for self-enumeration; this is done to avoid the biases introduced by interviewers. Or a part of the information may be collected through an interview. In family budget surveys, for example, a standard procedure is to interview the household in the evenings and take down details of the expenditure incurred by the household as noted in the household account book. Indeed a variety of procedures may be used.†

† Another procedure is the telephone interview method used mainly for radio and television research in the United States (Sec. 20.4). In this method contact is established by telephone,

7.9 DESIGN OF FORMS

The problem of the design of forms—questionnaires or recording schedules—should be discussed as soon as the planning of the survey starts. This is one of the most important aspects of the survey. The layout will depend upon the type of form. If it is a questionnaire to be answered by the respondent unaided, the form should be attractive looking. The questions should be simple and clear. The number of questions should be reduced to the barest minimum. There should be proper space for recording the answers, especially when the "other" category is to be specified. In a recording schedule, attractiveness is not a major consideration; field handling is more important. In either case there should be proper space for entering the codes, say on the right-hand corner, for facilitating the job of the punch operators. If different parts of the form are to be answered by different categories of the population, this should be spelled out very clearly. Experience shows that it is very helpful to divide the form into three parts: one giving identification particulars of the respondent, the other giving classificatory information, and the third the topics of the survey.

7.10 QUESTION ORDER

Careful consideration should be given to the problem of the order in which the questions should appear on the form. In order to guard against confusion and misunderstanding questions should be arranged logically, one question leading to the next. Specific questions should always follow general ones. The opening question should be very interesting; this will ensure that the respondent cooperates in the survey. The respondent is likely to be put off if you start badly with a dull and difficult question. As far as possible the questions should admit of simple answers of the yes or no type. If there is any question which is likely to affect adversely the response on the succeeding questions, it should be relegated to the end. The first few questions should be so designed that the interviewer can establish rapport with the respondent. When this has been done, other questions of a more intimate nature such as the income of the respondent or his knowledge of a certain subject can be asked. By and large the sequence of the questions should be determined keeping in view the convenience of the interviewer who has to bear the main brunt of the job in the field.

7.11 QUESTION CONTENT

The next step is to decide what should go into the questionnaire keeping in mind the needs of the survey and the information to be collected. Some of

---

and respondents are asked a set of questions relevant to the study. When applicable, the method has the advantages of greater speed, lower cost, and wider geographic spread of the sample.

the questions will be factual such as the number of persons in the household or the expenditure for bread last week. Other questions may relate to opinions or attitudes such as whether yesterday's radio talk by the President was helpful or what the respondent thinks about widows marrying again.

The study director must make sure whether respondents are in a position to answer the questions being considered, whether they are sufficiently informed on the subject of investigation. We may ask a farmer in a developing country the value of his production last year. But can he answer this question reliably? Part of his production was used at home, another part went to the local money lender, and the balance was retained for seeds. He cannot figure out numerically the value of his total production. If such a question is to be used at all, the question should be subdivided into several subquestions to help the farmer make the right sort of accounting. With opinion questions it is all the more difficult to make sure whether the respondents can give reliable information. Whether informed or not, the respondent may decide to express some opinion in order to terminate the interview.

The second point to check is whether the respondents will be willing to answer the questions correctly. There is generally some resistance to answering questions of a confidential nature or questions that are embarrassing. Thus a question that is likely to offend may have to be omitted or modified suitably. And then there is the prestige bias. Households, for example, are known to overstate expenditure but understate income. Old people are prone to overstate their age and so on. Whenever such a bias is likely to arise, efforts should be made to check or corroborate the information collected by employing different methods of approach within the same questionnaire and by reference to outside sources wherever possible.

The question content should not be biased or loaded in one direction. The questionnaire should not unduly favor one side of the issue. The questions should not introduce unwarranted assumptions about the subject matter. Consider, for example, the following questions in a labor force survey: "Do you work for a living? If not, how do you manage to exist?" These questions seem to put the respondent on the defensive to explain why he does not work for a living, assuming that every decent person should work to earn a living. Surely these questions are loaded in one direction. A different sequence is needed.

7.12 QUESTION WORDING

The choice of the language used in expressing a question is of the greatest importance. It is too often assumed that the respondents must be aware of the concepts and definitions used in the questionnaire since these are obvious to the survey team. Actually the survey team is a group of specialists while the respondents are nonspecialist in that field. If the terminology is ambiguous, the respondents will have to use their own judgment and different persons will judge differently. This causes confusion and errors. Consider, for example, the apparently simple question, "Who are the members of this family?" In the first place the term "family" has to be defined. Does it mean a group of persons who are related by blood? Is a cousin or a nephew included? What is meant by membership? Are only those to be included who ordinarily live at this address or those who spent the previous night at this place? What about the son at school in the neighboring district, the daughter in the hospital, and the husband on assignment in a foreign country? To take another example, let us examine the following question asked in a labor force survey: "Are you at present employed, or unemployed, or outside the labor force?" Such a question allows the respondent to decide whether he should be classified as employed or not. But the respondent is not supposed to know the exact implications of the statistical concepts such as employment or labor force. The result is that the data obtained by asking this question can be considerably biased. Furthermore, it is difficult to interpret the classification obtained thereby because each respondent will have a different concept of what is meant by "employment."

The basic principle in good question wording is to use the simplest words that will convey the exact meaning. The meaning of the question becomes clear when the words used are well known and mean the same thing to everyone. The question "Do you operate land?" used in some agricultural surveys is poor. It is not clear whether the person is supposed to be a cultivator or an owner of the land. The question "At what time do you go to bed?" should be preferred to the question "At what time do you consign your limbs to sleep?" Again, consider the question "What kind of a house do you have?" The answer may be, beautiful or three-storied or cement-painted or commodious and so on.

Leading or loaded questions should be avoided. Such questions are slanted toward a particular kind of answer and therefore can mislead the respondent. If you ask a person, "Do you not work for a living?," the obvious implication is that normally everyone is supposed to work for a living and that the respondent is not expected to be an exception. The answer is very likely to be "Yes, of course I work for a living," although the respondent may be a housewife looking after the children. The same type of remarks apply to the question, "What brand of cigarettes do you smoke?" This question assumes that the respondent is a smoker and that the only further information needed is the brand of cigarettes. Similarly, questions of a hypothetical nature should be avoided. The answer to "Would you accept a better job, if offered one?" is obvious. No useful purpose will be served by tabulating answers to such questions. Almost everyone will say "yes."

Furthermore, the investigator will have to decide whether to ask a more personalized or less personalized question to produce better results. As an example, the choice may have to be made from among the following:

"Is the teaching load at this university considered to be excessive?"

"Do you feel that the teaching load at this university is excessive?"

"Are you satisfied or dissatisfied with the teaching load at this university?"

The answer to the last question will express the individuals' own feelings while the answer to the first the general feeling of teachers at the university.

7.13 QUESTION TYPES

With respect to the form of response, the questions may be divided into two categories. A closed or fixed-response question is one in which the responses are limited to stated alternatives. Examples are

Do you own a car? ☐ yes ☐ no

What is the marital status of this person? ☐ single ☐ married ☐ widowed ☐ divorced

The box actually checked is the answer to the closed type of question. On the other hand, an open question is one in which the respondent is free to decide the form the answer should take. The job of the interviewer is to write down the response as it is given. Examples of this type of question are

What is the total area of this holding?

Do you think the present government is doing a good job?

In the case of the closed or fixed-alternative questions, the responses are easy to analyze since they are automatically coded. And the answers are given in a frame of reference that is relevant to the purpose of the inquiry. This form should be preferred when the range of possible answers can be adequately visualized. But care must be taken to ensure that the list of alternatives is exhaustive. The usual practice is to make the list of alternatives end with the "other" category to make it exhaustive. The main disadvantage of this system is that answers may be forced into a category to which they do not properly belong. This happens when the respondent does not know the answer to the question and is forced by the interviewer to fall in line with one of the alternatives. The interviewer may not like to put the respondent in the "do not know" class. When the respondent wants to give a qualified answer, the interviewer may not accept it since there is no room on the questionnaire to record such an answer. Another disadvantage is that the respondents tend to follow the middle course when they are given a choice between a number of answers.

With the open-ended questions the respondent is given the opportunity to answer in his own terms and in his own frame of reference without being encumbered by a list of alternatives. The coding of these answers can be done later in the office in a standardized manner. Experience, however, shows that it is not always easy to code such answers. Sometimes a reference may have to be made to the respondent to find out the exact code to be used.

To summarize, closed questions should be preferred when the possible alternatives are limited, known, and clear-cut. This is so when you are dealing with factual information such as age, marital status, or level of education. When the issue is complex or the relevant dimensions are unknown (as with many questions relating to opinions or attitudes), the open type of question may be used with advantage. Indeed, one may begin with an open type of questionnaire and gradually develop a closed type of questionnaire when greater information is available regarding the form of response.

7.14 EXAMPLES

We shall now present examples of a questionnaire and of a schedule. In order to save space, just a section of the form is exhibited. The questionnaire relates to a survey of health and a public opinion survey, and the schedule to a demographic survey.

Individual Questionnaire (Health Survey)

1. Were you sick at any time during the last 14 days?

   ☐ yes ☐ no

2. If yes: Did you talk to a doctor about your illness?

   ☐ yes ☐ no

3. If yes: What did the doctor say the illness was?__________________
4. If no: What was the matter (nature of illness)?__________________
5. When did it start?__________________
6. Are you now free from all symptoms relating to the illness?

   ☐ yes ☐ no

7. If yes: When did you consider yourself free from all symptoms?____________
8. Did this illness cause you to cut down on your usual activities?

   ☐ yes ☐ no

9. If yes: How many days of the last 14?__________________
10. Did you have any other illness during the last two weeks?

    ☐ yes ☐ no

Individual Questionnaire (Public Opinion Survey)

1. Do you think there is a real danger that atomic bombs will ever be used against the United States? Why do you feel that way?
2. Do you think the United States will be able to work out an effective defense against the atomic bomb before other nations could use it against us?

   ☐ yes ☐ no ☐ do not know

3. Which one of these statements do you most agree with?

   ☐ The most important job for the government is to make it certain that there are good opportunities for each person to go ahead on his own.

   ☐ The most important job for the government is to guarantee every person a decent and steady job and standard of living.

4. Do you think most people who are successful are successful because of ability, luck, pull, or the better opportunities they have had?

   ☐ ability ☐ luck ☐ pull

   ☐ better opportunities ☐ do not know

Recording Schedule (Demographic Survey)

Deaths during the last two years ended on________________

					Time of death							Recall period		
Serial number	Relation to head	Sex (male—1, female—2)	Age at death (years)	Age at death in weeks for "0" in column (4)	Date	Calendar month	Calendar year	How determined	Place of death	Attendance type	Relation of deceased to informant	Last year—1 Year before last—2	Weeks	
(1)	(2)	(3)	(4)	(5)	(6)	(7)	(8)	(9)	(10)	(11)	(12)	(13)	(14)	

REFERENCES

1. Mahalanobis, P. C. (1944), On Large Scale Sample Surveys, *Phil. Trans. Roy. Soc.*, (*B*)**231**.
2. Moser, C. A. (1958), "Survey Methods in Social Investigation." Heinemann, London.
3. Raj, D. (1968), "Sampling Theory," McGraw-Hill Book Company, New York.
4. Raj, D. (1964), "Handbook of Household Surveys," United Nations Statistical Office, New York.
5. Yates, F. (1960), "Sampling Methods for Censuses and Surveys," Charles Griffin & Company, Ltd., London.

8
Sampling from Imperfect Frames

8.1 INTRODUCTION

For the selection of the sample there must be a frame—a map or list or other acceptable description of the material—from which sampling units may be constructed and selected. The nature and accuracy of the frame available will determine the structure of the survey. We shall call a frame perfect if every element in the universe occurs once and only once on it separately, there are no other elements shown, and the information is accurate. Perfect frames are rare in actual sampling work. There are always some defects present. The frame may not contain some of the units in the population (incompleteness) or certain units may occur more than once (duplication). If some of the information contained in the frame is not correct, the frame is inaccurate. If it has not been kept up-to-date since it was first made, the frame may become obsolete or inadequate at the time it is to be used. It is obvious that the defects in the frame will introduce errors in the results. Thus a frame has to be very carefully examined before it is used. Procedures will have to be devised by which the defects present in the frame are not allowed to vitiate the results of the survey. Some of these methods will be considered in the subsequent sections.

8.2 TARGET AND FRAME POPULATIONS

In every inquiry there is the target population: population about which it is desired to collect the information. The target population, for example, may be the totality of dwelling units in a city at a given point of time. Suppose the dwelling units are $U_1, U_2, \ldots, U_N$. The purpose is to reach this population through a readily available frame. Let us assume that the frame was made two years back when all dwelling units in the city were listed at the time of the census. Thus the frame lists the dwelling units $U'_1, U'_2, \ldots, U'_M$. The basic problem is to make rules of association between the units in the frame and the units in the population. The rules should be such that the selection of a unit U'_i with a known probability from the frame leads to the selection of a unit U_j in the target population. That is, it should be possible to calculate the probability with which a reporting unit U_j in the target population is selected and this probability should be nonzero. If this criterion can be satisfied, the frame is serviceable (4, p. 218).

It should be pointed out at the outset that it is not a simple matter to produce such rules of association. The selection of a unit from the frame may not lead to any reporting unit in the population, as when the dwelling counted in the census has been demolished. Furthermore, a new story may have been added to a dwelling or the name of the street may have been changed and so on.

8.3 EXTRANEOUS UNITS

Suppose the frame contains a certain number of units which are not in the target population. For example, some of the establishments in the frame (based on an industrial census) may be no longer in business; a number of dwellings listed in the frame may have been demolished. When a sample is selected from the frame, some of these out-of-scope units may appear. But no great trouble is involved if we follow the rule that the value of y, the variate of interest, is zero for each of these units. Thus their selection in the sample will not vitiate the results. But the effective sample size will diminish somewhat, and the sampling error may increase (see Examples 3.8 and 3.9). Therefore, as far as possible such units should be eliminated before the sample is selected. But if the number of such units is believed to be small and if it is expensive to identify them in advance, one should not be much concerned about this aspect of the frame.

Sometimes a substitution is made for the extraneous unit in the sample by taking the unit next to it on the list. Although this procedure preserves the original sample size, the probability of selection of those units which happen to follow extraneous units is increased. Since the extraneous units are not known in advance, it is impossible to calculate the exact probability with which the units in the ultimate sample are selected. In the absence of information on these probabilities, the data collected cannot be analyzed properly. Therefore such procedures should not be used.

If an estimate of the proportion of extraneous units is available from a previous survey, the initial sample size can be increased correspondingly in order to arrive at the desired sample number at the end. If such an estimate is not available, a small pilot study may be conducted for the purpose. But it is high time to consider the possibility of revising the frame, if the number of superfluous units is known to be fairly large.

8.4 DUPLICATION

Sometimes two different units listed in the frame may lead to the same reporting unit in the target population. For example, the telephone directory may list the same hotel as Hilton Hotel and Hotel Hilton at two separate places. This increases the probability of selection for the duplicated units when a sample is selected. Sometimes these probabilities can be ascertained by scanning the list or by asking an additional question at the time of interview. We can then make use of these probabilities for analyzing the results. Suppose, for example, that a sample of establishments has been selected from the frame for getting information on companies (the target population). We can find from each sample establishment the company to which it belongs and the names of other establishments on the list which belong to the same company. The sample of establishments then leads to a sample of companies selected with probability proportionate to size (the number of establishments in a company). Such a sample can be analyzed without any difficulty.

Another procedure is to associate each reporting unit in the target population with only one entry in the frame. Thus in the case of the establishment sample, only those establishments will be recognized which are the seat of the company. Since a company has only one head office, the probability of selection is the same for each company. Similarly, a random sample of school children will give a random sample of school families if the family is selected when and only when the eldest child at school is in the sample.

In some of the methods discussed here, probabilities of selection have to be recalculated when duplicates are present in the frame. If it is considered undesirable to recalculate these probabilities, it is best to take steps to "unduplicate" the frame before selecting the sample. This is a tedious job especially when the number of units in the frame is large. One procedure is to transfer the frame to cards, there being one card for each unit listed. These cards are then sorted in various ways, e.g., by name, by address, etc., and duplicates are looked for among groups of adjacent cards and removed.

It should be mentioned that no substitutions need be made for the duplicates found in the sample. Nor is the problem solved by simply ignoring duplications in the sample.

8.5 INACCURACIES

Suppose there is a frame containing a list of blocks along with the number of persons enumerated in them at the time of the last census. The frame is to be used for carrying out a household survey in the area. For this purpose the blocks are stratified on the basis of census population and a sample of blocks is selected from each stratum. During the course of the survey it may be discovered that a block assigned, say, to the lowest stratum has in fact a very large population at the time of the survey. This is so because considerably new construction has taken place in the block since the last census. Thus the size of the block as given in the frame is entirely inaccurate. The question of whether this block should be retained in the lowest stratum or removed to the correct stratum arises. The answer is that the block should remain where it was selected. If this rule is followed, the results will be unbiased. The variance will, however, increase considerably since the strata made are not homogeneous. This situation can be avoided if such blocks with considerably new construction are identified beforehand. We may either assign these blocks to the proper strata before sample selection or sample them separately.

If it is a continuing survey, the problem of the occurrence of an unusually large or small unit in the stratum (due to inaccuracy of the frame) can be tackled more effectively. Suppose it is a monthly survey of retail trade based on a sample of establishments selected every month. We decide upon a suitable cutoff point y_0. All sample establishments which are found to be above this cutoff in all the months are assigned to a separate stratum (called the surprise stratum). Information for a particular month is collected on the establishments selected for that month as well as on the establishments in the surprise stratum. But the weight of the surprise stratum is correspondingly diminished to allow for the higher probability of their selection.

Sometimes it may be found that the frame available for use omits partially or completely a section of the population intended to be covered by the survey. In that case a supplementary frame will have to be constructed listing the units omitted by the main frame. For example, the frame available may be the list of blocks which were inhabited at the time of the last census. A sample taken from this frame will represent those blocks which were inhabited at the time of the census. To take in new construction in the uninhabited blocks, a supplementary frame listing all such blocks must be prepared and a sample of blocks selected from it. The two independent samples can be combined to provide estimates for the entire population.

There may be another type of inaccuracy in the frame. Suppose the purpose is to conduct a household survey. The frame available is a list of dwellings with their addresses. A random sample of addresses is taken from the frame. When we visit the sample addresses we may find that some of the addresses contain several dwellings and not just one as listed in the frame. The proper procedure to follow in this case is to collect information from all the dwellings at the sample address.

8.6 REDEFINITION OF TARGET POPULATION

Quite often we find that it is difficult to get at a class of units belonging to the target population. It may not be impossible to take them but it is believed that their inclusion will raise the cost of the survey substantially and may even delay it. In that case we may agree to redefine the target population by excluding this class of units, provided that the purposes of the survey can still be adequately served. For example, it may be discovered during the planning stage of a population survey that it is extremely expensive to take in the transient population—persons with no fixed address. The reason may be that different methods are needed to cover this part of the population and these methods are expensive. In that case the inquiry may be limited to settled persons only—those who can be reached by listing households in the selected areas. The results of the inquiry will, of course, apply to the sampled population only. Some idea, however, may be given of the magnitude of the part of the population excluded from the survey.

8.7 SAMPLING FROM LISTS

For many types of inquiry the frame available is a list of the units into which the population can be divided. Such lists are, however, rarely up-to-date. But this does not mean that these lists cannot be used as a frame for the selection of the sample. When judiciously used, these lists help in reducing the cost of the survey. One great advantage of sampling from a list is that there is considerable choice in determining the size of the cluster. We may take clusters of two units or of three units and so on. Furthermore, the sample may be a simple random sample of clusters or a systematic sample of elementary units or of clusters. If the addresses of the units are shown on the list, one can conduct the main survey by mail, thus reducing the total cost of the survey considerably. At the same time the data obtained in this manner are supposed to be more accurate because the biases of enumerators do not come in (Sec. 7.7). A list sample is particularly efficient when the population is skew and the large units are all listed. By taking all these units or a large proportion of them in the sample, it is possible to reduce the variance of the estimate substantially (Example 4.5).

The main difficulty with lists is that they can be seriously deficient.

When this is so, the list will have to be supplemented by another frame which contains the units not on the list. Generally the supplementary frame is based on a list of geographic areas from which a sample of areas is selected. All units in the sample areas which are not on the list form the supplementary sample. This subject is considered in the next section.

8.8 AREA SAMPLING

We shall define an area sample as one in which the ultimate sampling units are land areas and the reporting units (that is, the units that furnish information) can only be identified by geographic rules associating them with land areas in the sample (2). By contrast a list sample is one that does not meet these conditions. For example, we may select a few towns in the sample and a few blocks from each town. Lists of addresses are made in each block from which a sample of addresses is taken. The households found at these addresses are asked to give the information wanted. This is an area sample. On the other hand, a sample of 50 trucks selected from a list of 1,000 trucks is a list sample. The same is true of a sample of factory workers picked up from the payrolls of a group of factories.

Suppose the purpose is to design an inquiry relating to retail trade in a province. A fairly complete list of large establishments (employing five or more persons) is available from a recent census. A sample of large establishments can be selected, using varying sampling fractions in the strata, from this list. Then a sample of land areas is taken and all small establishments found in these areas are taken in the sample. The first sample is a list sample while the second one is an area sample. The results of the two samples can then be pooled to provide figures for the totality of retail establishments in the province. This is an example of the joint use of list and area samples. In this method great care has to be taken to see that all those establishments which have a chance of selection through the list sample are excluded from the area sample.

8.9 THE PREDECESSOR-SUCCESSOR METHOD

Another method of taking into the sample units which are not on the list is the following. Suppose the various units in a population can be considered to be ordered geographically so that it is possible to determine unambiguously the successor of each unit. A simple random sample of units is selected from the list. For each unit in the sample we determine its successor to see if it is on the list. If the successor is on the list, discard it. If it is not on the list, include it in the sample; then identify its successor and discard it if it is on the list but include it in the sample if not on the list, and so on. If the original sample is, say, a 10 percent sample from the list, this procedure gives a 10 percent sample of units not on the list. The two samples put together give a 10 percent sample from the entire population. The success of this method depends on the fact that the units in the population can be ordered geographically.

8.10 FRAMES USED IN SAMPLING WORK

An account will now be given of the diverse sources of frames usually used in sampling surveys.

POPULATION AND HOUSING CENSUSES

Most countries of the world carry out a population census every 10 years. Quite often this census is preceded by a housing census in which all places of abode are listed locality by locality in the various subdivisions of the country. The two censuses provide a useful frame for sampling inquiries during the intercensal period. Such a frame will have to be brought up-to-date in order to take in the new construction that has taken place. This is generally rather easy to achieve in the rural areas. A greater amount of new construction takes place in the urban areas, especially at the periphery of the towns. One has to look for these areas and list the new houses that have come up. The dwelling appears to be the proper unit of listing in the frame and not the individual or the household occupying it at the time of the census. It is useful to have lists of dwellings by block in the towns and by enumeration district in the other areas. The usefulness of the census as a frame can be enhanced if a master sample of areas is selected from it at the time the census is taken. Further subsamples of dwellings may be taken from the master sample from time to time depending on the needs of the survey. The master sample could be transferred to cards and used profitably by government and other organizations long before the final census tabulations have been released. The selection of the master sample is important when we take into account the considerable time lag between the date of the census and the final publication of the results.

The trend nowadays is to carry out a sample survey in conjunction with the census by using the census as a frame. Thus basic facts such as the number of persons and their sex and age composition and marital status may be collected in the census while data of a more delicate nature such as occupation and employment status may be collected through a sample survey. The latter type of questions are asked of only a fraction of the total population, say 5 or 10 percent. In order to make the sample a success, the rules for the listing of households and individuals in the census must be so framed that there is no room for the census enumerator to use his discretion in determining the order of listing. Suppose, for example, that every tenth household is to be taken into the sample. If the rules of listing are not rigid, there is a tendency on the part of the enumerator to change the listing so that a desired household is automatically selected in the sample. In this way systematic errors are likely to creep into the results (Sec. 19.12).

TOWN PLANS

Even when a nationwide census has not taken place in a country, it is not uncommon to find maps for individual towns. These maps are generally prepared by municipal authorities or tourist organizations. The maps show the main streets, the by-lanes, the roads, and the built-up areas. It is possible to divide the city into blocks with the help of these maps. Where there is any ambiguity, the map can be corrected easily by visiting the area. If it is considered important to have an idea of the relative sizes of the blocks, one can visit the blocks and count the number of dwellings or make eye estimates of them. Such a frame listing the blocks along with their measures of size can be used for two-stage sampling of the town. The blocks are selected with probability proportional to the estimated number of dwellings. Household lists are made for the selected blocks from which a sample of households is taken. With this procedure some of the smaller blocks may have to be amalgamated with the neighboring ones before the sample is selected. This is done in order to have a reasonable number of households in the sample from the selected blocks.

LISTS OF VILLAGES

The rural areas present a more difficult problem in countries where no census has taken place. In this case the best thing to do is to make accurate lists of villages by district or subdistrict. At the same time local people may be encouraged to undertake administrative counts of the number of people in the village. The cooperation of the village headman is essential in such an undertaking. Where such counts are available, it is possible to make use of this information for improving the precision of sample surveys. For carrying out a household survey a sample of villages can be selected from which households may be taken after careful listing.

Some difficulty may be encountered in allocating the few scattered houses outside the villages listed. No problem is involved if the households living in these houses owe allegiance to a particular village; these households are appended to this village. If this system does not prevail, the village to which these houses are nearest may be said to contain them. However, the rules of association will have to be framed before the fieldwork begins (5).

DIRECTORY OF ECONOMIC INSTITUTIONS

In addition to the population census many countries take a census of non-agricultural establishments every 5 or 10 years. This census is a valuable frame for surveys of establishments—industrial and distributive. If such a census has not been taken, one has to fall back on registration lists issued by government departments, tax records, etc. Although these lists do not usually provide all information needed for the directory, at least the larger establishments are expected to find a place on the list. The smaller establishments can be taken from an area sample (Sec. 8.8).

For current use these directories have to be kept up-to-date. Establishments that go out of business need to be eliminated and those starting new business must be added. Because of the high turnover involved, this type of activity must continue all the time. The administrative lists should be examined periodically to take in the new establishments.

There are advantages in gathering data on distribution and industry at the same time. Certain establishments carrying on mixed kinds of activity do not belong exclusively to the one or the other sector. A combination of the two inquiries facilitates the collection of data from such multiactivity enterprises.

FRAMES FOR AGRICULTURAL INQUIRIES

Broadly speaking there are two types of data collected in agricultural surveys. In the first type are included items such as the population dependent on agriculture and the distribution of holdings by ownership. Examples of the other type are areas under crops and crop yields. The frames provided by population and housing censuses should be adequate for collecting information on the first type since the reporting unit is a household in these inquiries. Thus this type of information on agriculture may be usefully collected along with a population survey. Another advantage of using this system is that errors of coverage and of classification can be reduced.

In the underdeveloped countries where a large proportion of farmers are illiterate and have no numerical idea of the areas operated by them, information on areas and yields can be better collected by taking a sample of areas rather than of households. For this purpose it is best to have a frame based on cadastral maps of villages or other similar areas. Countries not possessing such maps should be encouraged to make long-term plans for instituting cadastral surveys and prepare maps of areas.

The problems become more difficult when cadastral maps are not available. If a list of villages is available along with some auxiliary information such as cultivated area or population, a sample of villages may be selected and all fields associated with the sample villages may be enumerated. A rough sketch of the village showing the boundaries of the fields should help in making sure that no parts of the village have been missed.

Sometimes it is discovered that aerial photographs exist for the major part of the country although no cadastral surveys have taken place. In that case the possibility of using such photographs as a frame for agricultural surveys should be explored. It should be borne in mind that it is not easy to read an air photograph; trained persons are needed for this job. If the photographs can be enlarged to the extent that major roads, streams, and other landmarks shown can be identified, it should be possible to select areas of a reasonable size on the basis of these photographs and identify them on the ground. Enumeration of these areas can provide valuable statistics of land utilization.

8.11 EXAMPLES OF SAMPLING FROM FRAMES

In this section we shall present some actual problems of sampling from frames and show that their solution depends on the type of frame available for use.

SAMPLE OF MANUFACTURING ESTABLISHMENTS

The purpose is to take a sample of manufacturing establishments from an area in order to estimate the total employment, the production, and the cost of raw materials, etc. A characteristic of such populations is that a small proportion of the units accounts for a very large proportion of the total (employment, production, etc.). That is, the distribution of the population is skew with respect to the variable under consideration. In such a situation it is extremely hazardous to base the results on a simple random sample of all establishments. The proper plan is to stratify the establishments by some measure of size and take disproportionate samples from within strata. The stratum containing the largest establishments may be taken with certainty; the sampling fraction in the next largest stratum may be, say, 60 percent and so on, the sampling fraction in the stratum of the smallest establishments being the lowest. Such a sample is expected to produce a much more precise estimate than a random sample or a stratified sample with proportional allocation.

The selection of such a sample will be governed by the nature of the frame available. If an up-to-date frame of the establishments exists and the information is on cards, we can make the strata on the basis of previous employment, calculate the best sampling fractions to use, and select the sample from within strata. If information on employment is not available, we may decide to collect it by mail rather inexpensively and use this for purposes of stratification.

Now suppose that a list of establishments is not available or that the available list is known to be very incomplete. In that case we will make all efforts to prepare a list of the very large establishments by asking government licensing departments, chambers of commerce, and other bodies likely to know the whereabouts of large establishments. In this connection it is important to note that such a list need not be 100 percent complete. All these establishments will be taken in the sample with certainty. However, no such list can be easily made for the smaller establishments because these establishments are numerous and are scattered over the area. To sample this part of the population we shall have to take an area sample. Suppose we have a map of the town from which the sample is to be selected. The town is divided up into blocks. Using local knowledge we may divide the blocks into three strata: those known to contain very many units, those containing moderate numbers, and those likely to contain very few units. A fairly large proportion of blocks is taken from the first stratum, lists of establishments made, and a relatively smaller proportion of units taken in the sample from each sample block to make the overall sampling fraction equal to, say, 10 percent. The sampling fraction for blocks is a little less in the second stratum but that for establishments is more to achieve the same overall sampling fraction of 10 percent. In the third stratum a very small proportion of blocks is taken, say 10 percent, and all units in the selected blocks are taken with certainty. In this manner the list and area samples can be used together to produce a satisfactory sample of units from the population (1).

SAMPLE OF HOUSEHOLDS FROM AN AREA

The objective is to draw a sample of households from a city in order to collect information on employment, cottage industry, and household expenditure. The main characteristic of such a population is that the variates under study are somewhat evenly distributed over the population and their distribution is not highly skew.

One simple method of selecting the sample is to secure an up-to-date map of the town and divide it into blocks which can be identified on the ground. The blocks are numbered in a serpentine fashion—that is, neighboring blocks have contiguous numbers—and a systematic sample of blocks is taken which gives an even spread of the sample over the whole area. The precaution is taken that as far as possible the blocks are not very variable with respect to size. If the blocks are not large, complete lists of households are made in each block and a random or systematic sample of households is selected. If the blocks are large, they are divided into segments on the map. A segment is selected at random from each block and household lists made to select a sample of households (3).

Sampling theory suggests that in surveys of this type it is convenient as well as efficient to have a self-weighting sample. That is, the weight of each household is made the same at the time of calculation. This can be achieved by taking the same overall sampling fraction from each block. Thus a 2 percent sample of households can be obtained by taking 10 percent of the blocks and every fifth household in the sampled blocks. It may be noted that there are certain characteristics such as rent or income which are somewhat correlated within blocks while there is very little correlation between households belonging to the same block with respect to other variables such as proportion unemployed. In the former case a smaller number of households may be taken from a block and a relatively larger number in the latter situation.

The sampling procedure is different if a list L of dwellings is available, say, from a recent housing census. As a first step we use local knowledge to locate those areas which are newly inhabited or in which considerable new construction has taken place since the census. These areas are assigned to one stratum, say, A. This stratum is divided up into small areas, a sample of such areas is selected, households are listed within the selected areas, and the sample of households is taken. The entire town minus stratum A may be called stratum B. This stratum is supposed to be divided up into the substrata B_1 and B_2. The substratum B_1 comprises all the dwellings in B and listed in L and the substratum B_2 comprises those dwellings in B which are not on the list L. A sample of dwellings selected from L gives at once a sample from B_1. In order to take a sample from B_2 we divide stratum B into identifiable blocks and take a sample of blocks. All dwellings included in these blocks and on the list L are struck off. The dwellings that remain form a sample from the substratum B_2. Information is collected from the households in the sample taken from the strata A, B_1, and B_2 and the results are suitably pooled to provide estimates for the entire town.

SAMPLING FOR RARE ITEMS

Suppose the problem is to estimate the proportion of persons suffering from a rare disease D. It is believed that the proportion is quite small and that the persons suffering from it are not confined to a single small area. No list giving the names and addresses of persons suffering from the disease is available. If a sample of such persons is to be taken from the entire population, it is best to divide the population into large clusters and canvass the clusters completely in the hope of contacting some persons suffering from the disease. If, for example, the sample is to be taken from an area comprising villages, a sample of villages may be taken and enumerated completely. The sampling procedure will be different if a partial list of the patients suffering from the disease is available from hospitals situated in the area. In that case a sample of names will be taken from the list and contacted wherever they are. This sample will be supplemented by an area sample in which whole areas are canvassed in search of persons suffering from the disease. All persons found in the area sample which are also on the list obtained from the hospitals will be disregarded since they have a chance of selection through the list sample.

SAMPLING FROM A FILE

The simplest situation is that in which the units comprising the population are on cards, there being one card for each unit. As an example, the population may be the totality of employees in a large manufacturing establishment. If there is a card for each employee, the file of cards becomes the population of interest. It is very simple to take a sample from this file.

Suppose there are 10,000 employees and a sample of 500 employees is needed. If the cards are numbered serially, a 1-in-20 systematic sample can be taken by selecting the first card at random and taking every twentieth card thereafter.

It may be found that some of the cards in the file belong to persons who have left the company. If the purpose is to study present employees only, the initial sample will have to be larger than 500. Suppose the proportion of persons on the file who have left the company is 1/10. Then the sampling interval to be used is $20 \times (9/10) = 18$. When the ineligible cards are removed from the sample, the number left is expected to be 500. In case the proportion of ineligible cards is not known, a small sample may be taken to estimate this proportion.

To take another example, suppose there is a land register in which all fields are listed. There is a card for each field giving identification particulars of the field and the name and address of the owner. A person may own several fields located in more than one commune. If a random or a systematic sample of cards is taken, it will not be a random sample of owners. In this sample persons owning more fields have a higher chance of selection as compared with those who own fewer fields. In order to estimate, say, the proportion of absentee farmers from the sample, the probability with which a farmer is selected in the sample must be calculated. If it is found inconvenient to calculate these probabilities, the register should be unduplicated so that each owner occurs once only.

In case the only information on the cards in the land register is the geographical identification of the field and the cards are arranged geographically, it may be better to select the sample in clusters in order to reduce travel costs. If the cards are numbered serially, we may define our clusters as groups of cards numbered 1–4, 5–8, 9–12, ... A 10 percent sample of clusters will produce a 10 percent sample of fields. It should be cheaper to collect information from this clustered sample than from a random sample of fields from the area.

REFERENCES

1. Hansen, M. H., et al. (1953), "Sample Survey Methods and Theory," vol.1, John Wiley & Sons, Inc., New York.
2. Hansen, M. H., et al. (1963), The Use of Imperfect Lists for Probability Sampling at the U.S. Bureau of the Census, *Bull. Intern. Statist. Inst.*, **40**.
3. Kish, L. (1967), "Survey Sampling," John Wiley & Sons, Inc., New York.
4. Raj, D. (1968), "Sampling Theory," McGraw-Hill Book Company, New York.
5. Yates, F. (1960), "Sampling Methods for Censuses and Surveys," Charles Griffin & Company, Ltd., London.

9
The Design of Samples

9.1 INTRODUCTION

We shall assume that the question of the frame to be used for the survey has been settled. The next task is to design the sample for the survey. This is a technical undertaking in which considerable use is made of sampling theory and practice. The design of samples is always subject to administrative constraints and the type of auxiliary information available. In this chapter we discuss the considerations involved in the choice of different sampling techniques.

9.2 STRATIFICATION

This technique is frequently employed in sample design. On the basis of information available from the frame, the units are allocated to strata by placing within the same stratum those units which are more-or-less similar with respect to the characteristics being measured. If this can be reasonably achieved the strata will be internally homogeneous, that is, the unit-to-unit variability within a stratum will be small. This means that the sampling error of an estimate based on a probability sample selected from the stratum is expected to be small. This is how stratification brings about increased precision of the estimates made.

Another reason for stratification is that estimates of a specified precision may be needed not only for the entire population but also for certain subdivisions of it. For example, data on employment may be needed for each of the regions in the country in addition to national estimates. In that case the regions become the major strata. The regions may be further stratified for purposes of sampling.

Sometimes it is found that different parts of the population do not allow the same sampling procedure. In an employment survey, for example, the private households, the institutional population (living in hotels, jails, etc.), and members of the armed forces require different methods for the selection of the sample. In this situation the three categories become the major strata of the survey. In the field of industry, information from the large establishments can be collected by mail as these establishments have records and resources to fill out the questionnaire. The small establishments, however, have no records and only an enumerator making a personal visit can collect the information. Thus the large and the small establishments become the strata of the survey.

9.3 MODE OF STRATIFICATION

If there is only one characteristic of interest, say, the average number of cattle per village in an agricultural survey of villages, the best plan is to stratify the villages on the basis of the number of cattle as obtained from a previous census. In case this information is not available, a correlated variable may be used for stratification, such as the population of the village. In either case the strata will be formed by cutting at a suitable number of points the frequency distribution of the units for the characteristics to be used.

Quite often a large number of variables are to be investigated from the same survey. But it is not practicable to use different stratifications of the same population for different characteristics. With different stratifications we will have to take an excessive sample size which may be well beyond our resources. What is needed here is a compromise stratification carried out by making use of information on a number of variables. This method goes by the name of *multiple stratification.* Some applications of this method in actual surveys follow.

In the United States Current Population Survey the purpose is to collect information on the labor force. In this survey counties are stratified on the basis of degree of urbanization, geographic location, migration, type of industry (in urban areas), and type of farming (in the rural areas). Considerable judgment is exercised and numerical information, too, is used in setting up the strata. In the United States Annual Survey of Manufactures the units (companies) are distinguished by industry group and structure (multiunit or single-unit). Within each group the companies are stratified by number of employees as obtained at a previous census, there being 5 strata ranging from 4 or fewer to 250 or more employees. The data collected relate to employment, wages and salaries, cost of materials, and value of shipments. In the Greek survey of family budgets the communes are stratified by population and altitude as it is believed that the level of living is correlated with the size of the locality and the physical features as judged by altitude. In the multisubject Indian National Sample Survey the bases for the stratification of villages are population, per capita expenditure, and geographical features.

9.4 THE NUMBER OF STRATA

Ordinarily we like to make as many strata as we can. The reason is that a heterogeneous stratum can always be broken up to bring about greater homogeneity. But a large number of strata can be created only if we have sufficient information in numerical terms on a number of characteristics for each unit in the population. When the information available is meager, we cannot go very far in the matter of stratification.

The size of the sample, which is dictated by the total cost of the survey, puts an upper limit on the number of strata that can be made. The reason is that at least one and preferably two units must be selected from each stratum in order that we may calculate the sampling error from a probability sample. Suppose, for example, that no more than 200 villages can be investigated in a survey. This means that, with two units per stratum, we cannot make more than 100 strata.

In nationwide surveys it is customary to have the number of strata larger than 50, preferably 100 or even more. In addition to increased precision, the use of a large number of strata permits a stable estimate of the sampling error. If only one unit per stratum can be taken, the device of collapsed strata may be used for the estimation of error (Example 5.11).

It is of interest to note that the United States Current Population Survey started with 68 strata but now as many as 330 strata are being used. There were in all 250 strata in the Indian National Sample Survey when it was started in 1950. The Greek population survey of 1962 used 49 strata with two primary sampling units per stratum.

9.5 ALLOCATION OF SAMPLE TO STRATA

Sampling theory suggests that, if units are to be selected at random from within strata, the best allocation of the sample to a stratum is in proportion to the size of the stratum and the standard deviation of the variate y in the stratum. This holds when there is a linear relationship between the cost of the survey and the sample sizes within strata. Exact use of this principle is difficult in practice for a variety of reasons. The standard deviation of y within strata is not known. Second, the optimum allocation for one characteristic will generally be different from the optimum allocation for another characteristic. But we cannot afford to use different allocations within the same strata. Thus usually the use of a compromise allocation is called for.

It is a curious fact that very often we decide upon the sample allocation first and then divide the population into strata. That is, we decide to take two primary sampling units per stratum and then begin making the strata accordingly. If for example at most 200 primary units can be taken, 100 strata will be made and two primary units are selected per stratum. A procedure commonly used is to make the strata of equal aggregate size. This means that the total value of the variate x, with which other variates are correlated, is made about the same for each stratum. This is a convenient procedure and efficient, too, if the strata thus formed are internally homogeneous and differ considerably from each other with respect to x.

In surveys of economic institutions such as manufacturing establishments, it is found that the distribution involved is highly skew. A small proportion of the units accounts for a large proportion of the total value of y for the population. In this case it is extremely important to stratify the population by size of unit. Only a small fraction of the units is taken into the sample from the lowest stratum. The sampling fraction is progressively increased from the lower to the higher strata and is unity for the highest stratum (containing the largest establishments).

9.6 GAINS FROM STRATIFICATION

The gains from stratification will be large if the strata differ markedly from one another. Except in the case of highly skew populations, we are generally not in a position to make all the strata very different from one another. We do not ordinarily possess the data needed for making such an efficient stratification. Thus the gain from stratification is rarely expected to be dramatic. Occasionally geographically contiguous units are placed in the same stratum in the hope that such units are similar. But this does not carry us very far.

Another point to be noted is that stratification is not very effective for subgroups of the population. For example, a stratification based on income may produce a precise estimate of the average rent per household but may give a poor estimate of, say, the average rent paid by physicians. A further point to note is that stratification often does not help in improving the precision of proportions calculated from the sample. The reason is that in a multisubject survey one cannot usually create strata which differ markedly with respect to a proportion. The gains from stratification are therefore always small in this case.

9.7 SYSTEMATIC SAMPLING

The selection of every kth unit from a list, the first selection being random, constitutes a systematic sample. This method of selection is widely used for its convenience. Another advantage of this method is that it is easy to check whether sample selection has been done according to instructions. Furthermore, there are certain situations in which systematic sampling, apart from convenience, is likely to give higher precision than simple random sampling. This will be so when there is trend present in the list with respect to the characteristic y of interest. In this case centered systematic selection (the selection of the central unit from the first k and every kth unit thereafter) can be still better. Suppose, for example, that at the listing stage we can collect information from households regarding the amount of land operated by them. The households can then be arranged with respect to this characteristic and a systematic sample selected. Such a sample is expected to contain different types of households in their correct proportions and should therefore be quite efficient.

Systematic sampling is at its worst when there is periodicity in the data sampled and the sampling interval has fallen in line with it. When this happens most of the units in the sample will be either too low or too high, which makes the estimate very variable. Great care has to be taken in the use of this method when periodicity is suspected. The risk can be reduced by taking not one but many random starts, giving rise to a number of systematic samples each of a small size. When a number of random starts is planned, it is useful to take them in complementary pairs of the type $i, k + 1 - i$. This is particularly important when the stratum size is not an integral multiple of the sampling interval (4, p. 228).

A rigorous estimate of the variance cannot be obtained from a single systematic sample of units. Some assumptions will have to be made regarding the type of population sampled. These assumptions are, however, not needed if a number of independent systematic samples can be selected (Example 3.16). Furthermore, systematic sampling is usually used at the last stage of the sampling process. In that case the variance can be estimated, provided that the necessary precautions have been taken when sampling at the first stage (4, p. 119).

9.8 CLUSTER SAMPLING

Ordinarily the sample is selected in clusters in order to reduce costs. It is not practicable, for example, to make a list of all the 30,000 households in a city in order to select a sample of 300 households. The cost of making a complete list of households for sample selection is prohibitive. Even if such a list exists, we may decide to select the households in clusters—say by selecting whole blocks. If a simple random sample of households is taken, the sample will be spread all over the area and this will add to the cost of collecting the data. This cost can be reduced considerably by restricting the inquiry to a few chosen blocks. The blocks selected may be enumerated completely or a further sample of households may be taken from each block. In the latter case we are using two-stage sampling.

The variance of the estimate of the mean in simple random sampling of clusters depends, apart from the sample size and the population variance, on the correlation of y between units within the same cluster (4, p. 109). If the units within a cluster are more similar than units belonging to different clusters, the estimator is subject to a larger variance; thus the smaller the intracluster correlation the better. Suppose, for example, that we want to estimate the sex ratio in a population. We may use the household or the individual as the sampling unit. Suppose we use the household as the unit of sampling. If one member of a household is male, the other is likely to be a female giving rise to negative intraclass correlation. Thus a sample of households is likely to give a better estimate of the sex ratio than a sample of persons (Problem 4.10).

In actual practice clusters are made by putting together geographically contiguous households, fields, factories, and so on. The intraclass correlation coefficient is likely to be positive in such a case. If the size of the cluster is small, the correlation may be high; if the size is fairly large, the correlation is expected to be lower. In either case the variance of the estimate is likely to increase. But then the cost of collecting the information may be considerably reduced by sampling in clusters. The question of cost is considered in the next section.

9.9 COST FUNCTIONS IN CLUSTER SAMPLING

Suppose the objective is to collect information on the living conditions of persons in an area covering 2,500 miles and containing 300 villages of moderate size. We want to select n villages in the sample and collect information from about m households selected in each village. How will the cost of the survey depend on n and m? It is clear that a part of the cost—the overhead cost—is independent of n and m. An office is to be established and properly equipped to run the survey. A number of lists and maps relating to the area are to be acquired and handled. Let us denote the overhead cost c_0. A major cost of the survey is the time taken by the investigators to travel from one village to another. This will depend on the total distance involved in traveling through the sample villages. This distance does not increase proportionately with an increase in the number of villages, n, selected in the sample. We know from theory (3) that a lower bound for the expected travel among n random points in a region measuring A square miles is

$$\sqrt{A}\left(\sqrt{n}-\frac{1}{\sqrt{n}}\right) \simeq \sqrt{An}$$

If the average cost per mile is c_1' dollars, the total cost of travel from one village to another can be approximated as $c_1'\sqrt{An} = c_1\sqrt{n}$.

Then there is the cost of selecting a village in the sample, making a list of households there, and selecting a sample of m households from the list. If this means c_2 dollars per village, the total cost of sample selection and listing is $c_2 n$. The major component of cost that remains is the collection of information. If c_3 is the cost of collecting information from a single household and traveling to it, the total cost under this heading may be taken as $c_3 nm$. Altogether, the total cost of the survey (excluding statistical analysis) may be represented as (1)

$$C = c_0 + c_1\sqrt{n} + c_2 n + c_3 nm$$

A more realistic approach to the development of a suitable cost function is to conduct a pilot inquiry using different values of n and m in different strata. Alternatively we may conduct the first survey using different values of n and m and ask investigators to keep daily records of the time taken for the various operations of the survey. This data can then be graduated by a cost function of the form developed here, and the coefficients c_0, c_1, c_2, and c_3 determined. There is an opportunity to revise the values of the cost coefficients if the survey is conducted over several occasions. In this case we can find more stable values of the cost factors involved. A joint use of the cost function obtained in this manner and the variance function will provide an answer to the question of how many villages to select and how many households to interview from a village.

9.10 MULTIPHASE SAMPLING

It is sometimes convenient and economical to collect certain items of information on all the units in the main sample and other items of information on only some of these units selected at random from the original sample. The items investigated in the main sample (first-phase sample) may be cheaper to enumerate than those in the subsample (second-phase sample). Or, some basic items of information may be enumerated in the first phase leaving the second phase for the collection of information on the other items. This procedure is called two-phase or double sampling. The second-phase sample may be enumerated at a later time, in which case information collected in the first phase may be used for selecting the second sample (5).

In all cases, the first-phase information may be used for improving the precision of estimates made for the variables measured in the second phase (see Examples 6.1 to 6.4). As another example, information obtained in a complete census may be used to improve the estimates made from a sample, in which case the complete census and the sample are analogous to the first- and second-phase samples respectively

Household expenditure surveys provide a good example of the use of multiphase sampling. In these surveys it is found that the variability of expenditure on items purchased regularly such as bread is small while that on items purchased infrequently such as furniture or clothing is large. Suppose a sample of 1,000 households is taken for collecting this type of information. The relative precision of the estimates made for the regularly purchased items will be unnecessarily high while that for the infrequent items will be very low. This imbalance can be remedied by using the multiphase technique. Information on items which are extremely variable may be collected from the full sample and a subsample of a moderate size may be used for items which are moderately variable. A further subsample of households may be taken for the items which are the least variable. In this manner we can collect information of comparable precision for different items in the survey and at the same time reduce the burden of response. The reader is referred to Sec. 15.9 for the problems involved in the use of this method.

Another use of multiphase sampling in household surveys is the following. Suppose we want to estimate y, the amount of money spent by households on food. Since a detailed list of items is to be prepared and investigated over a period, enumeration costs per household will be high. Thus only a small number of households can be investigated. Now suppose that the money spent on food is highly correlated with the number of persons in the household—a variate which is inexpensive to enumerate. Then we may decide to take a fairly large sample and collect information on the distribution of households by the number of persons per household. Within each size group, a small subsample is taken to get data regarding expenditure on food. The two samples are then combined suitably to make a precise estimate of y.

Multiphase sampling may be combined with multistage sampling. For example, a two-stage sample of villages and households may be taken for the collection of information on the size of household and income. A subsample of households is then taken for daily records of detailed expenditure during a two-week period. This is two-phase sampling at the second stage. It may be noted that there is a difference between multiphase and multistage sampling. In multiphase sampling the same sampling units are taken in each phase. In multistage sampling a hierarchy of sampling units is selected at each stage.

9.11 CONTROLLED SELECTION

When the population has been stratified and the probabilities of selection of the primary sampling units within strata have been decided upon, we may achieve further gain in precision by assigning a higher probability of selection to preferred samples of psu's and consequently a lower probability to nonpreferred samples. The preferred samples may be those which are geographically well spread over the area and the nonpreferred are those which are not well spread. This method of selection with restrictions is called the method of *controlled selection* (2). The preferred and the nonpreferred samples are formed in such a way that each unit in the population appears in one or more samples. The number of samples in which each unit appears must be exactly proportional to its assigned probability of selection. After the complete set of samples has been established, the random selection of one of them constitutes a controlled probability sample.

Another example of controlled selection is the following. Suppose we want to select a sample of 5 cities from a province containing 250 cities. Information on census population and altitude is available for each city. It is desired to select the sample in such a manner that each of the five population groups and the five altitude classes is represented adequately in the sample. A simple random sample of five cities may deviate completely from this requirement. We may then think of stratifying the cities with respect to population and altitude group, making 25 substrata in all. Since the number required in the sample is five only, the sample cannot be selected from each substratum. We may then arrange the 25 substrata or cells in the form of a 5×5 square, the rows representing the five population groups and the columns the five altitude groups. From the first column we select one of the five cells at random. The row containing the selected cell is deleted. From the second column a cell is selected at random from the four that remain. The row containing this cell is deleted and so on. In this manner five cells are selected such that each row and each column of the square is represented. And finally a psu is selected at random from each of the selected cells. This gives a sample of five primary sampling units in which the various population and altitude groups are all represented (4, p. 81).

9.12 NATIONWIDE SURVEYS

We have considered in some of the previous sections the problem of designing a sample for an area containing primary sampling units of a moderate size. This is far different from designing nationwide surveys, in which the problem is to devise a small sample from a very large population. For the design of such surveys the question of the sample design cannot be considered in isolation from other factors, such as the organization of the survey and whether it is to be a one-time survey or a continuing inquiry. We shall discuss these matters first, with particular reference to the needs of developing countries.

SURVEY ORGANIZATION

In many developing countries the economy is mainly agricultural, the labor market is unorganized, farming is on a subsistence level, illiteracy is widespread, very few records are kept, and the household and business activities are interwoven. As a result, conceptual, definitional, and measurement problems are very complex. Special studies are needed to solve these problems. Unless an organization is charged with this task on a full-time and continuing basis, the lessons learned from one inquiry will be forgotten and no improvement made in a future survey. The problem of building up the necessary human agency over the years is important. This points to the establishment of a permanent survey organization.

MULTISUBJECT SURVEYS

Developing countries need all types of statistics for making plans for national development. Thus population statistics, agricultural statistics, industrial statistics, and so on are all needed. It is extremely expensive to maintain separate survey organizations for each type of statistics. A realistic approach is to give the same organization the necessary training and experience to handle statistics in different fields. In addition to reducing costs, this arrangement will improve the quality of the data collected since one type of information on a unit will throw light on the validity of the other type of information on the same unit. This points to the need for conducting multisubject surveys rather than single-subject surveys.

CONTINUING SURVEYS

Suppose we wish to determine the employment pattern in an area. This can be done in two ways. An army of investigators is employed and information collected with reference to one week in the year. Or, a small number of trained enumerators is used to collect the information over the whole year. The second procedure should be more useful as the survey is now spread over all the seasons. With the first procedure, data for just one week are collected without regard to the fact that the characteristic of interest is seasonal in character and varies considerably over the year. The idea of staggering the survey over time also ties in very nicely with the need to establish a full-time sampling organization capable of collecting information in a wide variety of fields. This points to the need for establishing a continuing national sample survey for collecting the basic facts for the country and studying the changes that take place over time.

Another advantage of continuing surveys is that the cost of the survey can be spread over a long period, thus allowing greater effort to be directed to initial sample selection and the selection and training of field staff. In a one-time survey the cost of an elaborate sample design and the thorough training of field staff may be found to be excessive relative to the total cost of the survey.

FIELD STAFF

The field staff for the surveys may be selected centrally, trained at headquarters, and sent out to the field for the collection of data. Since the staff will be in travel status all the time, its traveling and subsistence allowances will have to be paid. If this is found to be expensive, some of the staff may be recruited from the primary sampling units in the sample with the understanding that they can be asked to collect information in the neighboring primary sampling units as well. Another procedure consists in selecting all the field staff from the primary sampling units, each person working in the psu of his residence. No subsistence costs need be paid in this case. But greater supervision is needed. Second, the sample will have to be clustered considerably in order to justify the use of one full-time investigator per area.

9.13 DESIGN OF NATIONWIDE SAMPLES

A nationwide survey such as a labor force survey or a household survey of family budgets must be based on a multistage sample. The first stage or primary sampling units will be large areas, preferably administrative subdivisions of the country with well-defined boundaries, for which auxiliary information based on a previous census or on a large-scale survey is available. This information can often be used with considerable advantage for improving the precision of the estimates made. The number of primary units selected in the sample should be fairly large, normally 100 as a minimum. The total number of primary sampling units in the country should be many times the number selected in order that the full benefit of economies derivable from cluster sampling be obtained. If, for example, there are 50 major subdivisions of the country, these will not be used as primary sampling units; ordinarily all of them will have to be taken in the sample.

A major consideration in deciding upon the psu is the degree of its heterogeneity with respect to the characteristics to be measured. Ideally a psu should be a miniature of the entire country so that, if selected, it can tell the full story about the country. Another consideration is that it should be large enough to be an efficient sampling unit but not so large that travel between second-stage units within it is excessive or that supervision of the operations of the survey becomes difficult to organize. Some of the small psu's can be amalgamated with neighboring psu's in order to achieve a reasonable size.

STRATIFICATION

When information is available, it should be possible to group the primary sampling units into a large number of strata. As far as possible the units within strata should be similar and the strata should differ as much as possible from each other in order that full advantage of the gains due to stratification be taken. With two psu's in the sample per stratum the number of strata will be one-half the total number of psu's in the sample. If, however, it is felt that greater precision can be achieved by further splitting the strata, one should not hesitate to make as many strata as the number of psu's to be taken in the sample. When an equal number of psu's is to be selected per stratum, it is advisable to make the strata of about equal aggregate size (and not necessarily of equal number of psu's). This will imply allocation of the sample to strata in proportion to their measures of size.

When the number of strata has been decided upon and the average measure of size calculated, it might be found that there are certain large primary sampling units which are bigger than the average measure of size per stratum. In that case, all such psu's should be treated as strata and taken into the sample with certainty.

SELECTION OF PSU'S

It will ordinarily be found that the psu's within a stratum differ considerably with respect to an important measure of size x. In this case it is advisable to select the sample of psu's with unequal probabilities, giving a higher chance of selection to the larger units and a lower chance to the smaller units (as judged by x). The method of sampling with probability proportionate to size (pps) will be used. If two psu's are to be taken from a stratum, sampling should be without replacement, especially when the number of psu's in the stratum is not large. A number of methods are available for this purpose. The psu's in a stratum can be listed at random and a systematic sample taken with probability proportional to x (Example 3.22). Alternatively, the psu's in a stratum can be randomized into substrata and one psu selected with pps from each substratum (Example 3.24). Or, the first psu is selected with pps and the second with pps of the remainder (Example 3.23).

It is clear that the sample will ordinarily contain a greater number of large primary sampling units when the selection is made with probability proportionate to size. This should not increase the cost of subsampling since it is not proposed to make a complete list of all elementary units within a selected psu. On the other hand, there may be better transport facilities in large psu's and the density of population may be higher, which reduces the cost of travel within the psu.

SUBSAMPLING

Let us assume that n primary sampling units have been selected from each stratum. The selection is without replacement and the chance of selecting the psu U_i is p_i'. Further, let a psu be broken up into four substrata—based on two size groups and two altitude groups—and lists of second-stage units, say communes, are made within each substratum. Suppose the substrata are of about equal population. One commune is selected from each substratum with pps and a list of all households in the selected commune is made. If the number of households is M_{ijk} (ith psu, jth substratum, kth commune), what should the sample number m_{ijk} be for the collection of information on y? One solution which is convenient and efficient is to choose m_{ijk} in such a manner that the sample becomes self-weighting for a given overall expected sampling fraction. Suppose the expected sampling fraction for households is desired to be 1/200. Then an unbiased estimate of the stratum total for y is (4, p. 124)

$$\hat{Y} = \mathop{S}_{i=1}^{n} \frac{1}{p'_i} \sum_j \frac{M_{ijk}}{p_{ijk}} \frac{Sy_{ijkl}}{m_{ijk}}$$

where p_{ijk} = probability of selecting kth commune in jth substratum
p'_i = probability of selecting ith psu in sample from stratum

The purpose is to choose the number of households m_{ijk} in such a manner that

$$\hat{Y} = 200 \mathop{S}_{i} \sum_{j} \mathop{S}_{l} y_{ijkl}$$

Thus the sampling interval M_{ijk}/m_{ijk} for the selection of a systematic sample of households in the selected commune is simply given by

$$\frac{M_{ijk}}{m_{ijk}} = 200\, p_{ijk} p'_i$$

It is clear that the number m_{ijk} is about the same everywhere if the measures of size, that is, the population data in this case, are good. Provided the same expected sampling fraction is used in all the strata, the estimator will be completely self-weighting.

LISTING

Having selected, say, a number of communes from each primary sampling unit, it is extremely important to make an accurate list of the third-stage units, dwelling units in the present population survey, for further selection of the sample. For this purpose the field workers should be given simple, clear instructions, instructions which can be translated into practical procedures. The instructions should be more extensive and detailed and the training more thorough if the enumerator is far removed from control, as in widespread national samples. In the case of a survey confined to a small area, the enumerator can always ask for clarifications when handling an unusual situation; in a large-scale survey he has to be trained beforehand in handling such situations. In writing the instructions the different field tasks should be broken up into simpler units, to be performed one by one.

When the communes are large, it should be possible to select the sample by introducing another stage in the randomization process. A good map of the commune may be used for dividing it into blocks with well-defined boundaries. The blocks may be assigned measures of size and stratified suitably. A sample of blocks is taken and lists of dwelling units made in each sample block in order to select a sample of dwelling units.

RANDOM NUMBERS

While the primary sampling units and the second-stage units can be selected in the office, the selection of blocks and dwellings has to be left to the enumerator. Therefore it becomes important to devise selection procedures which are easy to control. In one method the random numbers are selected in the office and given to the enumerator in a sealed envelope. The enumerator is required to list the units in a fixed order. When the list is completed, the envelope containing the random numbers is opened by him to select, say, a systematic sample of units. In another procedure the enumerator is required to relist the units following a definite method. The total number of units in the list determines the page to be used from the book of random numbers. Both procedures are easy to apply and easy to check (2).

ROTATION OF THE SAMPLE

Some kind of rotation of the sample is needed to avoid an undue reporting burden on the survey respondents if they are to constitute a permanent panel in a continuing survey. Asking the same respondents to furnish information every time might increase the rate of refusals substantially. To avoid this difficulty, a system of partial replacement of the sample may be adopted (Sec. 6.2). For examples of this technique, see Secs. 14.12 and 18.8.

ESTIMATION PROCEDURE

The method of forming estimates of population totals, means, ratios, or proportions will be governed by the type of sample used. The method indicated in Example 5.6 may be used when the primary sampling units are selected with replacement. When the selection is without replacement with unequal probabilities, the methods of Examples 5.8 to 5.10 may be employed.

REFERENCES

1. Hansen, M. H., et al. (1953), "Sample Survey Methods and Theory," vol. I, John Wiley & Sons, Inc., New York.
2. Kish, L. (1967), "Survey Sampling," John Wiley & Sons, Inc., New York.
3. Marks, E. S. (1948), A Lower Bound for the Expected Travel among *m* Random Points, *Ann. Math. Statist.*, **19**.
4. Raj, D. (1968), "Sampling Theory," McGraw-Hill Book Company, New York.
5. Yates, F. (1960), "Sampling Methods for Censuses and Surveys," Charles Griffin & Company, Ltd., London.

10
Execution of Surveys

10.1 PILOT SURVEYS

It is clear from the preceding three chapters that there are many aspects of the survey on which a rational decision can only be made if a pilot survey is conducted before the main survey is begun. This is especially important if no previous experience is available on the subject of the survey.

One of the points to be decided is the cost function of the survey. In order to utilize the resources in the best possible manner it is important to know the cost components: time spent on travel from psu to psu, length of interview, time spent on listing a psu, analysis of the data in the laboratory, and so on. A pilot survey can be a great help in estimating the cost components.

Some information on the variability of different sampling units is needed in order to form a rough idea of the precision expected from the survey. The data collected in the pilot survey can be used for making the estimates needed.

The questionnaires prepared for the survey must be tested thoroughly in order to decide upon the final version. This can only be achieved through a series of small-scale pilot surveys. If a decision is to be made on two different wordings of a question, it is best to direct the two forms to random samples selected from the area. An analysis of the responses should help in deciding upon the better wording. In more complex situations, factorial designs may be used for economizing the effort to reach a definite decision.

As a rule the pilot survey should be a miniature of the main survey or census in which all the operations intended to be used should be tested to see how they work and what modifications are needed to make them work more efficiently and harmoniously. A question usually asked is whether the pilot survey should be based on a random sample. When the objective is to test the different procedures of the survey, a random sample is not essential. What is important is to select areas or units which are supposed to present a variety of situations and see how the proposed procedures work under different conditions. A random sample may not present this variety. However, usable estimates—say, of variance functions—can only be made from probability samples.

Many surveys have problems peculiar to them. For example, the size of the sampling unit in agricultural surveys of crop yields appears to be crucial.† (There has been a lot of controversy on the subject especially in India.) In that case considerable pilot work is needed to find an appropriate size of the sampling unit to be used in the main survey. If, however, a survey has already been conducted on the subject, no new pilot survey is needed. The present survey can be used as a pilot for the subsequent inquiry.

An important consideration in pilot surveys or census tests is the clustering of the sample. If the sample is spread all over the area, supervision of the operations involved becomes difficult. And it may not be possible to test all the allied procedures on the same material. Thus, whole blocks may be selected in urban household surveys for carrying out the test or the entire village may be enumerated in a rural survey. When all the operations are performed at the same place, their interrelationships become more apparent. The usual method of selecting the sample in clusters is multistage sampling.

It should be pointed out that the essential conditions under which a pilot survey is conducted should not be materially different from the conditions under which the main survey is intended to be taken. If this precaution is not taken, the usefulness of the pilot as a guide for the main survey is likely to be impaired.

10.2 SELECTION OF ENUMERATORS

In most surveys the data are to be collected through enumerators who work part time or full time. The nature of the enumerator's job is such that great

† See Sec. 12.6.

care has to be exercised in his selection. The enumerator should be able to create a friendly atmosphere and put the respondent at his ease. He should be able to ask the questions properly and intelligently and record the response accurately and completely. If the respondent has any legitimate questions, he should be able to answer them. By his friendly and courteous manner he should be tactful enough to discourage irrelevant conversation and elicit response from those who are apparently unwilling to cooperate.

What criteria should be employed for the selection of enumerators to ensure that the right sort are picked? This is a difficult question to answer. We can say in general terms that the enumerator should be honest in his work. If he is dishonest and prone to inventing figures, there should be some way of detecting and dismissing him. He should be accurate in recording respondents' answers, following instructions, and using definitions. Unless he is interested in the work, he cannot give his best. Since he may be called upon to take part in widely different types of inquiries, such as a household survey on expenditure or an establishment survey of value added, he should be a person who can readily adapt himself to different situations. A reasonable degree of intelligence is required for following instructions, and the enumerator should also have attained an adequate level of education to ensure his ability to follow directions.

An initial intelligence test may help in weeding out the undesirable applicants. This may be followed by a further test in elementary arithmetic. Then an interview with the applicants to sort out problems such as their availability for the survey and the salary they expect. Once selected and found to be qualified, the enumerators should be encouraged to remain in the profession. Since fieldwork is a strenuous job, there should be adequate provision for rest, and the salary should be attractive. Lest they become bored with the job, some variety may be provided by transferring them to the main office temporarily where they can help in editing the data. This will also give them an insight into the purposes of the work they are engaged in.

10.3 TRAINING OF ENUMERATORS

When enumerators have been selected for the job through appropriate tests, their training is a matter of the greatest importance. The enumerators should know the purposes of the survey and how the results may possibly be used. The manner in which the data are to be collected and the interview to be conducted should be explained at length with examples. They should know the definitions of the terms used in the questionnaire or schedule and the intricate problems involved in using them in the field. Mock interviews can be used in the classroom, the instructor and the student assuming the roles of respondent and interviewer respectively. The instructor brings up difficult situations and the whole class fills out the questionnaires, which are later discussed at length. The enumerators are then sent to the field to test their questionnaires. The data collected under supervision are examined by the instructors, who determine whether reasonable standards have been maintained. Those who measure up to the standards can be asked to begin their job.

The training is best carried out with the help of instruction manuals. These manuals explain clearly the job of the enumerator at each step. How to make a list of the sample units, say, households, in the selected area, the selection of the sample in the field, the information to be collected, the application of definitions, the handling of borderline cases are all explained there. Extra copies of the questionnaires are made for them to practice with. The main purpose of the training program is to bring about uniformity in the procedures of the survey. Unless all enumerators understand a definition in the same way, it is difficult to add up the data collected by different enumerators.

In continuing surveys the training of enumerators should never stop. The enumerators should be required to come to headquarters once or twice a year for group training sessions, for clarification of difficult situations encountered by them, and for exchange of views and special demonstrations. A training program should be arranged whenever a new schedule is to be added to the survey. Sometimes the enumerators can be given short-term assignments for editing and coding the questionnaires. This is done to impress on them the value of collecting accurate and complete information in the field.

10.4 SUPERVISION OF ENUMERATORS

Supervision of the work of the enumerators is absolutely essential. The mere presence of supervisors in the field has a wholesome effect on their performance. The supervision should be carried out by superior staff, better paid, better qualified, and more experienced. From time to time the supervisory staff should themselves undertake fieldwork in order to appreciate the difficulties involved. Wherever possible, they should be specially trained for the task of supervision.

Although the details may vary, a supervisor is usually in charge of a certain number of enumerators in an area. He is required to give assignments to the enumerators and keep them informed of changes in the instructions. All queries and difficulties are referred first to him. He receives the completed schedules from the enumerators and makes a rough check for obvious inconsistencies and omissions. He keeps an eye on the work of the enumerators. He is the person to make sure that instructions are being followed.

10.5 CONTROL OF QUALITY OF FIELDWORK

In addition to the training program and the provision of supervisory staff, a number of positive steps will have to be taken to ensure that the survey is under statistical control; that is, the errors to which it is subject are random and no assignable causes of variation are present. The control of quality can take several forms. Some of the well-known procedures generally used will be discussed here (2, 5).

10.6 OBSERVATION

The supervisor accompanies the enumerator to the field for a few hours once a month or once in two to three months as the need may be. He watches the enumerator as he asks questions and interprets the responses. He observes the enumerator's adherence to survey procedures, efficiency in planning his work, and tact in answering questions raised by respondents. Following the observation, the supervisor discusses strong points and deficiencies with the enumerator. At this point a card may be made out showing his rating, which determines the frequency with which the enumerator should be observed under this program.

10.7 REINTERVIEWS

The supervisor selects a sample from the production of the enumerator and collects his own information on the sample units involved. In surveys involving physical measurements or observations collection of a duplicate set of data presents no problems. In interview surveys it is more difficult to collect the same kind of information again but experience shows that the method works. By comparing the two sets of records and by reconciling the differences observed, gross differences will be apparent. If it is found that the enumerator is not honest and is not following the instructions, all his returns are reviewed. Such enumerators are removed from the field. There is an advantage in asking the supervisor to reinterview a part of the sample for which he is not provided with the records of the original interview. These data can be matched at headquarters by the superior staff to find out whether the supervisor is doing his job well.

It is worth pointing out that the purpose of this program is to keep the quality of data under control. If the sample shows that the enumerator's work is acceptable, that is, the error rate is within reasonable limits, he is allowed to go on with his work. If not, he is either removed or retrained. In this manner the quality is controlled within specified standards; it does not mean that all errors are corrected.

10.8 FIELD EDIT

When the questionnaires or schedules are handed over by the enumerator to the supervisor while in the field, the supervisor scrutinizes these to check omissions, inconsistencies, illegible writing, and other errors before they are passed on to headquarters. This editing serves a number of useful purposes. Unless the questionnaires are edited on the spot, the need for further information to correct some of the wrong entries may only be discovered when the team has moved to another area. Second, if the errors are discovered at this stage, the enumerator can be instructed not to make such errors in the future. Most of the obvious errors can be corrected at this stage without making a reference to the respondent since the interview is still fresh in the mind of the enumerator. Third, the job of the coder is facilitated when the supervisor has made sure that the entries are legible.

10.9 FOLLOW-UP OF NONRESPONSE

It is inevitable that, in spite of best efforts, some of the respondents will refuse to cooperate. Definite instructions must be given to cover situations of this type. Normally the enumerator informs his supervisor, who then tries to contact the respondent and secure the information. If the supervisor does not succeed, he should as a minimum get some information by which the respondent can be classified properly. In a household survey, for example, this information may relate to the number of persons in the household, the type of dwelling, profession, etc. This fragmentary information can be used later to determine the effect of refusals on the characteristics of participating households. In all cases the reason for nonresponse should be recorded.

One method of dealing with the nonresponse problem is to make a list of the nonrespondents and take a small subsample of them. Then all attention is directed toward the subsample for securing a response with the help of the supervisory staff of the survey. When this method is to be followed, rigorous instructions for taking the subsample should be given. Systematic sampling can be used with advantage. (See Example 6.8.)

In some surveys the major cause of nonresponse may be nonavailability at home when the interviewer has called. If recalls are expensive, the probability of availability at home may be estimated from those available by asking an additional question. The data obtained can then be adjusted to allow for those not available. (See Example 6.9.)

It should be mentioned that enumerators must not be allowed to make substitutions for those not found. If this practice is followed, the enumerators will not take pains to persuade the nonrespondents to cooperate. There will be a tendency to substitute for any one who is not considered to be a good respondent. This will introduce bias in the survey results.

10.10 INTERPENETRATING SUBSAMPLES

The internal consistency of the survey can be checked by using the method of interpenetrating subsamples. In this method the entire sample in a stratum is broken up into two (or more) parts, each part being allocated to a team of enumerators. A comparison of the results by team will reveal discrepant work. The disadvantage of the method is that the amount of traveling may substantially increase when both (or all) the teams have to work in the entire stratum. Furthermore, the discrepant work of an individual team can only be revealed if the differences are large relative to the errors of the survey. Suppose, for example, that the standard error of each team (for the estimate under discussion) is 5 percent and that 25 teams are involved. Thus the standard error for the survey estimate is 1 percent and that for the difference between two teams is 7.1 percent. If one of a pair of teams has a bias of 15 percent, there is a 50 percent chance that the difference between the two teams will be significant. (See Example 6.6.)

10.11 TIME REFERENCE AND REFERENCE PERIOD

The point in time at which the survey is to be put into the field and the period to which the data will pertain are two of the questions which should be decided in view of the purposes of the survey. In some countries it is impossible to do fieldwork during the rainy season due to impassable roads. If the government has announced certain plans which are likely to make people suspicious, this is not the right time to start a new survey.

The reference period is the period for which the respondents provide information. For example, in a survey of family budgets, households may be asked to give details of expenditure relating to the previous week or the previous month. The week or the month is then the reference period. The survey director has to decide what this period is going to be. If the reference period is long, respondents may not be able to recall the purchases accurately. When the item is one that is purchased frequently, such as bread, it may not be possible to recall the expenditure for longer periods than a week. If it is an item that is purchased infrequently, such as furniture, respondents may be able to report expenditure for a considerably longer period. On the other hand, the effect of the tendency to include expenditures not incurred during the reference period is more pronounced when the reference period is short than when it is long. Considerable research is needed to decide upon the proper reference period. The pilot survey can be a good vehicle for carrying out this type of research.

10.12 PROCESSING OF THE DATA

After the data have been collected the scene shifts from the field to the office. The data are to be given a thorough check, coded, transferred to cards or tape, and tabulated. These operations are as important as the collection of the data. Errors are likely to arise at every step and hence the need to be watchful. The different steps in the analysis of data will now be explained in greater detail.

10.13 EDITING

It cannot be taken for granted that the data coming from the field are free of all errors. Even the best of enumerators makes mistakes; a question may not be asked or the response may be recorded wrongly. It becomes important to check that there are no missing entries in the completed questionnaire or schedule. Sometimes, it may be possible to deduce from other data in the form what the missing entry should be. Otherwise, a reference may have to be made to the enumerator. Apart from checking for completeness, we have to make sure that the entries are consistent and accurate. If, for example, the questionnaire shows a woman of 58 giving birth to a child, the age of the mother is probably reported incorrectly. If a university professor is shown to be working 6 hours a week, the concept of "number of hours worked" is not in line with the accepted definition used in the survey. Inconsistencies of this sort are to be cleared up before the questionnaires are passed on for analysis.

The editing of schedules is best done by preparing a written manual of instructions. This is done by going through the first few batches of questionnaires received from the field in order to discover the types of problems arising and the items needing clarification. The instructions can be revised as more experience is acquired. This is especially true of continuing surveys in which the instructions based on an earlier survey can be used with minor modifications.

There are two ways of editing a schedule. The schedule may be edited as a whole or page by page by different persons. The first procedure is generally preferable since answers to other questions can be used for checking consistencies. With the second procedure the task of editing can become very boring.

10.14 CODING

The responses in the edited questionnaires are now to be translated in numerical terms in order to facilitate the analysis. This is done by setting up a list of codes for the possible responses to a question. For example, the sex of the respondent may be coded as : male—1, female—2. In an employment survey the entries against the item "reason for being economically nonactive" may be coded as: housewife—0, student—1, retired—2, too young—3, too old—4, pensioner—5, *rentier*—6, other—7. In this case it may be suggested that codes 5 and 6 should be combined or that the category "too old" be broken down further. All this cannot be done arbitrarily. The question of how detailed the list of codes should be must be considered in the light of the purposes of the survey and the way in which the results are to be used. As a general rule, the list should be fairly broad to start with. Some of the codes can be amalgamated later if this is considered advisable, but they cannot be broken down without going through the entire material again.

The usual procedure for setting up the code list is to examine a number of completed schedules and check the range of answers given. If some of the categories are too small and unimportant, these may be relegated to the "other" class. The most frequent categories may be further subdivided if this can throw greater light on some of the questions the survey is expected to answer. The list of codes should be neither too long nor too short.

Having decided upon the codes to be used, a program is arranged to train the prospective coders in the art of coding. The codes are explained and illustrated with examples from the material to be categorized. The coders then practice on a sample of the data and the problems that arise are discussed in order to bring about uniformity and consistency in the procedure. As a result of this exercise some of the categories may be dropped. Those trainees who are found to be unreliable coders are retrained. When their performance is judged to be satisfactory, they are put on the job. The coders are instructed to see the supervisor whenever a new problem arises. Periodic checks are made to ensure that the coders have not become careless. In this manner uniformity in the coding of the material can be maintained.

There are several methods by which codes can be entered in the schedules. Sometimes boxes are provided for codes either adjacent to the item to be coded or down the margin of the form, linked to items by numbers. In this case the original data do not have to be transcribed; this saves time and copying errors. A disadvantage is that sometimes, as a result of the editing process, too little space is left for entering the code.

Another variant of the method is to print the code numbers on the questionnaire opposite the possible answers to each question. The enumerator checks the item which represents the response and thus the proper code is automatically indicated. The full response is not written down. The method saves time and money; no copying errors are involved. This method will work when the range of possible answers has been visualized beforehand, as in a repetitive survey. However, it becomes more difficult to detect errors of reporting when the full response is not known. Second, the procedure is very rigid. If a code is to be split up or a new one is to be added, it is too late to incorporate these changes. The questionnaire along with the codes has already been printed.

In a repetitive survey the enumerators may enter most of the responses in terms of codes shown in the "instructions to enumerators." Considerable saving in time can be made in this case since the full response is not to be recorded. Alternatively, a special card (giving the responses in codes) may be made out by the enumerator from the original questionnaire. This card can be used for hand tabulation when facilities for mechanical tabulation do not exist.

10.15 TABULATION

When the material is edited and coded, it is ready for analysis. One of the operations is to sort the cards for preparing frequency distributions—the number of units belonging to each category. Another operation consists in aggregating a quantitative variate such as income, number of persons in the household, or total cultivated area. Both operations can be performed by hand or by machines. The mechanical method of analysis is faster and more accurate.

If machines are available, the coded information is transferred to punched cards before actual tabulation begins. The operator punches holes in the cards at the appropriate places as he runs his eye over the codes in the questionnaire. The punched cards are verified on a similar machine, which will stop when the two punchings disagree. The work of the punch operator is facilitated if the codes are all in one column or in a row. When punching is over, the punched cards go to the machines for sorting and tabulating.

If the survey is an isolated inquiry, one must decide whether the data should be tabulated by hand or by machine. If the volume of work is small, many of the simple one-way and two-way tables can be prepared by hand. If, however, the number of cross classifications is large, machine tabulation will be quicker and more accurate. When the operations involved are tedious, monotonous, and repetitive, the human brain is much more prone to errors than the machine. Another advantage of mechanical tabulation is that machines have built-in devices for comparing, checking, and transcribing data. But they have to be programmed by specialists. When such persons are available and the organization can afford to hire them, machine tabulation is preferable.

In addition to supplying estimates of the characteristics of interest, it is the duty of the organization to provide the users of the data with the margin of uncertainty to which the figures are subject. Thus the calculation of standard errors should form an integral part of the tabulation program. One method is to decide upon the key items in the survey and calculate the standard errors of their estimates. This calculation can be done by hand or by mechanical means. Since the rigorous calculation of sampling errors can be a laborious process, approximate methods may be used. For example, the entire sample may be broken up into a convenient number of random groups. The standard error can then be calculated simply by following the method of interpenetrating samples (3, p. 194).

10.16 CONTROL OF DATA PROCESSING

When the information on questionnaires is prepared for machine readability it is important to take steps to limit the quality loss in processing at a level consistent with cost and data utility. For this purpose quality control techniques are employed to ensure that the errors of editing, coding, and punching are within acceptable limits. These techniques take the form of verification of the processed data. Verification is of two types—dependent or independent. Verification is said to be *dependent* when the inspector or verifier reviews the work of a producer (a coder, or a puncher) and approves or disapproves of it. For example, an experienced and trained coder may be asked to inspect a sample of codes assigned by production coders. If the work is rejected, a corrector examines all the items and corrects those found in error. The chief difficulty with this type of verification is that the inspector is influenced by the work of the original producer (coder, in this case) and thus often fails to identify and correct a substantial proportion of defective items. The corrector of rejected work is similarly influenced by the work of the original producer and the changes made by the inspector. The result is that the risk of accepting work of poor quality increases and the quality of accepted work declines. Therefore it becomes necessary to control the work of inspectors and correctors too. In one method, deliberate errors are planted in the material prior to verification. The proportion of defectives undetected by the verifier gives an estimate of the error rate of the verifier. Experience shows that the error rate is high when the work involves considerable judgment and interpretation such as industry and occupation coding in labor force surveys.

In the process of inspection called *independent verification*, the work is repeated by one or more verifiers operating independently of the producer, and the outcomes are compared. For example, codes are assigned to the same item by two coders working independently and compared to the code assigned by the production coder. If all three agree, the production coder is right. If the two independent coders agree but disagree with the production coder, the production coder is wrong. The argument used is that it is highly unlikely that coders in agreement will be incorrect when coding is done independently. Another variation of the plan is to match the code assigned by the production coder to a precoded item. If there is agreement, the code is correct. If not, an adjudicator decides who is correct (1).

10.17 USE OF COMPUTERS

A good deal of survey work is nowadays tabulated by computers. The great advantage of computer tabulation is not so much in the saving of time, although this may be considerable, as in the range and flexibility of processing made possible by computers. A large number of variables can be studied simultaneously. By means of such techniques as computer simulation, it is possible to simulate and analyze the operation of extremely complex systems, which cannot be studied economically by other means.

An advantage of tabulation by computers is the possibility of using mark sensing or electronic scanning of marks to transfer information on schedules into machine-readable form without the need for handpunching of cards. Computers can be made to perform many of the tasks formerly carried out by clerks. Computers are able to apply a large number of complex criteria to determine when a return is unacceptable and when missing data can be properly imputed. In a population survey, for example, the imputation can be made simply by selecting systematically from a subgroup of units with a prescribed set of similar characteristics. A by-product of the application of the computer to tasks of imputation is a count of the number of cases in which such action is taken. This information provides the users of data a definite measure of the extent to which data have been supplied by the editing method.

Computers also greatly reduce the burden of coding. For example, if the return for a unit includes street and number information, it is possible to code this mechanically to a predetermined statistical subdivision such as a block. Another advantage is that detailed and complex cross-tabulations can be carried out, which would not be feasible by the use of conventional equipment. This is made possible by the ability of the computer to store vast amounts of information and work at high speed. Furthermore, complex weighting and estimating procedures can be used which can often improve considerably the reliability of sample results.

10.18 REPORT WRITING

We have now come to the end of the journey. The data have been collected and analyzed. It is now time to write a report on the survey. Two kinds of reports may be presented: either a general report giving a description of the survey for the use of those who are primarily interested in the results, or a technical report giving details of the sample design, computational procedures, accuracy and allied aspects. In this connection, it will be useful to follow the recommendations issued by the United Nations Statistical Office on the subject of the preparation of reports. These recommendations are outlined below (4).

10.19 GENERAL REPORT

STATEMENT OF PURPOSES OF THE SURVEY

A general indication should be given of the purposes of the survey and the permissible margin of error, and the ways in which it is expected that the results will be utilized. In this connection it is useful to distinguish integrated, multisubject, continuous, and *ad hoc* surveys. In an integrated survey, data on several subjects (or items, or topics) are collected for the same set of sampling units for studying the relationship among items belonging to different subject fields. Such surveys are of special importance in studies on standards of living. Integrated surveys of consumption and productive enterprises are also of special importance in developing countries where the related activities are frequently undertaken in an integrated manner in the household.

When in a single survey operation several subjects, not necessarily very closely related, are simultaneously investigated for the sake of economy and convenience, the investigation may be called a multisubject survey. The data on different subjects need not necessarily be obtained for the same set of sampling units, or even for the same type of units (e.g., households, fields, schools, etc.).

The most usual example of continuous surveys occurs where a permanent sampling staff conducts a series of repetitive surveys which frequently include questions on the same topics in order to provide continuous series deemed of special importance to a country. Questions on the continued topics can frequently be supplemented by questions on other topics depending upon the needs of the country.

DESCRIPTION OF THE COVERAGE

An exact description should be given of the geographic region or branch of the economy or social group or other categories of constituent parts of a population covered by the survey. In a survey of a human population, for example, it is necessary to specify whether such categories as hotel residents, institutions, persons without fixed abode, military personnel, etc., were included and to indicate the order of magnitude of the categories omitted. The reporter should guard against any possible misconceptions of the coverage of the survey.

COLLECTION OF INFORMATION

The nature of the information collected should be reported in considerable detail, including a statement of items of information collected but not reported upon. The inclusion of copies of the questionnaire or other schedules and of relevant parts of the instructions used in the survey (including special rules for coding and classifying) is of great value, and such documents should therefore be reproduced in the report if possible.

The information may be collected by direct investigation, or by mail or telephone. Direct investigation may involve objective methods of observation or measurement. The method of collection should be reported, together with the nature of steps taken to ensure that the information is as complete as possible (e.g., methods of dealing with nonresponse). The extent and causes of nonresponse, etc., should be stated.

It is of importance to describe the type and number of investigators, e.g., whether full or part time, permanent or temporary, with particulars of their training and qualifications.

REPETITION

It is important to state whether the survey is an isolated one or is one of a series of similar surveys. Where the survey is repetitive and some of the sampling units reappear in the successive stages, this should be stated.

NUMERICAL RESULTS

A general indication should be given of the methods followed in the derivation of the numerical results. Particulars should be given of methods of weighting and of any supplementary information utilized, for example, to obtain ratio estimates. Any special methods of allowing for nonresponse should be described.

DATE AND DURATION

There are two periods of time which are important for any survey: (1) the period to which data refer, or the *reference period*, and (2) the *time reference*, that is, the period taken for the fieldwork. In order to minimize memory lapses, the length of the reference period is sometimes fixed but not the end points, as in "preceding week" questions (the time reference being comparatively long). In such cases, the reference period may be called *moving* in contrast with a *fixed* reference period when the end points are fixed. The reference period, if properly selected, enhances the use and value of the information collected in the survey. Sometimes different reference periods are used for different topics in the same survey. This is done with a view to eliciting more accurate information from the respondents, as in a family budget survey, where questions on food items may be asked for the preceding week or month, but information on clothing or furniture or some durable goods may be asked for the whole year. In health surveys, different reference periods are used for different items, e.g., illness during the preceding two weeks, or hospitalization during the year. When the fieldwork is conducted on a continuous and successive basis, the parts of the survey which are operationally separate are called *rounds*.

ACCURACY

A general indication of the accuracy attained should be given and a distinction should be made between sampling errors and nonsampling errors.

COST

An indication should be given of the cost of the survey, under such headings as preliminary work, field investigations, analysis, etc. Resources used in the conduct of the survey but not included in the costs should be stated.

ASSESSMENT

The extent to which the purposes of the survey were fulfilled should be assessed.

RESPONSIBILITY

The names of the organizations sponsoring and conducting the survey should be stated.

REFERENCES

References should be given to any available reports or papers relating to the survey.

10.20 TECHNICAL REPORT

SPECIFICATION OF THE FRAME

A detailed account of the specification of the frame should be given; this should define the geographic areas and categories of material included and the date and source of the frame. If the frame has been emended or constructed *ab initio* the method of emendation or construction should be described. Particulars should be given of any known or suspected deficiencies.

DESIGN OF THE SURVEY

The sampling design should be carefully specified, including details such as types of sampling unit, sampling fractions, particulars of stratification, etc. The procedure used in selecting sampling units should be described and if it is not by random selection the reporter should indicate the evidence on which he relies for adopting an alternative procedure.

PERSONNEL AND EQUIPMENT

It is desirable to give an account of the organization of the personnel employed in collecting, processing, and tabulating the primary data, together with information regarding their previous training and experience. Arrangements for training, inspection, and supervision of the staff should be explained, as also should methods of checking the accuracy of the primary data at the point of collection. A brief mention of the equipment used is frequently of value to readers of the report.

STATISTICAL ANALYSIS AND COMPUTATIONAL PROCEDURE

The statistical method followed in the compilation of the final summary tables from the primary data should be described. If any more elaborate processes of estimation than simple totals and means have been used, the methods followed should be explained, the relevant formulas being reproduced where necessary.

It frequently happens that quantities of which estimates are required do not correspond exactly to those observed; in a crop-cutting survey, for example, the yields of the sample plots give estimates of the amount of grain in the standing crop, whereas the final yields will be affected by losses at various stages, such as harvesting, storing, transport, marketing, etc. In such cases adjustments may have to be made, the amount of which is estimated by subsidiary observations, or otherwise. An account should be given of the nature of these adjustments and the ways in which they were derived.

The steps taken to ensure the elimination of gross errors from the primary data (by scrutiny, sample checks, etc.) and to ensure the accuracy of the subsequent calculations should be indicated in detail. Mention should be made of the methods of processing the data (punched cards, hand tabulation, etc.), including methods used for the control of errors.

In recent years electronic computers have been applied to the making of estimates from sample surveys. In addition, they have been programmed to tasks of editing with consequent improvement of consistency of the primary data and possibly of their accuracy. Their use goes beyond the mere speeding up of existing methods and lowering the cost. They have made possible some forms of estimation (e.g., the fitting of constants in analysis of variance with unequal numbers in the subclasses) on a scale which would have been impracticable with hand methods. In some instances regression on previous surveys of a series with overlapping samples can substantially improve the precision of estimates. The use of these computers makes possible changes in the allocation of resources between the collection of the data and the processing.

The amount of tabular matter included in the report, and the extent to which the results are discussed, will depend on the purposes of the report. If a critical statistical analysis of the results embodied in the final summary tables has been made it is important that the methods followed be fully described. Numerical examples are often of assistance in making the procedure clear. Mention should be made of further tabulations which have been prepared but which are not included in the report, and also of critical statistical analyses which failed to yield results of interest and which are therefore not considered to be worth reporting in detail.

The inclusion of additional numerical information which is not of immediate relevance to the report but which will enable subsequent critical statistical analyses to be carried out should be carefully considered. If, for example, in addition to the class means of each main classification of the data, the subclass numbers (but not the means) of the various two-way classifications are reported, a study of the effects of each of the main classifications freed from the effects of all other classifications can be made (provided the effects are additive) without further reference to the original information.

ACCURACY OF THE SURVEY

Standard deviations of sampling units should be given in addition to such standard errors (of means, totals, etc.) as are of interest. The process of deducing these estimates of error should be made entirely clear. This process will depend intimately on the design of the sample survey. An analysis of the variances of the sampling units into such components as appear to be of interest for the planning of future surveys is also of great value.

Comparisons between independent investigators covering the same material should be given. Such comparison will be possible only when interpenetrating samples have been used, or when checks have been imposed on part of the survey. It is only by these means that the survey can provide an objective test of differential bias among the investigators. Any such comparisons or checks should be fully reported.

The existence and possible effects of nonsampling errors on the accuracy of the results, and of incompleteness in the recorded information (e.g., nonresponse, lack of records, whether covering the whole of the survey or particular areas or categories of the material), should be fully discussed. Any special checks instituted to control and determine the magnitude of these errors should be described, and the results reported.

Another source of error is incorrect determinations of the adjustments arising from observation of quantities which do not correspond exactly to the quantities of which estimates are required.

ACCURACY, COMPLETENESS, AND ADEQUACY OF THE FRAME

The accuracy of the frame can and should be checked and corrected automatically in the course of the inquiry, and such checks afford useful guidance for the future. Its completeness and adequacy cannot be judged by internal evidence alone. Thus, complete omission of a geographic region or the complete or partial omission of any particular class of the material intended to be covered cannot be discovered by the inquiry itself and auxiliary investigations often must be made. These should be put on record, indicating the extent of inaccuracy which may be ascribable to such defects.

COMPARISONS WITH OTHER SOURCES OF INFORMATION

Every reasonable effort should be made to provide comparisons with other independent sources of information. Such comparisons should be reported along with the other results, and the significant differences should be discussed. The object of this is not to throw light on the sampling error, since a well-designed survey provides adequate internal estimates of such errors, but rather to gain knowledge of biases and other nonrandom errors.

Disagreement between results of a sample survey and other independent sources may of course sometimes be due, in whole or in part, to differences in concepts and definitions or to errors in the information from other sources.

COSTING ANALYSIS

The sampling method can often supply the required information with greater speed and at lower cost than a complete enumeration. For this reason, information on the costs involved in sample surveys is of particular value for the development of sample surveys within a country and is also of help to other countries.

It is therefore recommended that fairly detailed information should be given on costs of sample surveys. Costing information should be given under such headings as planning (showing separately the cost of pilot studies), field-work, supervision, processing, analysis, and overhead costs. In addition, labor costs in man-weeks of different grades of staff and also time required for interviewing, travel, and transport costs should be given. The collection of such information is often worthwhile, since it may suggest methods of economizing in future surveys. Moreover, the preparation of an efficient design involves a knowledge of the various components of cost as well as of the components of variance. It must be emphasized that the concept of cost should be regarded broadly in the sense of economic cost and should therefore take account of indirect costs which may not have been charged administratively to the survey. Wherever possible, the costing data should distinguish the time and resources devoted to the various operations involved in the survey.

EFFICIENCY

The results of a survey often provide information which enables investigations to be made on the efficiency of the sampling designs, in relation to other sampling designs which might have been used in the survey. The results of any such investigations should be reported. To be fully relevant the relative costs of the different sampling methods must be taken into account when assessing the relative efficiency of different designs and intensities of sampling.

Such an investigation can be extended to consideration of the relation between the cost of carrying out surveys of different levels of accuracy and the losses resulting from errors in the estimates provided. This provides a basis for determining whether the survey was fully adequate for its purpose, or whether future surveys should be planned to give results of higher or lower accuracy.

OBSERVATIONS OF TECHNICIANS

The critical observations of technicians in regard to the survey, or any part of it, should be given. These observations will help others to improve their operations.

REFERENCES

1. Minton, G. (1969), Inspection and Correction Error in Data Processing, *J. Am. Statist. Assoc.*, **64**.

2. Moser, C. A. (1958), "Survey Methods in Social Investigation," Heinemann, London.
3. Raj, D. (1968), "Sampling Theory," McGraw-Hill Book Company, New York.
4. United Nations Statistical Office (1964), "Recommendations for the Preparation of Sample Survey Reports," ser. C, no. 1, rev. 2, United Nations, New York.
5. Yates, F. (1960), "Sampling Methods for Censuses and Surveys," Charles Griffin & Company, Ltd., London.

part four

Data Collection in Selected Fields

11
Surveys of Agricultural Area

11.1 INTRODUCTION

This part of the book deals with the problems involved in the collection of data in a number of fields. The fields of study include population, agriculture, industry, labor, trade, and transport. In this chapter the problem of the estimation of agricultural areas is considered. This topic is of great importance to the developing countries with an agricultural economy.

11.2 BASIC CONCEPTS

An insight into the basic agricultural structure of a country can be obtained by breaking down the total area of the country according to land utilization. In this manner it is possible to determine the part of the total land that can be used for agricultural production of different types. The broad categories of land utilization are arable land (normally used for growing rotation crops), land under permanent crops, land under permanent meadows and pastures, land under woods and forests, and all other land such as built-up areas, roads, parks, etc. The total arable land can be further broken down as land under temporary crops, under temporary meadows, under market and kitchen gardens, temporarily fallow (left idle for a brief period to recoup before cultivation starts again), and all other arable land. The land under temporary crops can be allocated to the important crops. The category of land under permanent crops includes crops such as coffee, fruit trees, and vines. Such crops remain on the land for a period longer than a year and do not have to be planted after each harvest. A useful breakdown of the land under permanent meadows and pastures is according to whether it is cultivated, that is, seeded and looked after, or uncultivated, that is, it exists naturally.

Table 11.1 presents a summary example of land utilization.

Table 11.1 Land utilization, France, 1962

Category	*Area (thousand hectares)*	*Percent*
Arable land	19,075	34.6
Land under grass	13,066	23.7
Vineyards	1,414	2.6
Wood and forest land	11,566	21.0
Trees	1,165	2.0
Uncultivated agricultural area	4,020	7.3
Nonagricultural land	4,831	8.8
Total	55,137	100.0

Another useful breakdown of the total area is by tenure (see Table 11.2 for an example). Information on agrarian structure is important as the nature of this structure has a considerable bearing on the efficiency of

Table 11.2 Land in holdings by tenure, Belgium

	Owned		*Rented*		
Year	*Area (hectares)*	*Percent*	*Area (hectares)*	*Percent*	*Total hectares*
1846	613,571	34.22	1,179,583	65.78	1,793,154
1856	628,292	34.32	1,202,225	65.68	1,830,517
1866	642,721	32.68	1,323,958	67.32	1,966,679
1880	713,059	35.95	1,270,511	64.05	1,983,570

production. The important categories are area owned by the holder (who operates the land) or held in ownerlike possession, area rented from others, area operated on a squatter basis, and area under tribal or communal tenure form.

A further useful indicator of the efficiency of production is the fragmentation of agricultural holdings. (A holding is all land operated by a single person.) The fragmentation is determined by the number of parcels on which the holding is based. (A parcel is all land operated by a person at one place surrounded by land of other holders.) Table 11.3 provides an example of this type of data in which the holdings are classified by size (total area in hectares).

Table 11.3 Fragmentation of holdings, Malaya, 1960

		Number of holdings containing				
Size group (hectares)	*All holdings*	1 *parcel*	2 *parcels*	3–4 *parcels*	5–9 *parcels*	10 *or more parcels*
Less than 1.2	203,572	138,148	45,538	16,744	3,002	140
1.2–2.0	99,152	39,238	32,908	22,600	4,146	260
2.0–4.0	100,752	25,140	32,640	35,226	7,300	446
4.0–10.1	40,900	5,192	8,442	17,094	9,714	458
10.1–40.4	5,134	490	512	1,414	2,066	652
Over 40.4	140	38	20	26	22	34
Total	449,650	208,246	120,060	93,104	26,250	1,990
Percent	100	46.3	26.7	20.7	5.8	0.5

11.3 CENSUS OF HOLDINGS BY INTERVIEW

Basically there are two methods of collecting information on areas. We may ask all agricultural holders in the country to provide information on the land operated by them. Or, we may visit every piece of land in the country to determine its utilization. The first method is called a *census of holdings* and the second one a *cadastral survey* of the country.

Censuses of holdings are carried out in many countries, especially in the West, every 10 years. Since only the government can require holders to make returns and in exchange give them the assurance that the information collected will be kept confidential, such censuses are sponsored by governments. Information is collected on the number of holdings and their principal characteristics such as size, tenure, and use of land. The interview method is used in which the census enumerator interviews the holder. The interview is held on the holding so that the enumerator can check the information, if necessary. The fact that the data are collected for each holding makes it possible to tabulate the information by any administrative unit, small or large. Another advantage is that the data can be cross classified at any level, thus making it possible to study the relationships between the variables.

The data collected from a census of holdings are subject to two important limitations. In the first place, these data do not give a picture of all land in the country. It is only the areas that belong to some holding that are taken in the census. Second, the data on areas collected by the method of interview can be substantially inaccurate; a holder may not disclose some of the parcels in the holding such as those far away from the place of interview. An example of this is provided by a study carried out by the author (7) in Greece. A year before the census of holdings, a cadastral survey of five communes was undertaken. In this survey all land was measured and the name of the holder was recorded. This information was matched against the data collected in the census relating to these five communes. The following were the important findings:

1. In the five communes investigated, 9 percent of the resident farmers did not report their land at all, 22 percent of the parcels were not declared, and 7 percent of the agricultural area was not reported.
2. Only 65 percent of the absentee farmers (not residing in these five communes) mentioned these communes in their reports and declared 64 percent of parcels and 71 percent of area.
3. As for all land in the five communes, 10 percent of the agricultural area was not reported and 23 percent of the parcels were missed.

The method of interview may be used on just a sample of holdings in order to reduce costs. In this case the accuracy of the results obtained will depend upon the procedures used in the survey. As an example (6), three different surveys were conducted in India to estimate the average size of an

Table 11.4 Average size of operational holdings in India

Region	*Survey* 1	*Survey* 2	*Survey* 3
North	3.9	5.3	6.7
Northwest	11.9	12.6	20.6
East	3.5	4.5	4.8
Central	10.9	12.2	14.6
West	11.1	12.3	14.9
South	3.8	4.5	6.2
All India	6.1	7.5	8.9

operational holding. In the first survey information was collected by the investigators in the course of a single visit. The duration of stay in the village was longer in the second and third surveys. In the second survey the size of the holding was not a part of the main inquiry while the third survey centered entirely around the holding and the holder. The results obtained are shown in Table 11.4. It is obvious that different methods of interviewing produced different results. The size of the holding increased progressively as greater effort was made to collect the data.

Another example comes from Japan (1). In this study the data on planted areas were collected by interviewing farmers. At about the same time, actual measurements were taken to check the accuracy of the interviewing method. The results (Table 11.5) show that farmers understate the areas planted. The understatement is more pronounced when the crop is of high marketability.

Table 11.5 Comparison of area estimates based on interviewing and measurement, Japan, 1959

	Area (thousand hectares)			
Crop	*By interviewing*	*By measurement*	*Ratio (%)*	*Marketability of crop (%)*
Paddy rice	2,750	2,951	93	61
Wheat	522	592	90	60
Sweet potato	294	333	89	62
Rapeseed	142	182	78	80
Mandarin	48	64	76	96
Apple	41	58	71	100

11.4 CADASTRAL SURVEYS BY PHYSICAL OBSERVATION

The situation is entirely different when all land in the country has been cadastrally surveyed. In this case there is a map showing the boundaries of the fields situated in each administrative unit such as the village. The fields are serially numbered on the map and the geographical area of each field is known accurately. Sometimes there is an elaborate government agency charged with the task of visiting each field once a year in order to determine the utilization of land. In this situation it is possible to get fairly accurate statistics of land utilization by adding up the areas of all those pieces of land which fall under a particular category of utilization. Cadastral maps of villages have been found to be extremely useful in India for purposes of land utilization studies.

The mere existence of a cadastre in the country is, however, no guarantee that the area statistics are accurate. Unless the operations are carried out under supervision and sample checks made regularly, the area statistics based on a complete enumeration of fields or farms may be very inaccurate. As an example, consider Table 11.6, which gives areas under different crops obtained in two ways (4). The original figures were provided by the normal government agency whose job is to visit each field and note down the utilization of land. A group of experienced investigators made visits to a sample of fields and noted the actual utilization in the presence of and with the agreement of the normal government agents. These are the "checked" figures shown in column 4 of Table 11.6. The discrepancies are quite large. There is unmistakable evidence that the complete enumeration records are far from accurate; the algebraic discrepancy ranged from –44 percent to 552 percent.

Table 11.6 Comparison of crop acreages as reported by government agents and checkers, India, 1950–51

		Acreage		*Discrepancy as percent of column 3*	
Crop	*Number of comparisons*	*Original*	*Checked*	*Absolute*	*Algebraic*
(1)	(2)	(3)	(4)	(5)	(6)
1	1,170	1,005.2	863.6	31	16
2	277	91.7	81.3	45	13
3	1,819	1,739.0	1,301.0	59	34
4	610	183.9	255.4	93	–28
5	357	105.5	189.9	61	–44
6	247	58.1	38.0	85	53
7	130	174.1	26.7	580	552

It is sometimes said that a complete enumeration of all fields by means of a cadastral survey is an essential prerequisite for the determination of areas under different crops. This statement should be judged carefully in the light of the considerable time, money, and expertise needed in order to carry out a cadastral survey. If the object is to obtain areas under important crops at the national level, sampling methods can be used effectively to achieve this even when a cadastral survey has not taken place. It is only when area statistics are needed for small administrative subdivisions of the country or for minor crops that complete enumeration of all fields has to be undertaken.

11.5 CURRENT AGRICULTURAL STATISTICS

In view of the costs involved, a census of holdings or a cadastral survey is conducted at infrequent intervals, say, once in 10 years. But the agricultural situation in the country changes year after year. For example, farmers may start devoting a greater area to commercial crops at the expense of food crops, or some areas may be abandoned if the weather is unfavorable, and so on. Thus it becomes important to gather current statistics of areas each year rather than depend on the results of the outdated census.

In many of the economically developed countries current statistics of areas are collected by asking the agricultural holders to report on the areas harvested by them. These reports are added up to find the total cultivated area and the areas under different crops. The method is inexpensive and works when the holders are well informed and are willing to give accurate information on their areas. When the number of holdings is not large, the data can be edited quickly and the results brought out in time.

This method does not work in the developing countries for the simple reason that many of the farmers are illiterate. They are usually not interested in the statistical program of the country. In such a situation, it is not uncommon for the government to appoint reporters in selected areas all over the country. These reporters make eye estimates of the areas under different crops in the regions under their jurisdiction. When aggregated, these reports provide information on areas under different crops. A major defect of this method is that some reporters may not go to the field to make the eye estimates. Since crop reporting is not their main job, they are unlikely to be interested in this type of work. Further, the eye estimates may be far off the mark. The advantage of the method is that it is inexpensive and that the analysis of data is simple. If the reports are biased in the same direction, the estimate of the level for a given year may be seriously biased. But the estimate of trend over the period may not be so poor.

Current statistics based on eye estimates can be considerably improved if a cadastral survey of areas has taken place in the country. In this case cadastral data can be prepared for the administrative areas for which eye estimates are to be reported. The administrative areas are broken down into smaller segments with distinguishable boundaries. The job of the reporter then is to estimate for each segment the proportion of area under each crop. When the proportions are established, a simple multiplication by the total area gives the areas under different crops. This method provides data that are verifiable if the size of the segment is not large.

The United States presents a good example of the use of census data for obtaining current statistics. Suppose, for example, that information is to be collected on the total area intended to be planted in a given month. Questionnaires are mailed to a large number of agricultural holders asking them to furnish information on the area intended to be planted and the total area of the holding. These reports are aggregated for conveniently chosen units to estimate the total area x and the area to be planted y for the unit. The variable y is plotted against x to determine the regression coefficient of

y on x. Since the total value of x is known from the census, the regression estimate can be used for getting an improved value of the total area to be planted. In the same way a forecast can be made of the area to be harvested under each crop. The method works because the holders are well informed and are willing to give the information needed.

India presents an example of a different kind. Here more than 70 percent of the area has been cadastrally surveyed and there is a revenue collector for each village or group of villages. As a part of his duties the revenue collector goes from field to field and records the crop grown on each field. Since the area of each field is known from the cadastral register, the total area under each crop can be determined very easily. This exercise is undertaken at least once a year. If sample checks show that the work of the revenue agent is satisfactory, it is possible to collect usable current statistics of agricultural areas by this method.

11.6 SAMPLE CENSUS OF AREAS IN DEVELOPING COUNTRIES

There are many developing countries in which no cadastral survey has ever taken place nor a census of agriculture taken. A large proportion of the population is illiterate and so the question of collecting information by mail does not arise. As resources are limited, it is not possible to carry out a cadastral survey. The number of qualified enumerators is small and transportation is difficult. This rules out the possibility of a census of holdings by the interview method. In such circumstances sampling methods can play a useful role. By restricting the volume of operations it is possible to collect usable information with the help of a few trained enumerators. When the data collected are of the census type, these inquiries are called *sample censuses*. These sample inquiries can be used for obtaining national totals with sufficient precision; regional totals, too, may be obtained when it is possible to increase the size of the sample considerably.

One great difficulty encountered in the sample census is that the farmer is not very helpful. He does not know how much area he operates. The units of area differ from place to place and their relationship to a standard unit is not known. Thus the enumerators have to make their own measurements of the parcels in the holding. But only those parcels can be measured which are declared by the holder. Quite often, for fear of taxation or otherwise, the holder is reluctant to declare all land operated by him. The second difficulty with a sample of holdings is that the holdings are found to be very variable with respect to total area and the areas under different crops. The result is that the sampling errors are often quite large when the sample size is reasonably small.

It should be pointed out that the method of measurement of areas is being increasingly used everywhere. It is now realized that the method of mail questionnaires is not an objective method of collection of data. Even an interview may be suspect. Thus there is the need to make an adequate number of measurements on the ground in order to adjust the data obtained by mail or interview.

11.7 SAMPLING BY ADMINISTRATIVE AREAS

We now consider the problem of estimating areas under different crops by sampling methods. A number of methods have been used in different countries. Such methods are always tailored to suit the conditions obtaining there. In most developing countries it is found possible to make a list of villages and stratify them on the basis of available information. From each stratum a sample of villages is taken with probability proportional to some measure of size. If a cadastral map of the village is at hand, a sample of fields can be selected and investigated for the crops grown. Otherwise, a list of all households in the village can be made and a suitable sample of households taken. The parcels operated by the sample households are measured and the area under each crop is determined. This method may be called *sampling by administrative areas*. An actual application of the method follows.

A sample survey of the area under winter paddy was conducted in the state of Orissa (India) in 1950–52 (2). The state contains eight districts, each of which was divided up into 10 to 15 strata. Groups of contiguous villages were the strata of the survey. From each stratum 10 to 12 villages were selected with replacement with probability proportional to the number of fields in the village. Within a selected village the fields were grouped into clusters of eight, out of which a simple random sample of four clusters was taken. Enumerators visited the selected clusters and recorded information relating to land utilization and the area under different crops. No ground measurements of areas were needed in the seven districts which were cadastrally surveyed. In the eighth district a simple random sample of villages was taken from each stratum. The area of a selected village was divided up into blocks. The dividing lines were roads, footpaths, etc. The block boundaries were shown on the village map prepared on the spot by the enumerator. Each patch of land in the block was measured and the name of the crop was stated. The results obtained are given in Table 11.7. The table shows that the area under this principal crop can be estimated correct to 5 percent by taking 1,142 villages in the sample. The margin of error is of the order of 12 percent for most of the districts. Certainly a fairly large sample size is needed for making good estimates of areas under even principal crops. With such a sample, estimates of the minor crops will be subject to very large sampling fluctuations.

The sampling errors of area estimates can be reduced in continuing surveys by replacing the sample partially when there is high positive correlation between successive occasions. In the present survey this device was tried in

Table 11.7 Area under paddy, Orissa, 1950–52 (in thousand acres)

District	*Number of villages in sample*	*Number of clusters in sample*	*Estimated area*	*Relative standard error (%)*
1	179	716	1,006	7.2
2	136	544	715	6.1
3	149	596	895	6.8
4	110	440	289	6.5
5	149	596	619	5.2
6	139	556	892	5.1
7	91	364	299	8.7
8	189	....	465	10.5
Total	1,142	....	5,180	2.5

the first district. A total of 195 villages was selected from this district, one half of which matched with the previous survey. The relative standard error dropped to 4.7 percent The method did not work in the last district, the relative error being 9.4 percent. The explanation given is that this district is inhabited by tribal people who practice shifting cultivation. As a result the correlation coefficient between successive occasions is low.

Another example of a sample survey of areas is provided by the National Sample Survey of India (5). In the 1958 survey stratification was geographical and in all, 289 strata were made covering the entire country. From each stratum a number of villages was selected with probability proportional to

Table 11.8 Net areas under important crops, India, 1958

Crop	*Estimated area (thousand acres)*	*Relative sampling error (%)*
Rice	74,907	2.63
Jowar	37,037	4.35
Bajra	27,588	3.65
Ragi	4,529	7.37
Maize	9,789	6.48
Wheat	29,009	3.24
Barley	6,797	4.47
All cereals	189,656	1.64

Total number of villages in sample 3,126

area. From a selected village 8 clusters of 10 plots each were selected systematically when village records were available. Otherwise, eight households were selected systematically from the complete list of households in the village and all land operated by these households was investigated. In each case the data were collected by actual measurement of the plots. Some of the results are given in Table 11.8. It is apparent from the table that a sample as large as 3,126 primary sampling units is needed to produce usable estimates at the national level. The reason is that the area under a crop fluctuates widely from place to place.

11.8 GRID SAMPLING

When detailed maps of areas are available, it is possible to select a sample from an area by taking a number of points on the map at random. Grids or clusters of a suitable size and shape can then be marked on the map using the point selected as, say, the southwest corner of the grid. The boundaries of these grids are then identified on the ground and all areas within the grid are enumerated for land utilization. This method should work when the maps available are so elaborate that the grids marked on them can be located on

Table 11.9 Area under paddy and jute, West Bengal, 1956–57 (in thousand acres)

	Paddy			*Jute*		
District	*Number of grids*	*Area*	*Relative sampling error (%)*	*Number of grids*	*Area*	*Relative sampling error (%)*
1	4,533	1,022	0.98	4,056	26	7.97
2	3,180	657	1.21	3,100	1	42.86
3	4,729	704	1.40	4,869	1	22.32
4	6,883	1,920	0.83	6,819	28	5.92
5	935	199	2.14	810	11	9.95
6	1,999	441	1.53	1,673	59	5.28
7	5,798	1,456	0.66	6,152	66	3.24
8	2,833	215	2.94	2,832	69	3.67
9	3,697	457	1.72	3,668	135	2.75
10	2,530	521	1.30	2,491	57	4.48
11	2,479	287	1.85	2,486	78	3.59
12	3,236	434	1.73	3,279	50	4.68
13	434	62	4.30	433	6	9.95
14	2,046	369	1.74	2,197	63	3.08
Total	45,312	8,744	0.35	44,865	650	1.24

the ground without any ambiguity. When the topography of the area is complicated, it is difficult to identify the grid boundaries with the help of the map. The method has the advantage of simplicity; sample selection is simple when detailed maps are available.

This method has been extensively used in India for estimating the area under paddy in the state of West Bengal where detailed maps are available (8). In the 1956–57 survey about 50,000 square grids of the size of 2.25 acres were marked on village maps. The field staff were supplied with the plot list of the grids to be surveyed along with the village maps. They were required to identify each plot in the grid and record the utilization of land in the plot. The utilization was indicated in percentage terms for each category. Each grid covered on the average 10 to 12 plots. If a plot was not entirely within the grid, the utilization in the part included was estimated. This was done by determining the utilization for the whole plot and multiplying it by the proportion of the plot included in the grid. Table 11.9 gives the estimates for paddy (a food crop covering 90 percent of the total area) and jute (a commercial crop covering 7 percent of the total area). The estimation method used was the simple unbiased estimate. The table shows that a very large number of grids is needed in order to make estimates with a low percentage error.

Table 11.10 Gross area under cotton, Egypt, 1958 (in hundred feddans)

	Simple average		*Ratio estimate*	
Province	*Estimate*	*Percent error*	*Estimate*	*Percent error*
1	871	6.17	702	2.36
2	1,338	4.73	1,369	2.13
3	1,330	4.08	1,277	1.21
4	1,853	3.66	1,798	1.49
5	791	5.35	849	1.80
6	940	4.63	966	1.39
7	340	7.49	338	1.89
8	602	4.41	577	1.70
9	883	3.24	900	1.49
10	2,126	2.81	2,114	0.97
11	1,797	2.49	1,748	1.25
12	2,639	2.09	2,789	0.90
13	1,187	3.61	1,401	1.49
14	2,094	3.28	2,089	1.19
Total	18,791	1.00	18,917	0.40

If there has been a plot-to-plot complete enumeration of areas, it is possible to improve the precision of the estimates made by the method of grid (or cluster) sampling. This can be done by making the ratio estimate, the denominator being the plot area obtained from complete enumeration. This approach was used in Egypt in 1958 for estimating the gross area under cotton (1). The cultivated land in each district was divided into clusters of about 2,000 feddans† by combining adjacent villages. The main consideration in making clusters was stability in rotational practices. The area under cotton for the year 1957 was obtained by complete enumeration of all the clusters in the country. For the survey of 1958 a simple random sample of 25 percent of the clusters was taken and the area under cotton was obtained by actual measurement. Both the simple average and the ratio estimate were formed. Table 11.10 gives the results obtained (1). It is clear that the use of the ratio method can give an estimate of the area under cotton with a coefficient of variation of 0.4 percent when a 25 percent sample is taken. The relative errors of provincial estimates lie between 1 and 2 percent.

11.9 JOINT USE OF LIST AND AREA SAMPLING

An imaginative use of the census for purposes of obtaining current statistics on agriculture has been made in the United States. Two samples were selected for the 1960 survey, one called the *list sample*, based on farms enumerated in the 1959 census, and the other called *area sample*, to take in farms existing in 1960 but not covered by the census (nonassociated farms). The two samples were selected in such a manner that there is the greatest possible overlap between them. This was done by selecting what is called a *segment list sample* (3) in the following manner. A one-to-one correspondence was established between each farm in the census and a point on a map. The sample was obtained by identifying those farms in the census for which the corresponding points fall within the boundaries of a sample of segments selected with known probabilities.

The area sample of nonassociated farms was selected in two stages. As many as 999 counties were selected with probability proportional to the value of farm products as obtained in 1954. In the second stage about 2,700 segments, with about 3 farms per segment on the average, were selected with variable sampling fractions. A check was made of all farms with headquarters in the sample segments to see whether they were associated with any farm enumerated in the census. In this manner the nonassociated farms were identified for inclusion in the area sample.

The list sample of associated farms consisted of three parts. All farms in the segment list sample with value of farm products below $40,000 as

† 1 feddan = 0.4208 hectare.

obtained in the census were included in the sample. The second part consisted of a systematic sample of farms located in the sample counties and with a value of farm products lying between $40,000 and $100,000. The third part was based on a systematic sample of the largest farms (with value of farm products exceeding $100,000) selected in all counties in the country. In all there were 8,200 farms in the segment list sample and 3,800 in the other two samples.

The survey enumerator collected information from all farms except those included in the second and third parts of the list sample. Information on these farms was collected by mail in the first instance, to be followed by an interview if necessary. As far as possible the census data were used as auxiliary information for improving the precision of the estimates. Some of the results obtained are presented in Table 11.11. It is clear from the table that the use of the difference estimate brings about substantial improvement in the precision of the results.

Table 11.11 Approximate sampling errors of estimates of selected items, United States, 1960

		Relative standard error (%)	
Item	*Estimated total* (*thousands*)	*Unbiased estimate*	*Difference estimate*
Number of farms	3,253	2.5	1.0
Land in farms (acres)	1,133,907	3.4	2.4
Cropland harvested (number of farms)	2,950	2.6	1.0
Cropland harvested (acres)	317,980	2.2	0.9
Cotton harvest (number of farms)	....	6.6	3.0
Cotton harvest (acres)	....	5.3	2.9

11.10 MIXED CROPS

A major difficulty in estimating area under different crops is that quite often a number of crops are grown on the same field. The crops in the field may be at different stages of development when the area is estimated. And the pattern of intercropping may not be regular. In such cases it becomes very difficult to allocate the area to each crop. No general solution of the problem is available which is satisfactory under all circumstances. A number of procedures usually followed will be outlined here.

In principle, the area under mixed crops should be distinguished from the area under pure stand. The area under mixed crops may be further subdivided as area in which crop A is predominant, area in which crop B is predominant, and so on. Table 11.12 is an illustration of this method of

Table 11.12 Area under crops, Senegal, 1960–61 (in thousand hectares)

Crop	*Pure stand*	*Principal mixture*	*Total*	*Secondary mixture*
Groundnuts	345	347	692	30
Millet	255	58	313	313
Rice	46	1	47	0
Niebes	2	0	2	88
Total	648	406	1,054	431

allocating areas to crops (1). In this case the total area under crops is 1,054,000 hectares. The area of all fields in which groundnuts stand pure is 345,000 hectares and the area in which groundnuts form the principal component of the mixture is 347,000 hectares. And in 30,000 hectares groundnuts are a secondary component of the mixture and so on for other crops. It may be said that the last column of the table should be omitted as it is difficult to interpret. If this is done, a minor crop will not find a place in the table when it does not stand alone or is not the principal component of any mixture.

Another solution to the problem is to present the area under each combination of crops separately. This is unsatisfactory when the number of combinations is very large. A third method is to determine the principal component of a mixture and to allocate the entire area to the principal component. In this case it is important to distinguish the area under pure stand from the area under the mixture (Table 11.12).

11.11 METHODS OF MEASUREMENT

When cadastral maps are not available, the area of the field selected in the sample has to be measured. There are many methods of measuring a field, all methods being expensive and complicated. The basic requirement is to draw a figure of the field according to a convenient scale and measure its area. For drawing the figure one has to measure distances and angles. Distances can be measured with the help of surveyor's chain. It is a cheap instrument which can be easily handled. But two men are needed to handle it. Another instrument is the metallic tape. It gives more accurate readings since no sagging is involved. But it is not as strong as the chain. A standardized cord provides another simple means of measurement. The precaution to be taken with the cord is that it should not get wet. When these instruments are not available, the method of pacing may be used. But then the steps of the pacers should be standardized to avoid systematic errors. An inexpensive method, which is better than pacing, is the use of Smith's wheel when the ground is flat. In this method the operator walks the distance pushing a wheel (circumference 1 meter); the number of revolutions is registered on a counter.

Measurement of angles is not needed when the method of triangulation is used for measuring areas. In this method the area is divided up into triangles and the three sides a, b, and c of each triangle are measured. The area of the triangle is calculated from the formula

$$\triangle = \sqrt{s(s-a)(s-b)(s-c)}$$

where $2s$ is the perimeter of the triangle. Quite often this method cannot be used since it usually involves trampling the crop. In that case angles need to be measured. The instrument commonly used for measuring angles is the compass. After the distances and angles have been measured, a sketch of the field can be drawn to measure the area. While drawing the sketch care should be taken that it closes up.

11.12 CONCLUDING REMARKS

The foregoing discussion shows that there is no method of estimating areas which is best under all circumstances. The best method depends very much on local conditions. As a general rule, any method proposed to be used should be tested in a pilot survey. Furthermore, a permanent organization charged with the task of collecting annual information on important crops is likely to be much more effective than an *ad hoc* organization built up to deal with a particular crop in a single year. If local data are needed or information on minor crops is to be obtained, complete enumeration of all areas together with independent checks on a sample basis appears to be the answer.

REFERENCES

1. Food and Agriculture Organization (1965), "Estimation of Areas in Agricultural Statistics," Rome.
2. Indian Council of Agricultural Research (1952), "Report on the Crop Survey by the Random Sampling Method for Estimating Acreage under Principal Crops in Orissa State," New Delhi (unpublished).
3. Jabine, T. B., R. Hurley, and W. N. Hurwitz (1965), Sample Design and Estimation Procedure for the 1960 Sample Survey of Agriculture in the United States, in "Estimation of Areas in Agricultural Statistics," Food and Agriculture Organization, Rome.
4. Mahalanobis, P. C., and D. B. Lahiri (1961), Analysis of Errors in Censuses and Surveys with Special Reference to Experience in India, *Bull. Intern. Statist. Inst.*, **38**.
5. National Sample Survey (1958), Some Results of the Land Utilization Survey and Crop-cutting Experiments, *Govt. of India Rept.* 38, New Delhi.

6. Panse, V. G. (1958), Some Comments on the Objective and Method of the 1960 World Census of Agriculture, *Bull. Intern. Statist. Inst.*, **36**.
7. Raj, D. (1962), "Farmers Reporting at the Census" (report submitted to the government of Greece), Athens.
8. West Bengal State Statistical Bureau (1956), "Report on the Sample Surveys for Estimating Acreage and Yield Rates of Crops in West Bengal," Calcutta.

12
Surveys of Agricultural Production

12.1 INTRODUCTION

The problem of estimating the annual production of food crops and other agricultural commodities is of great importance to all countries. It is more so with developing countries which are making strenuous efforts to feed their populations and raise living standards. The production of crops such as rubber, tea, or jute can be estimated somewhat accurately by consulting the marketing organizations dealing with the commodities. But crops such as wheat or paddy, which are grown by millions of farmers all over the country, present a formidable problem. One method of estimating the total production of such crops is to estimate the total area under the crop and the yield rate. The product of the two components is an estimate of the total production of the crop. In the previous chapter the problem of estimation of areas was discussed. This chapter deals with the more difficult problem of estimating yield rates.

12.2 CROP REPORTS

In most countries statistics of crop yields are based on periodic reports from crop reporters—who may be farmers or government officials. The reporters make their reports on the basis of their own assessment of the condition of the crop and the judgment of farmers whom they may happen to ask. The method is highly subjective and is therefore subject to unknown biases. Experience at some places shows that the reporters have a tendency to lean toward the normal; the yield is understated when the season is good and overstated when it is poor. Such a comparison is possible since we do have objective, although expensive, methods of estimating crop yields. Take, for example, the Swedish data presented in Table 12.1. The crop reports are consistently lower than the yield obtained by objective methods, the underestimation ranging from 8 to 27 percent (8).

Table 12.1 Crop yields, Sweden, 1957 (in kilograms per hectare)

	Objective methods			
Crop	*Biological yield*	*Yield adjusted for waste, etc.*	*Crop report*	*Underestimation (%)*
Winter wheat	3,340	2,930	2,400	18.1
Winter rye	2,870	2,510	2,070	17.5
Spring wheat	2,810	2,470	1,820	26.3
Barley	2,660	2,300	2,120	7.8
Oats	2,390	2,040	1,670	18.1
Potatoes	22,280	17,310	12,560	27.4
Tame hay	7,320	4,240	3,690	13.0

Table 12.2 Comparison between official and survey estimates of production of wheat (in thousand tons)

Year	*Official estimate*	*Survey estimate*	*Percent difference*
1943–44	2,472	1,901	30.1*
1944–45	2,593	2,356	10.1*
1945–46	2,253	2,282	−1.3
1946–47	2,278	2,237	1.8
1947–48	2,475	2,194	12.8*
1948–49	2,165	1,974	9.7*
1949–50	2,408	2,426	−0.8
1950–51	2,498	2,520	−0.9
1951–52	2,247	2,374	−5.4*

* Significant difference.

Our next example comes from the state of Uttar Pradesh in India (1). This state has a long established system of crop reporting. Beginning in 1943 the production of wheat was also estimated by more objective methods (based on crop cutting) as well. A comparison of the two methods is given in Table 12.2. It is clear from the table that the production figures obtained by the official method (based on judgment reports) are wide of the mark and that the bias fluctuates considerably around the survey estimate. As a result the state government in 1952 discontinued the official method of crop reporting.

12.3 THE METHOD OF CROP CUTTING

The objective method of estimating yield rate consists in selecting a sample of parcels or fields at random and using for observation the method of physical measurement of yield at harvest. The sample units actually harvested are plots of a prescribed dimension located and marked in the parcel (or field) according to clearly defined procedures. The usual method is to make a list of first-stage units, say, villages, in the area to be studied. A sample of villages is selected and a list of fields growing the crop in question is prepared for each village in the sample. A sample of fields is taken and a plot is marked at random in the selected field. The plot is harvested and the produce is weighed after it has been dried. Surveys in different countries have shown that this is a practicable method capable of giving yield estimates free from bias and possessing a high degree of accuracy. But it is an expensive method, requiring the use of a large number of trained enumerators. The method is delicate in the sense that various biases can creep into the results if attention is not paid to the details. Some of the difficulties involved in the use of this method are discussed in the next few sections.

12.4 SELECTION OF FIELDS

Sample surveys of crop yields are usually carried out crop by crop. The reason is that different crops mature at different times. The sample for a particular crop can only be taken when the crop is mature. This means that there is only a very short period before the actual harvest during which the field is fit for sampling. Consequently, if a random sample of fields is taken, there is great difficulty in ensuring that all fields are visited at the appropriate times.

A method used to meet this difficulty is to employ a local enumerator, who makes a preliminary visit to the selected fields and fixes the date of harvest in consultation with the farmers concerned. The farmers are requested not to harvest the selected fields until a sample of the crop has been taken. The enumerator visits the field on the date fixed for harvesting and takes the sample. In this way very few fields are missed. The disadvantage with this method is that the fields in the sample are known in advance. For fear of taxation or otherwise, the farmer may try to reduce the yield by stripping ears or by removing shoots, etc. Second, the date fixed may be inappropriate in that the harvest is delayed beyond its proper time; this may lead to loss of yield due to shredding of grain, etc.

There is another method in which the fields to be sampled are not known in advance to the farmers. A sample of clusters of fields is taken on the spot (say, by taking a sample of households and asking them about the fields they operate) and those fields are determined which grow the crop in question, on which the crop has not been harvested as yet and which are fit for sampling. A sample of two fields is taken and the yield is determined by subsampling the fields. The difficulty with this method is that the crops coming into harvest at different times will be unequally represented in the sample. For example, if the work is begun late after a proportion of the fields have already been harvested, the yield will be understated if the later maturing crops give lower yields. This cannot be avoided unless the sampling is so adjusted that its distribution over time corresponds to the distribution over time of the actual harvesting of fields. Second, the interval between the time when a crop is considered fit for cutting and the actual time of harvest is not constant. If this interval is small for some crops, the method will give a smaller proportion of these crops in the sample. The bias involved cannot be disregarded unless it can be shown that there is no correlation between the yield and this interval. In the third place, the selection of only two fields from all those judged fit for cutting at a given time will result in overrepresentation of the areas in which the proportion of the fields under the crop is low. This source of bias can be eliminated by taking a sample from all or a fixed proportion of the fields judged fit for cutting.

12.5 LOCATION OF SAMPLE CUTS

In order to estimate the yield from a selected field, a plot or a sample cut of a suitable shape and size is to be located at random in the field. The sample cut is usually located by taking a pair of random numbers, say x and y. The enumerator walks x paces along the length of the field and y paces in the perpendicular direction. The point reached is treated as the southwest corner of the plot to be marked if the plot is a square, a rectangle, or a triangle. In the case of circular plots the point reached is the center of the circle. It is clear from the description of this method that certain parts of the field which lie near to the borders have no chance of selection. The reason is that when the sample point falls in these areas, the plot put up on the basis of the sample point does not lie entirely within the field and therefore has to be rejected. Thus a bias is likely to creep into the results if the border areas are different with respect to yield from the other parts of the field. The bias so caused is called *border bias*. It is clear that the border bias is expected to be smaller if the size of the plot is small.

A number of studies have been made (especially in India) to determine the extent of border bias. The method consists in observing the yield rate on the basis of plots located at different distances from the border and comparing the results obtained. Unfortunately, the conclusions drawn by the different agencies in India appear to be conflicting. While one agency asserts that there is no danger of border bias, the other finds that the yield increases as one proceeds from the border to the center of the field (3, 6). It appears that the investigator will have to be guided by his own experiments conducted under the agricultural practices prevalent in the country.

12.6 THE SIZE OF SAMPLE CUTS

The question of the size of the plot and its shape has been considered at length in many countries. India provides the greatest amount of information on this subject. To cut the argument short, there is no doubt that the investigator prefers to use plots of a small size. Small plots have the advantage that they can be rapidly harvested; the produce can be threshed on the spot by the investigator himself and weighed. If the produce is to be sent to headquarters, a small bag will do. There is, however, an obvious danger of bias through faulty location or demarcation of the plots if the enumerators are not sufficiently skilled or if the instruments used for demarcation are unsuitable. It has been observed by Mahalanobis and Sengupta (3) that there is a tendency on the part of the enumerator to include plants on the border of the plot, which do not rightly belong to it. This results in an overestimate of the yield. The percent overestimation decreases as the size of the plot is increased. A number of experiments were made by them in which three concentric cuts with radii of 2 ft, 4 ft, and 5 ft 7½ in. were used to give cumulative areas of 12.6, 50.3, and 100.9 sq ft respectively. By recording the yield of each cut it was possible to compare the results obtained from cuts of different size. It was found that the smallest circle gave a bias of the order of 5 to 15 percent relative to the others and that the medium and the largest circles did not differ from each other. These results led them to the conclusion that a circular cut of 50 sq ft in area was about the right plot to use for estimating the yield of rice and jute under Indian conditions.

The experience of the Indian Council of Agricultural Research has been not very different. Working under different conditions (using the normal government agency and not enumerators especially trained for the purpose), they have found that very small plots could overestimate the yield by as much as 42 percent (see Table 12.3). A number of cuts of various sizes were tried on wheat and the whole field was harvested in a number of experiments in

Table 12.3 Overestimation of yield with small plots

		Irrigated wheat		*Unirrigated wheat*	
Shape	*Area of plot (square feet)*	*Average yield (pounds per acre)*	*Percent overestimation*	*Average yield (pounds per acre)*	*Percent overestimation*
Triangle	471.55	831.1	....	539.0	
	117.89	870.6	4.8	598.2	11.0
	29.47	961.9	15.7	664.9	23.4
Circle	28.29	954.5	14.9	618.8	14.8
	12.57	1,183.3	42.4	767.7	42.4

order to make proper comparisons. It is clear from Table 12.3 that there is appreciable bias with plots as large as 118 sq ft. Accordingly, they recommend the use of large plots, such as an equilateral triangle with side 33 ft (7).

The common denominator of the two investigations is that it is dangerous to use very small plots for estimating the yield of crops such as wheat, paddy, or jute under Indian conditions. However, a plot of a small size is very suitable where a mobile field staff has to cover a large area during the relatively short harvesting season by rapidly moving from place to place just ahead of the normal harvest. Thus the smallest plot which does not overestimate the yield is the right choice for crop-cutting experiments.

12.7 THE SHAPE OF THE PLOT

We shall now consider whether the shape of the plot has anything to do with the under- or overestimation of yield. For this purpose a number of studies have been made in India using plots of different shapes—square, rectangular, circular, triangular, etc. A direct comparison between triangular and circular plots of about the same size (28 to 29 sq ft) is provided by Table 12.3. It appears that circular plots are less subject to bias than triangular plots. The same type of result follows from the investigations of Mahalanobis (3). In one of his studies Mahalanobis used circular, triangular, and square plots of the same size of 12.5 sq ft. Another shape used was the fork (a rigid tool in the shape of a fork with two parallel prongs) of the same size. The average yield of the crop was obtained for cuts of each shape. The comparison is given in Table 12.4. In this table the circular plot is considered the standard and the yields based on other shapes are given in percentage terms relative to the circle. It follows from the table that the triangular plot appears to be considerably biased upward. The following explanation may be given for the fact that the circular cut is the least biased. The bias is caused by the

Table 12.4 Effect of the shape of plot on yield

	Average yield relative to circular cut			
Shape	*Place* 1	*Place* 2	*Place* 3	*Total*
Circle ○	100.0	100.0	100.0	100.0
Triangle △	115.8	125.0	123.2	123.5
Square □	93.0	109.3	107.9	103.5
Fork ┌┴┐	91.1	100.4	108.0	103.5

human tendency to include plants on the border of the cut, which do not rightly belong to it. Thus overestimation is caused by disturbances on the border. Now, for a given area, the circle has the smallest perimeter and therefore the disturbance is a minimum in this case.

It should be pointed out that the question of the shape of the plot cannot be considered without regard to other factors. The convenience of marking the plot is also important. Experience shows that there is no hardship involved in marking a circular plot on the ground. Having selected the random point, all that need be done is to erect a vertical rod at the point. By rotating the arm of the circle it is fairly easy and quick to decide which plants are inside the plot.

12.8 EXPERIMENTS ON CULTIVATORS' FIELDS

The yield obtained from a sample cut is usually obtained under controlled conditions. This is different from the conditions under which the farmer will harvest his fields. The result is that the estimate of the yield obtained from the survey will not be comparable with the yield actually obtained by the farmers. For example, the farmer will lose a part of the grain during the full scale harvesting and threshing but the experimenter will not. Thus the results from the survey need to be corrected to allow for these factors. One method commonly used is to ask the farmers to harvest the fields by using the tools and machinery that they would normally use for the purpose. This is done on a sample of fields from which cuts have been taken. A comparison between the two figures provides the correction factor. As an illustration, consider the data presented in Table 12.5 based on two studies in Sweden (5, 9). It is clear from the table that the loss may vary from crop to crop and from year to year. Sometimes commercial yields are available from figures on marketing for a subsample of the sampled fields. These figures can be used for making a comparison with the corresponding sample yields in order to determine the correction factor or to verify their accuracy.

In another approach the survey procedures are made to resemble as

Table 12.5 Estimated loss in yield (%), Sweden

	Estimated percent loss	
Crop	1958	1959
Winter wheat	2.9	6.9
Winter rye	6.1	4.9
Spring wheat	3.7	8.0
Barley	6.1	6.1
Oats	6.6	11.2

closely as possible the procedures actually used by farmers. Thus the investigators are asked to cut the crop as the farmers do, cut it at harvest time, dry it as the farmers do, and so on. The correction involved is expected to be smaller in this case but investigations are certainly needed to determine the amount of correction. The reason is that farmers' practices are not uniform over the entire country.

12.9 MIXED CROPS

In developing countries it is a common practice to grow a number of crops simultaneously on the same field. The result is that the area under a crop has to be broken down as (1) area under pure stand and (2) area in which the crop is the predominant constituent of the mixture (Sec. 11.10). If the two kinds of areas are treated separately in crop-cutting experiments, no great difficulty is encountered in estimating the yield rates. This means that crop-cutting experiments are to be conducted in fields selected at random from both the categories. With regard to the second category (mixed crops), the yield of the crop obtained from the sample cut will relate to the entire area

Table 12.6 Yield rates of major crops, India, 1958–60 (in pounds* per acre)

	1958–59		1959–60	
Crop	*Pure crop*	*Mixed crop*	*Pure crop*	*Mixed crop*
Rice	921	619	901	435
Jowar	619	608	644	541
Corn	1,193	928	1,163	861
Wheat	792	656	759	630

* 1 kilogram = 2.2 pounds.

of the sample cut. Thus two yield rates will be obtained: one for the area in which the crop is in pure stand and the other for the area in which the crop dominates the mixture. Table 12.6 presents an example of this kind of situation. The example is based on the National Sample Survey of India (4). Since the areas under the two categories are known separately, it is fairly simple to estimate the total production of the crop.

12.10 SAMPLE DESIGN FOR YIELD SURVEYS

The sample design of a crop-cutting survey can only be decided in the light of information on the administrative setup of the country, the field staff available, local agricultural practices, the peculiarities of the crops to be sampled, and the auxiliary information already available. Therefore it is not possible to lay down a sample design which can be used everywhere with success. We shall limit the discussion to pointing out the broad considerations involved in the design of surveys for estimating the yield rate.

12.11 STRATIFICATION

Ordinarily an estimate of the production of the crop is needed for the various administrative subdivisions of the country, such as the administrative districts in India or the crop estimating districts in the United States. In that case these administrative divisions should form the principal strata of the survey. The sampling plan within a principal stratum should be so chosen that the sample can provide an estimate for the stratum at a prescribed level of precision. Each principal stratum should be broken up into strata which are expected to differ considerably from each other with respect to yield rate. If there is no accurate information available on yield rate from previous inquiries, geographical stratification may be employed. Convenience of sampling points to the use of administrative subdivisions such as counties as the strata. Ordinarily sampling within one stratum is done independently of that in another.

Usually it is not possible to take a one-stage sample from each stratum. Thus lists of first-stage units such as villages may be made. A sample of villages is then selected from the stratum with probability proportional to the cultivated area of the village. When the total number of villages to be selected in the entire sample is known, the number to be allocated to a stratum may be based on the proportion that the area under the crop (or cultivated area) in the stratum bears to the total area under the crop (or total cultivated area). Such an allocation is found to be sufficiently close to the optimum allocation. It should be pointed out that the total number of villages to be assigned to a principal stratum must not be smaller than the number needed for making a reasonably precise estimate there.

In the developing countries the type of mapping material available is usually not such that it is possible to divide the stratum into a number of homogeneous units by drawing lines on the map in different directions. This is the reason that it is recommended to use administrative areas such as villages as the sampling units; their use facilitates the organization of the survey. The village can be further subsampled by making a list of fields growing the crop and selecting a certain number of them. Within a selected field one or more plots of a specified size can be located at random and the yield determined. The design thus used is a multistage one. The first-stage units are selected in the office while the second- and third-stage units are selected in the field. The selection of a random plot from the field is an extremely important part of the whole procedure. It is here that biases are likely to occur which may ruin the entire survey. As a safeguard, rigorous procedures should be outlined for the field worker for the selection of the sample plot. Usually the selection is done by taking a pair of random coordinates from a corner of the field. With this procedure the central part of the field is oversampled (see Sec. 12.5). The bias caused thereby is not important if there is no correlation between the position of the plot and the yield obtained from it. Indian experience on this subject is not conclusive.

12.12 SAMPLE SIZE

The number of villages to be selected from a principal stratum will be governed by the level of precision needed to make estimates for the stratum. When the cost of the survey is simply proportional to the total number of experiments (or sample cuts), it is clear that the number of sample cuts taken from a village should be as small as possible. However, this simple cost function will not apply when the work is done by a party of enumerators moving from one place to another and considerable traveling is involved. In that case, more than one sample plot should be taken from each village. This can be achieved either by selecting more fields or by locating several sample plots in a selected field. An analysis of data from Indian surveys shows that the variability between fields within a village is many times more than the variability between plots in the same field. Similar results have been obtained in the United States from surveys on wheat. When this holds, one should take more fields and just one plot from each field. (In the initial surveys two plots per field may be taken to study the variability between plots.) As an example, the following results were obtained from a survey on wheat in a district in India (6). Table 12.7 gives the number of fields per village and plots per field for estimating the average yield with a relative error of 5 percent. It is clear from the table that whereas the number of villages decreases appreciably as the number of fields per village is increased from one to two, a further increase in the number of fields does not bring down the number of villages to be selected in the sample. Furthermore, the increase in the number of plots per field has hardly any effect on the number of villages. Two to three fields per village and one plot per field is about the optimum allocation.

Table 12.7 Number of villages to be selected in the sample

Number of fields per village	*Number of plots per field*			
	1	2	3	4
1	88	74	70	68
2	63	56	54	52
3	54	50	48	47
4	50	46	45	45
5	47	45	44	43
6	46	43	43	42

The argument is simple if an unrestricted random sample of fields can be selected directly from the area to be studied. Suppose we want to estimate the average yield of a crop with a coefficient of variation of 2.5 percent; that is, the margin of error is 5 percent. Then the number of fields to be selected in the sample is 1,600 times the square of the coefficient of variation of yield in the population of all fields. Thus the sample size needed is 1,600, 400, or 100 according as the coefficient of variation of yield is 1.0, 0.5, or 0.25 respectively. When the crop is not sown in mixture with other crops, the field-to-field variation is not expected to exceed one-half the average yield, which gives a coefficient of variation of 50 percent. In this case a sample of 400 fields will do. If a two-stage sampling method is used, the number of fields to be selected will be more than 400. Thus the total sample size needed is not very large.

12.13 THE ESTIMATION PROCEDURE

We shall assume that a sample of n villages has been selected with replacement with probabilities p_i, $\sum p_i = 1$, from a stratum. The ith village has M_i fields growing the crop, from which m are selected at random. The area of a particular sample field is A_{ij}, from which a plot is taken and the yield y_{ij} determined as per unit area. Then an estimate of the yield per unit area is

$$\hat{R}_1 = \frac{(1/n)\underset{i}{S}(M_i/mp_i)\underset{j}{S}A_{ij}y_{ij}}{(1/n)\underset{i}{S}(M_i/mp_i)\underset{j}{S}A_{ij}}$$

This is a ratio estimate. The use of this estimate is explained in Example 5.6. A simpler estimate, which works well in practice when p_i is proportional to M_i, is

$$\hat{R}_2 = \frac{SSy'_{ij}}{nm}$$

where y'_{ij} is the yield of the plot taken from the jth field. This is the simple average of the sample plots. When the area A under the crop is known, a third estimator to use is

$$\hat{R}_3 = \frac{1}{An} \underset{i}{S} \frac{M_i}{p_i} \frac{1}{m} \underset{j}{S} A_{ij} y_{ij}$$

A comparison of the three estimators has been given by Panse (6) on the basis of a number of surveys conducted in India. In these surveys the villages were selected with equal probability. The results obtained are shown in Table 12.8. The simple average $\hat{R}_2$ comes out as the best.

Table 12.8 Coefficients of variation (%) of different estimates of average yield

Survey	$\hat{R}_1$	$\hat{R}_2$	$\hat{R}_3$
Wheat (1947–48)	4.7	3.7	14.0
Wheat (1948–49)	5.7	2.5	10.0
Cotton (1944–45)	11.3	5.5	15.0
Cotton (1945–46)	13.2	6.9	14.0

12.14 THE INDIAN SURVEY OF FOOD PRODUCTION

As an example of a large-scale survey of food production being conducted in a developing country, we shall discuss the land utilization and crop-cutting survey of India, which is an integral part of the National Sample Survey (4). In the fifteenth round of the survey the entire country was divided up into 218 strata. The strata were formed by grouping administrative units believed to be homogeneous with respect to population density, altitude, and type of food crops. The strata populations were about the same. Within a stratum 12 villages were selected systematically with equal probability, the total number of villages in the sample being 2,616 for the land utilization survey. From each village 6 clusters of 10 plots each were selected for land utilization studies. Crop-cutting experiments were done in a subsample of 872 villages. For this purpose six fields were selected with probability proportional to area under the crop from all fields growing that crop in the village. Within a selected field two circular cuts of radii 2 ft 3 in. and 4 ft were taken to harvest the crop when it was mature. Some of the results obtained are given in Tables 12.9 and 12.10.

Table 12.9 Estimated area and production of rice, India, 1959–60

	Gross area (thousand acres)		*Production (thousand tons)*		
Zone	*Area*	*Relative error (%)*	*Number of experiments*	*Production*	*Relative error (%)*
North India	2,173	17.58	504	969	18.68
Central India	22,306	5.21	820	6,157	8.10
East India	38,824	4.37	1,243	15,446	5.95
South India	17,920	4.74	1,151	9,184	6.49
West India	9,934	9.54	448	3,844	12.10
Total	91,157	2.68	4,166	35,600	3.66

Table 12.10 Estimated area and production of food crops, India, 1959–60

	Gross area (thousand acres)		*Production (thousand tons)*		
Crop	*Area*	*Relative error (%)*	*Number of experiments*	*Production*	*Relative error (%)*
Rice	91,157	2.68	4,166	35,600	3.66
Jowar	63,725	5.08	1,206	15,532	9.44
Bajra	40,885	10.18	750	5,640	11.45
Ragi	6,710	10.13	446	2,693	9.13
Corn	15,337	7.87	833	6,962	8.88
Wheat	45,993	3.85	1,552	13,764	6.02
Barley	14,212	6.16	612	3,671	8.91
Total	278,019	2.25	9,565	83,862	2.79

A number of conclusions can be drawn from the results. With about 4,200 experiments judiciously dispersed all over India it is possible to estimate the total production of rice, the most important crop, with a margin of error of about 7 percent. The regional estimates of the production of rice are subject to large sampling errors. The estimates of the production of other crops at the national level are not as precise as one would normally wish to see.

12.15 CONCLUDING REMARKS

The results of crop-cutting experiments conducted in different countries appear to be encouraging. It seems that the sampling method of estimating the yield of a crop by harvesting sample areas immediately prior to harvest works well provided the area under the crop can be estimated satisfactorily. This method should be increasingly used in the developing countries and in other areas where there is reason to doubt the existing estimates. It is, however, very important to conduct a series of pilot surveys for the development and testing of methods. This job can only be tackled effectively by a full-time permanent organization which can make use of the experience accumulated. There is no intention of suggesting that the existing system of yield estimation by crop reporters should be scrapped immediately. It is only when the accuracy of the results of sample surveys has been tested over a number of years by comparison with other data that the sampling method can replace the reporting method. To make the sampling method a success, the fields must be selected at random, the sample plot must be demarcated accurately, its location should not be influenced by the appearance of the crop, and the time of crop sampling must be sufficiently near to the time of harvest of the crop. The size of the sample plot should be large enough so that the danger of bias arising from errors of demarcation is appreciably reduced. Objective tests should be made from time to time against full-scale harvestings. The method of double sampling may be used for increasing the precision of the estimate. In this method eye estimates are made on a large sample of fields. Sample harvestings are done on a subsample to calibrate the eye estimates. The procedure, however, needs adequate testing.

REFERENCES

1. Department of Agriculture (1957), "Random Crop-cutting Surveys for Estimation of Yield of Food Crops in Uttar Pradesh (India), 1943–53," U.P., Lucknow.
2. King, A. J., et al. (1942), An Objective Method of Sampling Wheat Fields to Estimate Production and Quality of Wheat, *U.S. Dept. of Agr. Tech. Bull.* 814.
3. Mahalanobis, P. C., and J. M. Sengupta (1951), On the Size of Sample Cuts in Crop-cutting Experiments in the Indian Statistical Institute, *Bull. Intern. Statist. Inst.*, **38**.
4. National Sample Survey (1963), Some Results of the Land Utilization Survey and Crop-cutting Experiments, *Govt. of India Rept.* 79, New Delhi.
5. Nilsson, B. (1960), Pilot Crop Cutting Survey, VIII, *Statist. Tidskr.*, **9**.
6. Panse, V. G. (1954), "Estimation of Crop Yields," Food and Agriculture Organization, Rome.
7. Sukhatme, P. V. (1947), The Problem of Plot Size in Large-scale Yield Surveys, *J. Am. Statist. Assoc.*, **42**.
8. Zetterberg, O., and T. Soderlind (1958), Pilot Crop Cutting Survey, VI, *Statist. Tidskr.*, **6**.
9. Zetterberg, O., and B. Nilsson (1959), Pilot Crop Cutting Survey, VII, *Statist. Tidskr.*, **8**.

13 Demographic Surveys

13.1 INTRODUCTION

The census of population is the best vehicle for obtaining local data on the number of persons and their distribution by sex, age, and marital status. Such a census is taken once in 10 years in most countries of the world. In order to study the changes that take place in the intercensal years the system of registration of vital events (births, deaths) is used along with migration statistics. It is, however, well known that registration figures are usually defective; this is so even in countries with a long tradition of registration data. Consider, for example, the data presented in Table 13.1, in which a comparison on the number of births and deaths per 1,000 persons in India has been made (3, 4, 11) between (1) the registration system (1952–54), (2) the sample census (1952–54) using a sampling fraction of 1 in 200, and (3) the National Sample Survey (1953–54) using a sampling fraction of 1 in 7,000. It is clear from the table that the registration figures are too low and that the accuracy of the data appears to increase as the size of the sample is decreased. This

Table 13.1 Comparison of registration and sample survey data on births and deaths, India, 1952–54

	Rate per thousand persons		
	Registration	*Sample census*	*National Sample Survey*
Births	24.2	29.6	34.3
Deaths	13.6	15.1	16.6

means that higher-quality data can be collected from a smaller sample, which permits one to devote greater attention to the reduction of nonsampling errors which appear to be more important than sampling errors. This analysis shows that sampling methods have an important role to play in the collection of demographic data. Some of the problems involved in the collection of such data are discussed in this chapter.

13.2 THE HOUSEHOLD AS SAMPLING UNIT

In most demographic inquiries the sampling unit used is the private household. A private household may be one-person or multiperson. A one-person household is a person living on his own. A multiperson household is a group of two or more persons living together and eating together. The persons may or may not be related. A boarder sharing meals and accommodation for payment is treated as a member of the household while a lodger not sharing meals is treated as a separate household. When this definition of the household is applied in practice, borderline cases do arise. Such cases are usually handled by prescribing criteria regarding length of stay, number of meals taken during the previous month, etc. It is, however, always safe to make the definition more flexible and include information on all persons who are (1) normally resident and present at the time of the survey, (2) normally resident but absent at the time of the survey, (3) temporary guests. It should be possible to identify these categories from the questionnaires. When this is done, any category can be excluded at the time of analysis in order to use a particular definition of the household.

There is another definition of the household which is usually used when a *de facto* census is taken. In this case a household is defined as a group of persons who slept in the housing unit on the night of the census. This definition is clear-cut and allows no room for doubt as to whether a particular person should be considered a member of the household. Furthermore the enumerator can check whether the members are around and thus collect information of a better quality directly from each member of the household.

When applied to a small area, the two definitions may give different results. But the two should agree for the country as a whole since each person is accounted for in both definitions. As an example, consider Table 13.2, in which the population of Greater Athens in private households is estimated by the two methods. In the survey conducted two days after the census a complete list was made of all categories of persons associated with the households in the sample. It was thus found possible to estimate the number of permanent residents and those who slept in Greater Athens on two specified nights. The results show that the differences between the three estimates are well within sampling errors (8).

Table 13.2 Estimated population of Greater Athens in private households, 1961 (in thousands)

Category (*A*)	*Listed in census* (*B*)	*Not listed in census* (*C*)	*Total* (*D*)	*Standard error* (*D*)
Number of persons who spent the census night (Mar. 19, 1961) in Greater Athens	1,581	16	1,597	95
Number of persons who spent the survey night (Mar. 21, 1961) in Greater Athens	1,583	20	1,603	95
Number of persons usually resident in Greater Athens	1,585	20	1,605	

13.3 AGE-SEX DISTRIBUTION

Most of the demographic tabulations are broken down by age and sex. Age is defined as the interval between the date of birth and the date of the survey or the date of death. It is usually expressed as the number of completed years (age last birthday) for adults and children and the number of completed months for infants under one year of age. Information on age is collected by asking the head of the household the date of birth or the age last birthday of the different members of the household. When aggregated, the age-sex distribution of the population is obtained (see Table 13.3 for an illustration).

Experience shows that reports on age are subject to considerable errors. There is an overstatement of ages of very young children. Young men and women tend to report incorrectly that they have attained majority before they have actually reached those ages. There is a tendency to report age in multiples of 5 and to avoid odd numbers. Old persons tend to overstate their age and this tendency is found to rise with the degree of illiteracy in the community. In societies in which numerical age has no importance, it becomes very difficult to obtain correct information on age. The enumerator

Table 13.3 Population of Mauritius by age and sex, 1944

Age-group	*Males*	*Females*	*Sex ratio*
Under 15	73,764	72,867	101.2
15–29	60,538	61,345	98.7
30–44	41,323	37,482	110.2
45–59	24,570	23,116	106.3
60 or over	9,380	13,190	71.1
Age not stated	751	859	87.4
Total	210,326	208,859	100.7

has to make a reasonable estimate of the age-group to which the respondent belongs. This is usually done by preparing a historical calendar showing important events such as epidemics, famines, earthquakes, or significant political changes. The estimation of female ages is assisted by clues such as bodily changes associated with menarche, number and age of children, and a fairly well-defined upper limit to the ages of childbearing.

When the data are tabulated by single years of age, it is possible to make an analysis of age misreporting. There is no unique way of making adjustments for heaping at selected digits. Although the heaped numbers can be redistributed in some arbitrary manner assumed to represent the underlying facts, the general approach to adjustment is to group the data into convenient classes for subsequent study and analysis. The grouping is usually in five-year age periods, so chosen that the grouped totals would presumably correspond close to like totals of the underlying data. A number of tests have been devised to determine the best system of grouping. By one criterion the best grouping is that which produces the smoothest curve throughout life. In another method, the observed data are compared with those obtained by curve fitting. These tests do not point to any particular age-grouping to be best under all circumstances. The age-groupings usually used are 0–, 1–4, 5–9, 10–14, ...

13.4 MARITAL STATUS

The composition of a household is determined by finding the marital status of each member and the relationship with the head of the household. Information on marital status is vital for studies of population structure, fertility levels, and trends. Usually marital status is defined in terms of four categories: those who are single (never married), married, widowed, or divorced. The information is collected from all persons in the survey, irrespective of the age of the person or the national minimum legal age for marriage.

It is often not difficult to determine the marital status of a person. There may, however, be situations in which the persons are not formally married, although they are living together as husband and wife. In this case it is better to break down the married group into three categories: persons formally married and living with spouse, persons formally married and living separately, and persons in stable *de facto* unions. At the stage of analysis some of these categories may be excluded or included, depending upon the problem to be studied. For example, persons in *de facto* unions may be excluded if the purpose is to calculate legitimate marriage rates.

13.5 BIRTHS

Sample surveys are being increasingly used to collect information on births in order to estimate the birth rate or test the reliability of the registration system. The retrospective method is usually used. In this method the selected household is asked to declare the number of babies born to women in the household during the reference period (which may be a year). The sex of the child born is asked and the date of birth. Only live births are considered, that is, those births in which the child showed some evidence of life such as the beating of the heart or movement of the muscles. Multiple births are counted as two, three, or four births as appropriate. Whether or not the birth took place in the area where the household is situated, it must be recorded if the woman is a member of the household selected in the sample.

Experience shows that the reporting of births is subject to considerable errors. A major source of error is the inability of the respondent to recall the event that took place long ago. Consider, for example, Table 13.4, in which the number of births obtained from a demographic survey (13) are compared with the registered births. The figures relate to only those households for which registration data were available. It is clear from the table

Table 13.4 Number of births recorded in the survey and in registration lists, Mysore, 1951–52

Period of recall in months	*Number registered*	*Number in survey*	*Percent*
Less than 4			
4–6	15	15	100
7–9	58	54	93
10–12	67	66	99
13–15	77	73	95
16–18	88	78	89
More than 18	15	11	73
Total	320	297	93

that the proportion of births as declared in the survey declines as the period of recall lengthens.

Another source of error is the tendency on the part of respondents to allocate the birth to the reference period although it occurred at a prior date. This is called *telescoping effect* or *border bias*. An example of this type of error is provided in Table 13.5. The birth rates are calculated from two types of data. One is the survey method, in which the number of births is recorded by a retrospective inquiry. In the second method data on births are obtained by continuous observation on households. The effect of border bias is apparent.

Table 13.5 Comparison of birth rates by two methods

	Number of births per thousand persons	
Country	*Survey*	*Continuous observation*
Guinea (1955–56)	63	50
Ivory Coast (1954–55)	55	49

In view of these errors, great care should be taken in the collection of data on births. Every married woman in the household should be asked whether a baby was born to her during the reference period. The name of the child, its sex, and the month of birth should be asked in order to get as complete and accurate a response as possible. If a maternal death is reported in the household, it is important to ensure that the birth involved is not overlooked. Information on births should also be collected from women visiting the household at the time of the survey and from boarders living with the household. In order to get a complete picture the survey of births should be extended to institutional households such as hospitals and maternity homes. But the procedures should be streamlined to make sure that the events reported are not duplicated.

When a birth is reported, it is useful to ask for the order of birth and the age of the mother at the time of birth. Answers can then be tabulated by age and parity. Such a tabulation can provide a check on the consistency of the data collected. For example, the cumulation of age–specific fertility of zero parity women to the age interval 30–34 should equal the proportion of women at age 30–34 having at least one child, and a correction can be applied to the births reported for the preceding year to ensure such equality.

13.6 LIFETIME FERTILITY

In addition to collecting information on births taking place during the previous year, it is useful to ask every woman in the household the total number of live-born children she has brought forth during her entire lifetime up to the date of the survey and the number still alive. All live births are to be recorded, whether legitimate or not, whether born of the present marriage or of prior marriage. The sex of each child born should be noted. When this information is available, it is possible to calculate the average number of children born alive to women of childbearing age and over and the average number of children still surviving per woman.

Since this type of information relates to the woman's entire lifetime, the data collected are subject to errors of omission as a result of memory failure, ignorance of the facts, or unwillingness to give the information. The errors are likely to be larger with older women, who may not remember events that occurred in the long past. Children who died in their infancy are not likely to be reported. As an example of the errors involved, consider Table 13.6,

Table 13.6 Average number of children born per woman by age of mother, Thailand, 1960

Age-group of women	*Reported average*	*Adjusted average*	*Percent underreporting*
15–19	0.07	0.145	51.7
20–24	0.86	1.084	20.7
25–29	2.29	2.664	14.0
30–34	3.78	4.303	12.2
35–39	4.98	5.653	11.9
40–44	5.73	6.552	12.6
45–49	5.91	6.881	14.1
50–54	5.74	6.881	16.6
55–59	5.64	6.881	18.0
60–64	5.43	6.881	21.1
65–69	5.33	6.881	22.5
70 or over	5.16	6.881	25.0

in which the average number of children born alive has been tabulated by age of mother. The reported average increases as the age of mother in the reproductive period increases but then starts falling off. Normally the average should remain constant beyond the reproductive period when there is no significant association between fertility and mortality and fertility is stable in this age range. The adjusted figures given in the table are obtained by replacing the declining fertility curve (2) by a horizontal line passing through the peak corresponding to the end of the reproductive period. Assuming the adjustment to be valid, there is no doubt that there is considerable underreporting of births.

In view of these errors steps should be taken to improve the accuracy of response on the number of children ever born to a woman. Suitable probes can be used and the surviving children checked against the household members listed. If a particular member cannot be accounted for, the reason should be found out. The age of mother at the time of each live birth should be asked. This will help determine the order of birth. In addition to asking each woman how many children were born alive to her, it is also useful to ask separately for the children living with her, those living elsewhere, and the number born alive but now dead. With this accounting procedure it is possible to ensure that those children who had died or left the family are not overlooked. Interviewers should make an unambiguous entry for every respondent, especially to enter a zero for women with no children, rather than leaving a blank which might indicate "no response."

When these precautions are taken the data collected provide material for a number of tests of consistency. For example, omissions of children ever born are likely to be sex selective, and such a tendency is revealed by a trend in the sex ratio of the reported children ever born as the age of woman increases. Another significant type of probable error is to leave out higher proportions of dead than of surviving children, especially on the part of older women. This form of omission may also be sex selective, resulting in an implausible contrast in estimated child mortality by age for the two sexes. Furthermore, additional tests of consistency, such as differential mortality in families with different numbers of children, can be made by tabulating in each five-year age interval the number of women with no children, with one child, and so on.

13.7 DEATHS

In order to estimate the rate of mortality in the population from a sample survey, all households in the sample are asked whether any deaths occurred among the members of the household during the previous year. The age at which the person died and the sex are recorded. For recording this event, death is defined as the permanent disappearance of all evidence of life at any time after live birth has taken place. Thus fetal deaths are excluded.

The results of retrospective inquiries on the number of deaths show that the data collected are usually subject to large errors. Consider, for example, Table 13.7, in which the registered deaths are compared with those obtained from a household survey conducted in a part of India (13). The comparison is limited to only those households which registered a death.

Thus 10 percent of the deaths registered were not declared by the households at the time of the survey. The proportion declared declined as the period of recall increased. Therefore memory failure is one of the reasons for inaccurate reports on deaths. Another reason is unwillingness to report a death as it is not a pleasant occurrence. Deaths of household members away from the regular place of residence are often omitted. Furthermore, no information on deaths can be obtained from single-member households which

Table 13.7 Number of deaths recorded in the survey and in registration lists, Mysore, 1951–52

Period of recall in months	*Number registered*	*Number in survey*	*Percent*
Less than 4	1	1	100
4–6	10	10	100
7–9	36	35	97
10–12	40	36	90
13–15	24	21	88
16–18	25	21	84
More than 18	8	6	75
Total	144	130	90

have disappeared as a result of death if information is collected from existing households only.

A number of precautions can be taken to improve reports on deaths. In the first place the interviewer should establish rapport with the informant before touching on the delicate subject of deaths. He should record the name of the person dead, age, sex, and relationship to the head of the household. If the household composition reveals a widow or widower, the interviewer should ask where the spouse died and make sure that the death has not been overlooked. If it is found that there was another household in the sample dwelling unit which disappeared because of the death of all members during the reference period, this information should be recorded.

When information on lifetime fertility is being collected, one of the questions to be asked is, "Are there any children born alive who are now dead?" If one or more children are reported dead, it is important to ask whether any of these children died during the reference period. The response to this question can be used to correct any omissions in the list of deaths prepared by the investigator. Furthermore, for every live birth reported by the household during the reference period (of one year), the fact whether the child is alive now should be ascertained. Finally, the survey should be extended to institutional households, such as hospitals and maternity and nursing homes. A number of persons dying in these institutions have no residence other than the institution; their deaths will go uncounted and unmourned if the survey is restricted to private households only.

13.8 INFANT MORTALITY

The number of infants dying before reaching the age of one year provides a useful index of the level of health in the country. This number is usually expressed as a percentage of the number of live births and is called the *infant death rate*. Sample surveys in a number of countries have brought out the fact that the calculated infant death rate appears to be very low. The explanation given is that infants who are born alive and die within one year are underreported. The reason is that there is often a disinclination to speak of a recent death of a young infant; such infants are not considered to have been members of the household. In order to improve the reporting of infant deaths, the enumerator should ask whether the child is alive now in case the household reports the birth of a child during the preceding year.

One difficulty with the use of the infant death rate is that the deaths of infants during the reference period do not correspond with the births during the same period, if it is a one-time survey. Continuing sample surveys, however, provide a valuable opportunity to estimate the *infant mortality rate* which is closer to the life-table rate. This is done by asking the number of births B_1 during the calendar year and of these the number of those that died D_1' in the same year. From the survey in the second year we can find the number of infants D_2'' dying in this year before reaching the age of one year out of the births B_1 in the previous year. The ratio of $D_1' + D_2''$ and B_1 provides the infant mortality rate. In this calculation the numerator and the denominator correspond with each other but the rate does not refer to any calendar year.

13.9 ADJUSTMENT FOR RECALL LAPSE

It has been observed that recall lapse is an important source of error in statistics of births and deaths collected through retrospective inquiries. A number of studies made in India and elsewhere (12) have shown that the number of vital events observed in the survey is a function of the period of recall; the number falls off exponentially as the period of recall increases. The following procedure can, then, be used for making an adjustment for recall lapse. The sample of households is staggered uniformly over the year and information is collected every month on the number of events occurring during the year preceding the date of inquiry along with the month of occurrence of the event. Thus it is possible to determine the number of events declared in the survey in which the period of recall is k months ($k = 1, 2, \ldots,$ 12). This gives the cumulated average monthly number of events when the period of recall is one month, two months or less, three months or less, and so on. An exponential curve of the form

$$y = y_0 e^{-ak^2}$$

can then be fitted to the cumulated data and the best values of the constants y_0 and a determined. The goodness of fit can be tested by using appropriate statistical tests. If the fit is good, we can substitute $k = 0$ in the equation of the curve to obtain y_0 as the average number of events per month when no recall is involved. This gives $12\, y_0$ as the best estimate of the total number of events in the year in which the bias due to recall lapse has been corrected. When compared with the observed number of events, the proportion understated in the survey can be estimated and the correction made. Based on this analysis, Som (12) discovered an understatement of 17 percent in the birth rate in urban India (1957–58) and of 38 percent in the death rate, the unadjusted figures obtained from the National Sample Survey of India being 30.5 and 8.5 respectively.

13.10 SAMPLING FOR BIRTHS AND DEATHS: RETROSPECTIVE METHOD

That it is possible to make plausible estimates of birth and death rates from a carefully designed and executed sample survey is shown by the National Sample Survey of India, among many others. In the initial rounds of this survey the final sample of households was based on multistage sampling of administrative areas and the reference period for the collection of data was one year. The survey gave a birth rate of about 33 per thousand and a death rate of about 14 per thousand for rural India. Beginning in 1958–59 a number of improvements were made in the survey design. In the first place, the entire village was taken as the unit of sampling. Thus all households in the selected village were required to give information on births and deaths. This improved the quality of the data as household reports could be checked for accuracy with the help of neighbors. Second, an attempt was made to include households which disappeared as a result of death of all members. This was done by asking for such households from responsible persons in the village and from every tenth household in the sample. Third, two reference periods were used, one "last year" and the other "year before last." Thus it was possible to compare the results for the same period on the basis of two succes-

Table 13.8 Rates of birth, death, and growth per thousand persons based on two recall periods, last year R_1 and the year before last R_2, rural India

	Births		*Deaths*		*Growth*	
Reference period	R_2	R_1	R_2	R_1	R_2	R_1
Aug., 1956–July, 1957	31.6	....	10.1	....	21.5	....
Aug., 1957–July, 1958	28.2	38.7	10.0	19.3	18.2	19.4
Aug., 1958–July, 1959	....	38.7	....	17.1	....	21.6

Sample: 432 villages 38,000 households

sive surveys, in one of which this period is referred to as "last year" and in the other "year before last." Table 13.8 gives some of the results obtained from the survey (5, 7).

Two important conclusions can be drawn from these results. With the improved technique the birth and death rates were higher than those observed in the earlier surveys. And the recall period of "last year" produced higher and therefore better figures than "year before last." The recall period "year before last" underestimated the birth rate by as much as 27 percent and the death rate by 48 percent as compared with the recall period "last year." There was no appreciable effect of the differential recall period on the rate of growth of the population.

13.11 SUCCESSIVE SURVEYS OF THE SAME AREAS: FOLLOW-UP METHOD

Another method of collecting information on births, deaths, and migration consists in selecting a sample of areas and observing the vital events in these areas over time. The composition of the households in the sample areas is obtained in the first round of the survey. These persons are followed up in the later rounds of the survey. In this manner reliable information on births and deaths in the household as well as migration from and to the household can be collected. In fact the data collected in the earlier rounds can be corrected when more information becomes available from the subsequent rounds.

Such a survey was conducted in Morocco during the period 1961–63 (10). In the first round of the survey all households in the sample areas were listed and their composition determined. Information on deaths was collected by the retrospective method, the reference period being 12 months previous to the date of survey. Six months later a fresh list of the members of the households was prepared without using the first list made in the first round. The date of birth of all household members was recorded and questions were asked regarding the last death in the family and all deaths in the household since February, 1961. A year after the beginning of the first round the third round was launched in which all persons listed in the first round were checked for survival. At the same time any discrepancies in the first- and second-round records were noted. Furthermore, data on births and deaths occurring during the intervening period of one year were collected by the retrospective method. More than 60,000 households were involved in the survey. The data collected during all the three rounds were used for making estimates of births, deaths, migration, and growth of population. The one difficulty experienced was that a number of persons (15 percent) found in the second round could not be matched with those in the first round. The main conclusion drawn from the survey was that there would have been a gross error of 17 percent for births and 36 percent for deaths if the data had been collected using the single-round retrospective method. The net errors, however, would have been much smaller, and of the order of 3 percent overenumeration for births and 9 percent for deaths. Second, the errors of enumeration would have been considerably smaller if it was possible to use both the household composition follow-up and the retrospective mortality questionnaire in two successive rounds.

A similar approach was used by the National Sample Survey of India in 1958–59 (5). A sample of 423 villages was resurveyed completely and a person-to-person matching was done to look for births, deaths, and migrants. The sample was also used to check the accuracy of the retrospective method of inquiry. An additional question on the number of days sick during the last month helped to identify a few more deaths. This method produced a higher death rate—a rate which is believed to be more accurate.

13.12 INTERNAL MIGRATION

Household surveys can also be used for collecting information on internal migration in the country. This information can be used for estimating the rate of urbanization and for determining the characteristics of migrants. Internal migration is defined as change of residence from one geographical unit to another in the country. It is obvious that the volume of migration depends on the reference period used and the size of the geographical unit chosen.

The question usually asked of members of the household to elicit information on internal migration is "Where were you, say, five years ago?" The answer to this question establishes the place of origin of the person on a fixed date in the past vis-à-vis the present place of residence. This makes it possible to calculate rates of out-migration and in-migration for individual localities. Furthermore, the movement of the population from rural areas to urban areas and vice versa can be studied by identifying the place of enumeration and the place of origin as rural/urban.

It should be mentioned that the reference period chosen for assessing migration should not be too long. When it is long, the informant may not be able to recall the previous place of residence accurately; there may have been several moves during this period and some persons may have died in the intervening period. And the response may be inaccurate because of possible change of geographical boundaries during the interval. On the other hand, the reference period should not be too short. When the reference period is very short, the number of changes of residence may be too few to warrant a study of the problem. Hence a balance has to be struck between the two conflicting requirements.

Another question sometimes asked is, "What is your native place?"

If the place of enumeration is different from the native place (that is, where the parents live permanently), the person is considered to be a migrant. In a number of other studies the question has related to the place of birth rather than to the native place. Another variant of the question is, "Where did you live before moving to this place?" A limitation of all these questions is that they do not relate to a specific time reference against which the rate of migration may be calculated. Similarly, the question "How long have you lived here?" does not give an indication of the direction of movement.

To give the reader an idea of the kind of tabulations made from a survey of migration, Table 13.9 is presented (6). The data are taken from the

Table 13.9 Percentage distribution of persons in labor force by migration status, urban India, 1955

	Sample	*Males*	*Females*	*Persons*
Permanent migrants	1	32.47	18.54	30.07
	2	33.59	17.13	30.92
	1 + 2	33.02	17.87	30.49
Temporary migrants	1	8.58	4.70	7.91
	2	8.06	3.47	7.32
	1 + 2	8.32	4.11	7.62
Nonmigrants	1	58.95	76.76	62.02
	2	58.35	79.40	61.76
	1 + 2	58.66	78.02	61.89

National Sample Survey of India, in which two parallel samples were taken from the urban areas divided up into 94 strata. There were in all 1,054 blocks and 16,703 households in the sample. A constant number of households was taken from each block. The degree of agreement between the two samples provides information on the precision of the estimates.

13.13 USE OF REGISTRATION DATA

When there is a system of registration of vital events in the country, it is possible to improve the estimates made from a sample survey by using the registration figures. The following device can be used (1). Suppose a random sample of households has been selected from which information is collected on the vital event in question. It will be found that a number of events reported in the survey are included in registration records and some are not. Similarly, a number of registered events relating to the sample households will be found to be not reported in the survey. Thus the following fourfold table can be made in which the number d is unknown.

	Registered	*Not registered*	*Total*
Reported	a	b	$a+b$
Not reported	c	d	$c+d$
Total	$a+c$	$b+d$	n

Assuming independence in the fourfold table, the proportion not reported should be the same among the registered events as among the not-registered events. This means that

$$\frac{c}{a+c} = \frac{d}{b+d}$$

from which an estimate of d is obtained as bc/a. Thus an estimate of the total number of vital events is $a + b + c + bc/a$, from which an improved estimate of the vital rate can be made. This method is very useful when it is possible to match the two sets of records conveniently.

13.14 TOTAL POPULATION

There are two ways of using the sampling method for keeping track of the total population of a country during the intercensal years. In one method the sample is used for estimating the birth and death rates in the country. Together with data on out- and in-migration, the birth and death rates can be used for estimating the total population. Since, however, information on internal and external migration is usually not available in greater geographical detail, this method cannot be used for estimating changes in urban and rural populations and for regions.

In the second method the sample population counts are directly used for estimating the total population. Usually two samples are selected for making the estimate: one of private households and the other of institutional households (such as hotels, prisons). Within a selected private household a complete and accurate list is made of all members who are (1) normally resident and present at the time of the survey, (2) normally resident but absent at the time of the survey, and (3) normally resident elsewhere but present at the time of the survey. In the institutional sector all those persons who are present in the institution at the time of the survey are covered. By following the sample design employed, an estimate of the total number of persons in the country can be made (see Example 5.8).

It is well known that sampling methods can be used with greater success for estimating the population of a country when a census has already taken place there. By taking a sample of areas with probability proportional to census population and by enumerating a subsample of households, it is possible to estimate the total population of the country with a reasonable margin of error. As an example, let us take the population survey of Greece conducted in 1962 (9). The three principal strata were Greater Athens, other urban areas, and rural areas. A sample of 280 block clusters was taken from Greater Athens and about 0.5 percent of households were investigated. From the other urban areas nine towns were selected and about 0.5 percent of households spread over 126 blocks were enumerated. The sample from the rural areas was based on 160 communes from which a 0.5 percent sample of households was taken. The results obtained from the survey are given in Table 13.10 along with the census figures.

Table 13.10 Population in private households, Greece, 1962

	April, 1962, *sample*		*March*, 1961, *census* (*thousands*)
	Population (*thousands*)	*Relative error* (%)	
Males	3,940.8	1.7	3,867.3
Females	4,365.4	1.1	4,238.1
All persons	8,306.2	1.3	8,105.4

Thus the sample spread over 280 blocks in Athens and 169 primary sampling units elsewhere produced an estimate of the population in private households with a relative standard error of just 1.3 percent. This was made possible by selecting the primary sampling units with probability proportional to 1961 population as enumerated in the census.

If the objective of the sample survey is to measure population changes from year to year, a sample of a moderate size may not be able to do this. The sampling and nonsampling errors can be such that accurate measurement of changes as small as 1 or 2 percent may be difficult. In this case the use of overlapping samples can be particularly helpful in reducing sampling errors. When, however, the period is longer, say, five years, it should not be difficult to obtain significant changes on the basis of a sample.

REFERENCES

1. Chandersekhar, C., and W. E. Deming (1949), On a Method of Estimating Birth and Death Rates and the Extent of Registration, *J. Am. Statist. Assoc.*, **44**.
2. Das Gupta, A., et al. (1963), "Population Perspective of Thailand," Bangkok (unpublished).
3. Government of India (1955), Sample Census of Births and Deaths in 1953–54, *Census of India, 1951, Paper* 1, New Delhi.

4. Government of India (1955), Sample Census of Births and Deaths, 1952–53, *Census of India, 1951, Paper* 2, New Delhi.
5. Indian Statistical Institute (1963), The Use of the National Sample Survey in the Estimation of Current Birth and Death Rates in India, *Proc. Intern. Pop. Union Conf.*, New York, 1961.
6. National Sample Survey (1962), Tables with Notes on Internal Migration, *Govt. of India Rept.* 53, New Delhi.
7. National Sample Survey (1963), Fertility and Mortality Rates in India, *Govt. of India Rept.* 76, New Delhi.
8. Raj, D. (1962), "Post-enumeration Survey of the Population Census of Greece," Government of Greece, Athens.
9. Raj, D. (1968), "Sampling Theory," McGraw-Hill Book Company, New York.
10. Sabagh, G., and C. Scott (1965), An Evaluation of the Use of Retrospective Questionnaires for Obtaining Vital Data, *Proc. U.N. World Pop. Conf.*, Belgrade.
11. Som, R. K. (1959), On Recall Lapse in Demographic Studies, *Proc. Intern. Pop. Conf.*, Vienna.
12. Som, R. K. (1968), "Recall Lapse in Demographic Enquiries," Asia Publishing House, Bombay.
13. United Nations (1961), "The Mysore Population Study," Population Studies, no. 34, New York.

14
Employment and Unemployment Surveys

14.1 INTRODUCTION

Data on the size and composition of the working population are needed in every country for purposes of economic planning. Indeed, the total manpower in the country, its quality (whether skilled or not, educational attainments and occupational structure), and the amount of work done (number of hours per week or days per year) are the limiting factors which set an upper limit to the growth of the economy's output. As key economic indicators these data serve a very useful purpose. But the collection of such data presents very difficult problems. The concepts involved are elusive and the measurement problems are complex. At least in the developing countries, a census is not the right medium to collect data of so delicate a nature; it is only a carefully planned and well-executed sample survey that can produce satisfactory figures in this field. Consider, for example, Table 14.1, in which employment data on the Greek population aged 10 or more are presented. The figures in column 3 were obtained from a census while those in column 2 from a carefully executed sample survey (5, p. 277). The table shows that the survey gave many more employed persons than the census, the difference

Table 14.1 Population of Greece by activity status (in thousands)

Activity status (1)	*April*, 1962, *sample* (2)	*March*, 1961, *census* (3)	*Difference* *Value* (4)	*Difference* *Sampling error* (5)
Employed	3,862.2	3,343.0	519.2	85.9
Unemployed	221.2	237.2	−16.0	13.2
Not active	2,667.8	3,033.0	−365.2	50.8

being as much as 15.5 percent. The reason is that the concept of employment is difficult to define and use. The definition used can be applied with greater success in a sample survey than in a census.

14.2 BASIC CONCEPTS

The first task is to define the population to which the employment statistics should relate. The population in institutions (such as prisons) is usually omitted and so are all persons who are too young, say, 13 years or less. For reasons of security the armed forces are excluded from the jurisdiction of the survey. The part of the population that remains is the subject of further study. This part is divided into the categories of employed, unemployed, and not in the labor force. The sum of the employed and the unemployed is called the civilian labor force. One of the purposes of the survey is to determine the percentage of the population who are in the labor force, that is, those who are economically active.

It should be pointed out at the outset that the definitions to be used must be relevant to the area for which the data are to be collected. The concept of "having a job" is meaningful in the economically developed countries where there is an organized labor market and people seek employment on the basis of salaries or wages. In the rural areas of the developing countries, however, the conditions are entirely different. There is no organized labor market; there are very few jobs carrying a salary and the work is of a seasonal nature. Therefore, the rural areas of developing countries need different concepts for the study of the employment problem. We shall first limit our discussion to urban areas with an organized labor market.

14.3 THE LABOR FORCE

All persons who worked for pay or profit during the period in question are considered to be employed. Also included in this class are all those persons who had jobs but were temporarily absent from work for reasons of illness or other physical disability, bad weather, strike, paid layoff, or paid vacation. Furthermore, unpaid family workers who assisted in the operation of the family enterprise for a specified minimum amount of time are also included in this category.

Just as the concept of employment is based on the activity of working, the concept of unemployment is based on the activity of looking for work on the part of persons who did not work during the period in question. Thus all those persons who had no jobs and were actively seeking work are considered to be unemployed. Also included in this category are those persons who could not look for work due to illness or due to the belief that no work was available during the period in question.

All persons not classified as "employed" or "unemployed" are considered to be not in the labor force. They do not have a job and they are not looking for it. They may be students, housewives, pensioners, rentiers, or disabled persons who are outside the labor market.

14.4 TYPE OF ACTIVITY AND EMPLOYMENT STATUS

The distribution of the labor force by type of economic activity is closely related to the economic organization of the country and throws light on the quality of the working population. There are two ways of distinguishing the type of economic activity. One is by way of occupation and the other through industry. The occupation of a person is the kind of work he does such as grocery salesman, bus driver, post office clerk. His industry is the activity of the establishment in which he works, such as banking, agriculture, or manufacturing. The two kinds of classification are not the same. Some occupations can be performed in several industries. For example, a person may be a clerk in a manufacturing industry or in commerce or in construction.

Another indicator of the quality of the labor force is its distribution by employment or work status. Some persons are employers of others and some are employees. There are many persons who are own-account workers, that is, they operate their own enterprises unaided by any employees. Finally there are unpaid family enterprise workers who give a helping hand in the enterprise of the family. The distribution of the working population by work status indicates the different roles that the workers play in the economic activities of the nation.

14.5 USUAL VERSUS CURRENT STATUS

In order to decide whether a person is in the labor force or not, the basis of classification may be either the "usual status" as determined by the dominant pattern of activity over a long period such as a year or the "current status" as elicited from activity over a short period such as a day, a week, or a month. Experience shows that the usual-status approach is vague and subjective while the current-status method is more precise and objective. The former approach, in which a person is asked whether he usually works for a living, presents a static picture. It provides a poor gauge of the volume of employment and unemployment during times of rapid change. On the other hand, the current-status approach provides a dynamic picture of the size and composition of the labor force.

14.6 PRIORITY VERSUS PRINCIPAL STATUS

During the period in question the activity status of a person may vary from day to day. For example he may have been looking for a better job at the same time that he continues to work at the old job. Should such a person be classified as employed or unemployed? There are two approaches to this problem. In one approach, the person is classified as employed if he has worked for more than half the time during the period in question. If he has worked for less than half the time, he will have to be classified as unemployed. This classification is unfortunate because a person working less than half time is basically an employed person however inadequate the quantum of employment may be. It is for this reason that a second approach is usually preferred. In this approach a strict system of priorities is established to permit the classification of the population as employed, unemployed, or not in the labor force. This system puts working first, looking for work second, and not in the labor force third. Thus a person who both worked and looked for work during the period in question will be classified as employed, since any amount of working takes precedence over looking for work. A great advantage of the priority-status criterion is that it makes the labor force categories mutually exclusive and exhaustive. Every person must belong to one and only one category.

14.7 REFERENCE PERIOD

Let us assume that the current-status approach will be used and that the priority criterion will be employed for determining the status of a person during the reference period. The next matter to be decided is how long the reference period should be. The decision is based on a number of considerations. If the reference period is too short, the estimates are liable to be subject to large fluctuations due to sampling in time, and unemployment figures tend to become inflated. On the other hand, the longer the reference period the larger is the estimate of employment and the smaller the estimate of unemployment. Furthermore, there is the danger of recall lapse when the reference period is very long. Consider, for example, Table 14.2, in which data on the urban

Table 14.2 Population of urban India by activity status

Activity status	*Percent of population when the reference period is*	
	Week	*Day*
Employed	31.33	30.94
Unemployed	1.83	2.43
In labor force	33.16	33.37
Not in labor force	66.84	66.63
Number of sample persons	50,000	

areas of India are presented using the week and the day as the two reference periods (4). It is clear from the table that the number unemployed is larger when the reference period is shorter.

These results are typical. Surveys with the month as the reference period have tended to produce very low unemployment; on the other hand, surveys adopting the day as the reference period have tended to show comparatively high rates of unemployment. With the week as the reference period the results obtained are found to be intermediate. This is one reason that the week is usually used as the reference period in surveys of employment and unemployment. With this period, there is no great risk of recall lapse and the events can be recorded fairly accurately.

14.8 THE QUESTIONNAIRE FOR URBAN AREAS

The framing of a questionnaire for an employment survey is a very delicate task. Much depends on the type of questions asked and the order in which they are asked. The reason is that attachment to the labor force is not a fixed fact but an attitude on the part of many persons such as women and students. Thus the response may vary considerably according to small variations in the way the questions are asked or the combination of circumstances at the time of interview.

Experience shows that it is safe to start with the question

1. What did this person do most of the time last week?
   If the answer is attending school or keeping house or something else (other than working), the further question is asked,
2. During last week did this person do any work at all?
   If the answer is yes, the number of hours or days worked is noted. If no, the next question to be asked is

3. Although this person did not work last week, was he only temporarily absent from his job?
   If the answer is yes, the reason for temporary absence is noted. If no, the person is further asked,
4. Was this person seeking work last week?
   In case the person says that he was seeking work last week, he is to be classified as unemployed. When the answer is no, the person is asked,
5. Although not currently seeking work, did this person want regular work?

If the answer is yes, the person is asked why he was not seeking work, although he wants regular work. If he says that he was sick or that bad weather prevented him from looking for work or that he believed that no work was available, he is classified as unemployed. In case the answer to question 5 is no, the person is considered to be not in the labor force. He is then asked whether he is retired, whether he is a pensioner, a student, and so on.

In case the answer to question 1 shows that the person worked for pay or profit during the last week, or the answer to question 2 or 3 is yes, information is collected on work status (employer, etc.) in the present job or in the last job held, occupation, and industry. When it is established that the person is unemployed, he is asked how long he has been looking for work and the type of work he is in search of (6).

14.9 EMPLOYMENT IN RURAL AREAS

In the developing countries conditions in rural areas are very different from those in the urban areas. There is no organized labor market and there are no well-defined jobs. There is a preponderance of self-employed persons, and the work is of a highly seasonal and irregular type. Even the very young and the very old participate in economic activity. The concept of unemployment as defined for the urban areas does not apply strictly. A person who works on his own piece of land in the season but is without work in the off-season cannot be called unemployed in the off-season. It is more appropriate to say that he is "available for work" in the off-season and that he is "working" in the season (1). The aggregate of those who are working and are available for work gives the economically active population. Those who cannot be classified as working or available for work are economically nonactive. The week previous to the date of the survey may be taken as the reference period. Using the priority-status criterion, working comes first, available for work next, and not active third.

To define our terms, all persons who worked for pay or profit during the reference week are considered to be *working*. Also included in this class are those persons who, in spite of availability of work during the week, abstained from work for noneconomic reasons such as illness, injury, bad weather, etc. Furthermore, unpaid helpers who assisted in the operation of an enterprise on at least one day in the week are also considered to be working.

All those persons who had not worked even on a single day during the week owing to lack of work and were currently available for work are said to be *available for work*. This class includes persons who were seeking work through some channels or not seeking work but were available for work at current rates of remuneration.

It should be pointed out that no age cutoff is contemplated in applying these definitions to the rural sector. The concept of work is to be interpreted liberally in the case of own-account workers. Persons available for transacting business at their normal place of work or in their usual rounds should be considered as working even though no business is transacted for want of demand. And the unpaid family helpers need not belong to the same family. Since there are very few means of seeking work in the rural areas, a person may be available for work although he has not sought work. To summarize, the terms "employed" and "unemployed" are replaced by "working" and "available for work" and certain adjustments are made in the definitions to suit the conditions in the rural areas.

14.10 MEASUREMENT OF UNDEREMPLOYMENT

It is characteristic of the economies of the developing countries that there are very few persons who are wholly unemployed since people have to make a living somehow in order to survive. Many people appear to be working much less than they are capable of doing. At the same time those who appear to be working hard and for long hours nevertheless seem to be poverty stricken. Thus it becomes necessary to distinguish between full employment and underemployment.

It is difficult to give a precise definition of underemployment—a concept that is beautifully vague. Underemployment is said to exist when persons in employment who are not working full time would be able and willing to do more work than they are actually performing, or when the income or productivity of persons in employment would be raised if they worked under improved conditions of production or transferred to another occupation, with their occupational skills being taken into account (6). Underemployment appears in two major forms:

Visible, which involves a shorter than normal period of work and which is characteristic of persons involuntarily working part time

Invisible, which is characteristic of persons whose working time is not abnormally reduced but whose earnings are abnormally low or whose jobs do not permit full use of their capacities or skills (sometimes called *disguised underemployment*), or who are employed in establishments or economic units whose productivity is abnormally low (sometimes called *potential underemployment*)

It is clear from the definition that the problem of measuring underemployment is a complex one. Considerable experimentation is needed taking into account the social, economic, and labor market conditions prevailing in the country. In some countries workers do not think in terms of work or pay by the hour; time has no relevance for them. And reliable data on earnings are hard to come by. Then there should be some way of taking into account the desire of the worker for more work in determining whether he is underemployed. The society's view as to what constitutes full employment is another factor. Furthermore, underemployment appears to differ considerably between the various categories of workers; urban employees present problems which are different from agricultural employees and the own-account and unpaid family workers in agriculture. In the face of these difficulties we shall present some of the approaches that have been devised to measure underemployment in the developing countries.

When relevant, visible underemployment can be measured by asking all persons who worked during the reference week the number of hours worked and whether the person was available for additional work. This method has been used in the Indian National Sample Survey and the results in Table 14.3

Table 14.3 Percent of working persons by hours worked per week, rural India

Period	*Up to* 28 *hours*	29–42 *hours*	43–56 *hours*	57 *or more hours*	*Available for more work*
May, 1955–May, 1956	29.86	17.79	31.44	20.91	12.44
Aug., 1956–Aug., 1957	25.40	18.37	32.09	24.14	15.29

were obtained for the rural areas during the period studied. Thus 12.44 percent of the employed persons were considered to be underemployed during the first year and 15.29 percent in the second year. The trend of underemployment was found to be upward. In the earlier rounds of the survey information had been collected on the intensity of employment and the average number of days at work out of the previous 30 days. The results obtained are given in Tables 14.4 and 14.5.

Table 14.4 Percent of working persons by intensity of employment (October, 1953, to March, 1954)

Intensity	*Percent*	*Intensity*	*Percent*
No work	4.98	Half	10.69
Less than quarter	4.07	Three-quarters	7.08
Quarter	4.33	Full	68.85

Table 14.5 Number of days at work out of 30 days

Period	*Primary occupation*	*All occupations*
Apr., 1952–Sept., 1952	18.71	
Oct., 1952–Apr., 1953	18.33	22.86
May, 1953–Aug., 1953	17.54	21.69

A different approach for measuring underemployment was used in Puerto Rico (3). All those persons were identified whose replies to questions on the labor force indicated that they were employed and were not own-account workers or unpaid family workers in subsistence agriculture. They were asked the following additional questions.

1. In addition to working last week, did this person also look for work last week?
2. What was the main reason that this person did not work more hours last week?
3. Did this person desire to work more hours last week?

If a person looked for more work or wanted more work but could not get it, he was classified as underemployed. No upper limit on the number of hours worked was imposed on own-account and unpaid family workers in nonagricultural industries or in commercial agriculture. However, 35 hours was used as the cutoff point for employees since very few of those employees who had worked 35 hours or more reported desire for more work. Table 14.6 is an illustration of the kind of results obtained. Thus 72 percent

Table 14.6 Degree of job seeking, male employees, Puerto Rico, June and July, 1952

	Hours worked in survey week			
	1–29	30–34	1–34	35–
Total persons	2,142	967	3,109	4,609
Those who looked for work in addition to working	564	195	759	181
Those who wanted more work	1,043	450	1,493	685
Number underemployed	1,607	645	2,252	866

of the male employees were found to be underemployed. With regard to own-account workers and unpaid family workers in subsistence agriculture, the question asked was whether the farmer and his family consumed or sold the major part of the farm's produce. All those who replied that they consumed the major part of their farm produce were classified as underemployed. It was found that this single question gave results which correlated highly with size of farm and type of crop. Those farmers who consumed the major part of their produce had only small farms producing minor crops.

14.11 GREEK LABOR FORCE SURVEY

We shall now present some actual surveys to show how sampling methods are used for collecting information in this field. The first survey to be described relates to the noninstitutional civilian population of Greece who spent the night of Apr. 8, 1962, in a private household. In order to design the sample for the survey, the country was divided up into the principal strata of Greater Athens, other urban areas, and the rural areas. Based on the 1961 population census 20 strata were created in Greater Athens. Within a stratum blocks were listed along with their population in private households, and two independent samples of seven block clusters were selected with probability proportional to population. This gave 280 blocks in the sample. A systematic sample of households was selected from each block cluster, the overall sampling fraction being 0.5 percent. The 62 municipalities and communes comprising the other urban areas were stratified on the basis of proportion of population dependent on agriculture/industry and the rate of population growth. From each of the nine strata formed two municipalities were selected with probability proportionate to population. Two samples each containing seven block clusters were selected with probability proportional to population from each municipality. A sample of dwelling units was taken from each block cluster, the expected sampling fraction for dwelling units being 0.5 percent.

The rest of the country was divided up into 135 reasonably efficient primary sampling units which were allocated to 20 strata on the basis of population and per capita cultivated area. Two psu's were selected from a stratum with probability proportional to population. All communes within a sample psu were arranged by altitude and two independent systematic samples, each containing two communes, were selected with probability proportional to population. There were 164 communes in the sample. Four enumeration districts were selected from each commune. The sampling rate for households within an enumeration district was determined to achieve an overall sampling fraction of 0.5 percent. The same questionnaire was used in both urban and rural areas. The questions asked followed the sequence described in Sec. 14.8. Some of the results obtained are presented in Table 14.7*a* and 14.7*b* (5, p. 276). The sample in Greater Athens was well spread and so the relative sampling errors were small. There was considerable clustering in the other urban areas and this resulted in greater sampling errors. Since unemploy-

Table 14.7a Noninstitutional civilian population of Greece by activity status, 1962 (in thousands)

(1)	*All ages* (2)	*Aged 10 or over* (3)	*Active* (4)	*Employed* (5)	*Total unemployed* (6)	*New unemployed* (7)	*Not active* (8)
				All Greece			
Males	3,940.8	3,127.8	2,465.2	2,345.2	120.0	24.6	662.6
Females	4,365.4	3,623.4	1,618.2	1,517.0	101.2	51.8	2,005.2
Persons	8,306.2	6,751.2	4,083.4	3,862.2	221.2	76.4	2,667.8
				Greater Athens			
Males	823.4	682.0	495.0	446.4	48.6	12.0	187.0
Females	955.2	828.2	203.0	153.2	49.8	33.2	625.2
Persons	1,778.6	1,510.2	698.0	599.6	98.4	45.2	812.2
				Other urban areas			
Males	813.4	657.6	478.8	438.2	40.6	4.2	178.8
Females	880.6	736.4	180.0	144.6	35.4	11.6	556.4
Persons	1,694.0	1,394.0	658.8	582.8	76.0	15.8	735.2
				Rural areas			
Males	2,304.0	1,788.2	1,491.4	1,460.6	30.8	8.4	296.8
Females	2,529.6	2,058.8	1,235.2	1,219.2	16.0	7.0	823.6
Persons	4,833.6	3,847.0	2,726.6	2,679.8	46.8	15.4	1,120.4

Table 14.7b Relative sampling errors (%) of estimates shown in Table 14.7a

(1)	(2)	(3)	(4)	(5)	(6)	(7)	(8)
				All Greece			
Males	1.7	1.7	2.1	2.2	6.7	12.8	2.8
Females	1.1	1.0	3.0	3.3	7.3	7.5	2.1
Persons	1.3	1.3	1.9	2.2	5.8	7.4	1.9
				Greater Athens			
Males	2.0	2.2	2.2	2.1	7.6	17.1	3.9
Females	1.7	1.8	3.2	3.1	7.1	9.3	2.0
Persons	1.6	1.7	1.8	1.8	5.6	9.1	1.9
				Other urban areas			
Males	4.0	3.8	3.4	3.0	12.3	25.6	5.5
Females	3.0	2.8	9.5	10.1	16.3	9.8	3.0
Persons	3.5	3.2	4.6	4.3	12.2	10.2	3.2
				Rural areas			
Males	2.4	2.6	3.2	3.4	16.4	25.2	4.8
Females	1.5	1.2	3.7	3.9	18.2	28.6	4.4
Persons	1.8	1.8	2.5	3.0	14.6	22.9	3.6

ment was no major problem in the rural areas, the estimates of unemployment were low and were subject to large errors of sampling. For the country as a whole the total active population was estimated with a relative error of about 2 percent, the corresponding error for the number unemployed being 5.8 percent.

14.12 THE UNITED STATES CURRENT POPULATION SURVEY

This is perhaps the best organized labor force survey in the world. It is a monthly survey designed to provide information on the characteristics of the American labor force at the national level. The reference period is the week containing the twelfth of the month. The concepts used are broadly the same as outlined in Secs. 14.2 to 14.7.

ORIGINAL SAMPLE

The original sample design was based on 68 primary sampling units. Neighboring counties were amalgamated to form about 2,000 psu's. These psu's were grouped into 68 strata (7, 8) of about the same population. The basis for stratification was degree of urbanization, geographic location, migration, type of industry, and type of farming. One psu was selected with probability proportional to population from each stratum. All enumeration districts (EDs) in the selected psu were arranged according to whether they were in urban, rural nonfarm, or rural farm area. The measure of size of an ED was the number of segments of six households it was expected to contain. A number of EDs were selected systematically with probability proportional to this measure of size. The number of segments to be selected from an ED was determined to achieve a sampling fraction of 1 in 2,050 in each stratum.

SAMPLE ROTATION

An ingenious system of rotation of the sample over time has been devised. To describe this system, consider samples S_1, S_2, . . . each containing eight subsamples (1 2 3 4 5 6 7 8). Each of the subsamples is a probability sample of segments taken from the entire population covered by the survey. The survey starts with eight subsamples $S_1(1234)$ and $S_2(5678)$. Next month $S_1(1)$ is dropped and $S_1(5)$ is substituted and similarly for S_2 as shown in Table 14.8. Examination of Table 14.8 brings out the following important characteristics of the rotation system when it is in full operation.

1. In any month six out of the eight subsamples are common with the previous month; i.e., there is 75 percent overlap between two consecutive months.
2. Four out of the eight subsamples are common to any month and the same month a year ago; i.e., there is a 50 percent year-to-year overlap.
3. Every month two new subsamples enter the survey. Of these one is completely new and the other returns after an absence of eight months.

Table 14.8 Rotation of sample

Year and month		Sample and subsamples			
		S_1	S_2	S_3	S_4
1968	Jan.	1234	5678		
	Feb.	2345	678	1	
	Mar.	3456	78	12	
	Apr.	4567	8	123	
	May	5678		1234	
	June	678	1	2345	
	July	78	12	3456	
	Aug.	8	123	4567	
	Sept.		1234	5678	
	Oct.		2345	678	1
	Nov.		3456	78	12
	Dec.		4567	8	123
1969	Jan.		5678		1234
	Feb.		678	1	2345

This system of rotation (four months in, eight months out, four months in) gives good measures of month-to-month and year-to-year changes without having an unduly long period of inclusion for any particular household.†

1954 SAMPLE

In 1954 the survey was redesigned, the new sample being spread over 230 instead of 68 primary sampling units with approximately the same number of households. This was made possible by removing the restriction in the original design that there must be a full-time supervisor in each psu. The greater spread of the sample increased the precision of the estimates as well as that of the variance estimates. For most items, the between-psu contribution to the total variance was found to be less than 20 percent of the corresponding contribution from the 68-area sample. In Table 14.9 are presented the sampling errors of estimates of a few selected items. The composite estimate mentioned in the table is a weighted average of two estimates for a given month. The first estimate is the result of two stages of ratio estimation in which the unbiased estimate is improved by so weighting the returns as to make the sample population approximate the known distribution of the entire

† The use of this rotation system has led to the finding that some estimates based on a given panel are a function of the number of times that the panel has been observed in the past. In particular, the number of unemployed married women is considerably greater if based on a sample of households that are being interviewed for the first time than it is if the households are being interviewed for the second time. No concrete explanation has been found for this phenomenon, which may be called the *first-month bias*.

Table 14.9 Sampling errors of selected items, March, 1955

		Standard error	
Item	*Estimate (thousands)*	*Unbiased estimate (thousands)*	*Composite estimate (thousands)*
Total labor force	64,012	832	286
Employment in nonagricultural occupations	55,061	826	330
Employment in agriculture	5,876	270	229
Males employed in agriculture	5,098	229	193
Employed males	41,865	544	187
Females in labor force	20,013	360	252
Unemployed	3,074	114	136

Sample: 230 areas 22,000 households

population with respect to basic characteristics such as age, color, sex, and farm/nonfarm residence. The second estimate is the preceding month's final estimate adjusted by an estimate of change between the two months based on households included in the sample in both months. The two estimates are given an equal weight to form the composite estimate.

REVISED SAMPLE

In May, 1955, the survey was expanded from a 230-area to 330-area sample. The overall sample size was increased to about 35,000 households. The expansion increased the reliability of major statistics by around 20 percent. When data from the 1960 population census were available, the survey was updated and the number of sample areas was increased to 357. In most of the sample psu's area sampling has been replaced by the selection of units from census lists. The sampling rate in a stratum is 1 in 1,662 (1964). These changes have resulted in a further gain in reliability, of perhaps 5 percent or so, for most statistics. In January, 1963, two new items were added to the monthly questionnaire. The unemployed persons were asked whether they were seeking full- or part-time work. The other question related to the level of family responsibility of unemployed persons.

CONTROL OF QUALITY

A strong point of the survey is the stress on the control of nonsampling errors. This takes several forms. In the first place all questionnaires are edited in the field to catch omissions, inconsistencies, illegible entries, and other errors. Table 14.10 gives a frequency distribution by enumerator assignments of the average number of errors per schedule for a particular year (7).

Table 14.10 Distribution of errors per schedule

Errors per schedule	*Number of monthly enumerator assignments edited*
0.00–0.04	986
0.05–0.09	1,138
0.10–0.19	1,151
0.20–0.29	397
0.30–0.39	182
0.40–0.49	95
0.50–0.59	51
0.60–0.69	37
0.70–0.79	18
0.80–0.89	14
0.90–0.99	7
1.00 or more	30

The median is about 10 errors per 100 schedules. These error rates for individual enumerators are also used as a basis for retraining, reassigning, or withdrawing assignments from enumerators whose error rates are too high. Second, a subsample of the households is taken every month for reinterview to study the type of errors made by enumerators and to determine the causes of these errors. Table 14.11 is an example of the sources of error determined during a particular period.

Table 14.11 Distribution of causes of difference between original interview and reinterview

Cause of difference	*Percent*
Different respondent	27.3
Due to enumerator	22.5
Question asked improperly	3.1
Misinterpreted answer	17.1
Misrecorded answer	2.3
Due to respondent	47.5
Misunderstood question	12.8
Reported incorrectly	34.7
Other	2.7
Total	100.0

In addition to furnishing information on individual enumerators, the reinterviews provide a basis for determining the quality of the program.

Table 14.12 shows results of the original and check interviews over a certain period (2). It is clear from the table that the differences are quite large for some of the items. This is especially true for persons working part time, those who have a job but are not at work, and those unemployed. Thus the differences tend to be concentrated among groups with marginal attachment to the labor force.

Table 14.12 Comparison of original and check interview

Employment class	*Number of persons based on*		*Percent identically reported*
	Original interview	*Check interview*	
Labor force	1,221	1,242	93.1
Employed	1,149	1,168	93.5
Agriculture	158	164	94.5
Nonagriculture	991	1,004	93.3
Full time	666	663	96.5
Part time	253	266	86.9
With job but not at work	72	75	86.7
Unemployed	72	74	86.5
Not in labor force	945	924	98.7

RECENT DEVELOPMENTS

In January, 1967, a number of changes were made in the procedures of the survey. The lower age limit on labor force concepts was raised to 16 years from 14 years. The questionnaire was revised to include new probing questions in order to increase the reliability of information on hours of work, duration of unemployment, and the self-employed. Additional substantive questions on the potential availability for work of persons not in the labor force were also introduced. In order to be counted as unemployed, a person must be currently available for work and must report some jobseeking activity. Also the sample was expanded to a 449-area sample and the overall sample size was increased to about 60,000 housing units. The expansion increased the reliability of major statistics by about 20 percent and made possible the publication of greater detail.

14.13 THE JAPANESE LABOR FORCE SURVEY

This is a continuing monthly survey designed on the lines of the United States Current Population Survey. The population covered is all Japanese persons 15 years old or over who are normally resident in the country. The reference period is the last seven days of the month for dynamic questions and the last day of the week for static questions. The current-status approach is used for determining employment status. Unlike the United States survey, all unpaid family workers who worked as much as one hour during the survey week are included in the labor force. Another innovation is the count of employed persons who wish to change jobs or locate an additional job and the number of persons who want jobs but are not seeking work (8).

In order to select the sample the noninstitutional EDs in the census are classified into 25 strata and a systematic sample of EDs is selected from each stratum with equal probability or with probability proportional to size. About 2,000 EDs are selected every month. Within a selected ED the dwelling units are listed from which a systematic sample is selected to obtain about 25,000 households or 70,000 persons 15 years old or over. The sampling rates at the two stages are adjusted to obtain the overall sampling fraction as 1/920, 1/1,380, or 1/1,840. The rotation scheme is similar to that used in the United States survey. Within a selected ED two sets of dwelling units are selected. The first set is surveyed for the first two consecutive months and then replaced by the second set.

About 1,500 enumerators take part in the survey and their work is supervised by 180 supervisors. In contrast to United States practice, enumerators visit the household twice each month, once before the survey week to explain the survey and leave a record sheet to be maintained by members of the household, and a second time to prepare the household schedule after the end of the week. When the completed schedules have been edited, the ratio method is used for making the estimates. For this purpose information is collected from the census and other sources on the number of persons by sex, age, and by area (rural/urban). Table 14.13 gives an idea of the order of magnitude of monthly sampling errors.

Table 14.13 Standard errors of monthly figures, Japanese labor force survey, 1967

Magnitude of estimate (thousands)	*Standard error (thousands)*	*Relative standard error (%)*
100	14	14.0
200	20	9.9
500	31	6.3
1,000	44	4.4
2,000	63	3.1
5,000	100	2.0
10,000	140	1.4
20,000	200	1.0
50,000	320	0.7

Sample: 2,000 EDs 25,000 households 70,000 persons

14.14 CONCLUDING REMARKS

It should be stressed that the techniques of measurement of economic activity, employment, and underemployment are not entirely satisfactory. Considerable research is needed to devise the methods appropriate to a given set of conditions. In fact a multidimensional approach to the measurement of economic activity may be desirable in order to gain a deeper understanding of the employment structure and to avoid omitting persons on the border line of the labor force. In view of the fine dividing line between economic and other activities, a more complete recording of the respondents' activities throughout the day may yield higher dividends.

Because of the seasonal character of employment in developing countries, it is important that the time reference of the survey be the entire year. Within the year, the sample of households can be staggered uniformly, each household providing data for, say, one week. This scheme will provide an average figure over the year. In case the households enumerated in a quarter of the year form a subsample from the entire population, it is possible to obtain quarterly estimates from the survey. The American practice of obtaining monthly estimates is noteworthy but probably too expensive for many of the developing countries of the world.

REFERENCES

1. Central Statistical Organization (1961), "Standards for Surveys on Labor Force, Employment and Unemployment," Government of India, New Delhi.
2. Hansen, M. H., et al. (1955), The Redesign of the Census Current Population Survey, *J. Am. Statist. Assoc.*, **50**.
3. Jaffe, A. J. (1961), A Survey of Underemployment in Puerto Rico, in "Family Living Studies," International Labor Office, Geneva.
4. National Sample Survey (1962), Tables with Notes on Employment and Unemployment, *Govt. of India Rept.* 52, New Delhi.
5. Raj, D. (1968), "Sampling Theory," McGraw-Hill Book Company, New York.
6. United Nations Statistical Office (1964), "Handbook of Household Surveys," New York.
7. U.S. Bureau of the Census (1963), The Current Populations Survey: A Report on Methodology, Tech. Paper 7, Washington.
8. U.S. Presidential Committee to Appraise Employment Statistics (1962), "Measuring Employment and Unemployment," Washington.

15
Consumer Expenditure Surveys

15.1 INTRODUCTION

One of the ways of assessing the economic level of individuals or households in a population is to find out how the community uses its income, or more specifically, the expenditure pattern of households of differing social and economic characteristics. This can be done by collecting data on consumer expenditure through what are known as household (or family) budget surveys. The usefulness of budget surveys is enhanced if they also include information on number of members of the household, amount and sources of income, and savings. Such surveys are invaluable for the construction or revision of consumer price indices. The details supplied by these surveys on consumer purchases, together with the estimated value of items acquired without monetary payment, have proved useful in the computation of national accounts. They provide the basis for comparisons of consumption levels at different periods and between different population groups. Useful as these surveys are, they present some formidable methodological problems relating to concepts and methods of measurement. Some of these problems are discussed in this chapter.

15.2 CONSUMER EXPENDITURE

Family budget surveys are usually concerned with private households and so institutions such as hotels, schools, hospitals, and prisons are excluded. For purposes of these surveys a private household is defined as a group of persons who live together and pool their income to provide for essential living expenses, in particular for a common food supply. This definition is usually equivalent to the one used in demographic surveys.

Household consumption expenditure refers to all expenditure on goods intended for consumption and on services. In includes payments by the household for goods and services supplied or in connection with the use of particular goods and services (see Table 15.1 as an example). Thus payments for community taxes relating to house occupancy, driving license fee, etc., are all included, and so are payments for education, health, and legal services. Consumption expenditure excludes direct taxes, life insurance premiums and other social security contributions, and repayments of loans. Cash contributions to other households are treated as disbursements, not expenditures, on

Table 15.1 Consumer expenditure, Sweden, 1952

	Expenditure	
Item	*Amount (million crowns)*	*Percent*
Food	8,352	35.6
Wine and liquor	527	02.2
Housing	3,163	13.5
Clothing and footwear	3,096	13.2
Miscellaneous	8,333	35.5
Furniture	317	
Sewing machines	69	
Doctors	96	
Hairdresser	204	
Books, stationery	145	
Postage	34	
Trams, buses, railways	523	02.2
Purchase of automobile	549	02.3
Tobacco	561	02.4
Domestic help	236	
Lottery tickets	280	
Total	23,471	100.0

the part of donors. Gifts in kind purchased for members of the household are counted as expenditures but those purchased for outsiders are sometimes treated as disbursements. Purchase of extremely expensive jewelry, of rare paintings, or of antiques is sometimes disregarded since these articles are not intended for ordinary use or consumption.

When the household is the owner of the accommodation occupied by it, the imputed value of the rent is considered as part of the expenditure of the household. This value is sometimes calculated on the basis of current rental values for dwellings of similar characteristics and location. In addition to the imputed value of rent, the cost of repairs and taxes are included in the housing expenditure of the household.

Included in the concept of consumption expenditure is the value of goods or services received in kind by an employee. The valuation of these items is usually at retail price. In principle, some allowance should also be made for items obtained by employees at less than current retail prices such as purchases made at special rates by employees of a department store. However, in view of the difficulties in collecting information of this type, the concessionary privileges are usually ignored.

Food and other products taken from the farm operated by the household are treated as expenditures at retail value. The same procedure is followed for products taken for home use from a business operated by the household. Certain products produced at home, such as clothing and furniture, are usually disregarded.

15.3 MEASUREMENT OF EXPENDITURE

There are three alternative methods of quantifying expenditure. In one method, the value of goods and services actually *consumed* by the household during the reference period is considered. Thus the consumption expenditure relates to all purchases of consumption items which are not added to stocks; actual consumption also includes amounts taken from stocks existing at the beginning of the period. The method therefore involves the burdensome task of measuring inventories at the beginning and at the end of the period as well as purchases in between. It is for this reason that this method is rarely used except in connection with specialized surveys such as food consumption studies (Sec. 20.7).

In the second method, consumer expenditure is based on the total value of all consumption goods and services *delivered* during the period, irrespective of when full payment is made or whether they are fully consumed during the period. Thus the full price paid for the items delivered to the household during the reference period is recorded as consumption expenditure. On the other hand repayments of credit or hire purchase installments in respect of items delivered to the household at a date outside the reference period are not regarded as consumption expenditure. The method is practicable for almost all goods and for many types of services such as transportation, laundry, etc. There are other services, such as rent, for which it becomes difficult to define when the delivery took place. In such cases it is not uncommon to record these services at the time of payment.

Actual payments made during the reference period, irrespective of when the goods and services are delivered or consumed, form the basis of the computation of consumer expenditure when the third method (the *payment* principle) is used. Thus only actual payments made during the period for goods procured on the hire purchase system are considered as expenditures and so is any down payment made for a newly acquired item. It is clear that this method and the second one will agree when the delivery and the full payment are more or less simultaneous. And in the case of small items regularly purchased the two methods should give about the same results over a period of a few months. The method is easy to apply even when it is difficult to define when the delivery of certain services took place.

It is usual to collect sufficiently detailed information during the course of the survey in order to be able to use the second and third methods interchangeably. For example, auxiliary information on any installment paid during the reference period is recorded, although the second method based on the delivery principle is intended to be used.

15.4 METHODS OF COLLECTING INFORMATION

There are two principal methods of collecting expenditure data from households. In the *interview method*, investigators obtain and record information on household transactions over a certain period in the course of single or repeated interviews. Usually detailed and fully classified questionnaires are used for this purpose. The household is required to recall all expenditures made during the period in question. The *account-book method* provides for the current recording of household transactions over a certain period by the respondents themselves. Specially designed account books are provided for the purpose. The record books may be of the open or semiopen type, that is, providing for either a simple listing of consecutive expenditures on successive lines of a page for each day, or a similar listing in different sections of the page specifying broad classes of expenditures such as food, clothing, etc. Sometimes a single account book is provided for the household to be maintained by the housewife; in other cases separate books are issued to each member of the household to record individual expenditures.

15.5 THE ACCOUNT-BOOK METHOD

The account-book method has very limited application in the rural areas of the developing countries where illiteracy is widespread. In the urban areas, where it can be used, the refusal rate can be quite high owing to the strain of maintaining daily records of expenditure for a protracted period. As a result, the sample may have to be partly based on a list of volunteers or some other nonprobability method such as the quota method. To give an example (9), 50 percent of the final respondents in the 1963–64 round of budget surveys in the European Economic Community came from the basic probability sample, about 25 percent from the reserve probability sample, and the remainder from the volunteers. In the 1958 budget survey of Sweden (11) about 22 percent of the households selected for keeping account books did not cooperate. It was found possible to collect data from more than one-third of these noncooperating households by the method of interview, thereby reducing the nonresponse rate to about 14 percent.

A special source of error present in this method is the tendency on the part of households to change their purchasing habits once they start using account books. Even if there is no nonresponse and all purchases are correctly entered in the account book, the changed behavior of the households is likely to produce a bias. The very fact of record keeping renders the reporting households less representative of the community which they have been selected to represent. Some evidence of this phenomenon is provided by Table 15.2, based on a survey in the United Kingdom (8). In this survey each cooperating household was asked to keep daily records of all its expenditures for a period of two weeks. A separate account book was provided to each member working full time or over 16 years of age. The survey was staggered over the whole year, fresh households being visited each week. The results show that the average per household of the expenditure recorded during the first week is significantly higher than the average of that recorded during the second week. A part of the explanation for this abnormal situation may be that the fact of record keeping has changed the character of the participating households.

There are other possible sources of bias present in the expenditure data of Table 15.2. During the first few days of recording there may be a tendency on the part of respondents to include purchases made prior to the commencement of the survey, thus overstating the expenditure incurred during the first week. Again we have "border bias" or "telescoping of events." Another explanation is that some purchases in the second week may be omitted as a result of fatigue in maintaining detailed and accurate records of expenditure or due to a lowering of conscientiousness with which records of purchases are kept. The effect of fatigue is that the second-week average is lower than the first. Whether the abnormal situation presented in Table 15.2 is due to conditioning or telescoping or fatigue individually or jointly is a debatable point. Considerable research is needed involving well-designed experiments in order to identify the main effects and interactions of the three possible sources of bias: conditioning, telescoping, and fatigue.

Table 15.2 Average household expenditure on selected items, United Kingdom, 1959 (households with weekly income £20 or less)

Item	*Number of households*	*Average recorded expenditure per household (pence)*	
		First week	*Second week*
Tea	548	44.0	41.3
Bread	548	71.0	69.2
Potatoes	548	37.3	35.5
Eggs	548	47.6	42.2
Soap, etc.	548	30.0	28.4
Coal	548	69.5	36.4

The accuracy of the data collected by the account-book methods depends to a large extent upon the promptness of the recording. The records are expected to be accurate when the recording is completed on the same day or on the spot as purchases are made. If data are recorded hurriedly and carelessly for a period of many days at a time, less accuracy is to be expected. It is therefore important to check the respondents' books frequently in order to get worthwhile data.

Some difference in results may be expected according as the account books to be used are of the open or semiopen type. In a Canadian study (1) it was found that the itemized listing in the semiopen type achieved higher levels for the items listed but lower levels for the unlisted products. Since the itemization could serve as a shopping reminder or a prompting device which could condition purchasing behavior, the higher levels for listed items could not be regarded as necessarily reflecting increased accuracy. But there is a great advantage in using the itemized approach in that the items can be precoded, which can save a considerable amount of labor at the processing stage. Account books of the open type, in addition to lacking uniformity, may be illegible or inconsistent, requiring a greater volume of checking and coding. When some of the account books have to be rejected, the random character of the sample is impaired.

Finally, account books offer the basic material for a detailed study of purchasing habits, including the frequency and regularity of purchases and the preference, if any, as to day of purchase. Such studies are invaluable for designing family budget surveys.

15.6 THE INTERVIEW METHOD

The interview method possesses several advantages over the account-book method. The response rate is likely to be much higher when the interview can be made both interesting and stimulating, thereby creating a favorable atmosphere for winning the cooperation of the respondent. When fully classified questionnaires are used, the items can be precoded. This reduces the cost of manual coding. Since the interviewers are expected to make the entries uniformly in accordance with instructions, very few completed questionnaires are likely to be rejected for want of clarity. With regard to field costs, it is likely that fewer personal contacts per unit are required than under the account-book method. The net result is that the cost per unit with the interview method is likely to be lower than for the other method. Perhaps the strongest point of the method, as far as the developing countries are concerned, is that it can be made to work after the necessary experimentation has been done; the account-book method is out of the question when illiterate respondents are required to complete the books.

The principal weakness of the interview method is that the data collected may be subject to large response errors. The accuracy of the data depends considerably on the ability of the respondents to recall items of expenditure and place them correctly in time. Errors due to memory faults may considerably detract from the value of the results. As an example, consider Table 15.3, in which a comparison has been made between the record-book and the interview methods by using both methods on the same households (4). In the case of the interview method the question asked related to the amount spent in the previous week. It is clear from the table that the figures obtained in the interview were generally higher than those given in the record books of the same households. For all items combined, the interview figure was higher by about 29 percent.

Table 15.3 Weekly expenditure per person on selected items, United Kingdom, 1951

	Expenditure (pence)	
Item	*By record book*	*By interview*
Rent	41	42
Rates, mortgage payments	23	34
Fuel and light	48	62
Insurance	21	36
Time-payment installments	10	11
Telephone bill	4	3
Clothing and footwear	80	124
Tobacco, alcoholic drink	69	85
Entertainments	25	27
Fares	21	16
All items	342	440

The accuracy of the expenditure data obtained by the interview method depends on the period of recall vis-a-vis the frequency and regularity of purchases. Expenditure on items purchased frequently, such as bread, may not be recalled easily for periods longer than a few days or a week. Errors of reporting are likely to increase if the recall period for such items is long. Information on items purchased in bulk to meet the needs of a week or a month may be obtained with reference to a longer period. The period of recall can be considerably increased, without introducing serious reporting errors, for items purchased at long intervals such as furniture or major household equipment.

Apart from the memory faults of the respondents, differences in the practices and approaches of individual interviewers can bring about appreciable bias and variability in the data collected. The manner in which the questions are asked and the techniques of probing used can influence the answers of respondents, thereby introducing possible bias and increased variability. As an example, consider the data presented in Table 15.4. This table is based on a British survey (3) in which the method of interpenetrating subsamples was used for collecting expenditure data. The sample in each area was divided into two parts and an interviewer was allocated at random to each part. As a part of the study, a comparison was made with the account-book method. For every variable studied, the data were tabulated by interviewer, from which the between- and within-interviewer variances were computed. The significance of the variance ratio F indicates interviewer variability. It is clear from Table 15.4 that some subjective element is present in the data collected by the interview method. As many as 9 out of the 19 observed variance ratios were found to be significant at the 5 percent level when the interview method was used while none was significant with the account-book method.

Table 15.4 Comparison of account-book and interview methods, United Kingdom, 1965

Method	*Number of variables studied*	*Number of F ratios*	
		Below 1	*Significant at* 5% *level*
Account book	26	10	0
Interview	19	2	9

15.7 CHOICE OF THE REFERENCE PERIOD

The most important considerations in the choice of the reference period for the collection of expenditure data are the magnitudes of the response errors and the sampling errors. Broadly speaking, the level of reporting errors varies directly with the length of the period of recall. And, for a given period of recall, reporting errors are likely to be smaller for major items than for minor items. Furthermore, apart from a few exceptions, the minor purchases are made more or less regularly and the major ones rather irregularly and infrequently. Assuming that these statements are about right, it is apparent that the best reference period need not be the same for all items. Indeed, the reference period to be used for an item depends on the regularity and frequency with which the item occurs. Consequently, it is best to divide the items into a number of categories such as the following:

1. Daily purchases such as food and transportation
2. Irregular but not quite small purchases such as clothing and household utensils
3. Occasional big purchases such as furniture and major household equipment
4. Regular payments made at fixed intervals such as house rent and electric bills

For items in category (1) which are purchased in small quantities on a daily basis, an extension of the reference period from, say, a week to a month might result in a mere repetition of a cycle of identical purchases. Thus the sampling error is not likely to be appreciably affected when the reference period is made longer. On the other hand, a reduced reference period will presumably lead to some reductions in response errors as a result of the reduced period of recall when the interview method is used. Hence a short reference period such as a week is indicated in this case. About the same kind of argument holds in the case of items in category 4. As an example, consider Table 15.5, based on a pilot survey in Greece (7) in which information on expenditure on food was collected for three weeks. The last column of the table gives relative standard errors for one week, two weeks, and three weeks respectively. It is clear that the relative standard error of the estimate did not decrease appreciably by increasing the length of the reference period.

Table 15.5 Expenditure on food, pilot survey, Greece

Period	*Average expenditure (drachmas)*	*Standard error of estimate*	*Relative standard error of cumulative estimate (%)*
First week	271	13.5	5.0
Second week	254	13.5	4.7
Third week	272	13.5	4.6

With regard to items in category 2, the sampling errors are expected to be somewhat large if the reference period is too short. This is so because the number of households reporting any purchases in this category is expected to be small. Since the purchases are not quite small, the reporting errors are unlikely to increase when the reference period is increased from one week to, say, a month. Thus a somewhat longer reference period such as a month is indicated. By following the same argument a period of three months or even a year may be used for items in category 3.

Items such as medical expenses need special consideration. Because of serial association over the period, lengthening of the reference period may be of no avail. If no household member is sick during a week, the household is unlikely to report any medical expenses in the following week, too. In this case probably the best way of reducing the sampling error is to increase the number of households in the sample.

Some evidence of the undesirability of increasing the period of recall unnecessarily is provided by a British survey (3). In this survey expenditure on clothing was asked for the previous month and for the month before. The data were tabulated by interviewer to study the differences between interviewers. The results obtained are given in Table 15.6. It is easy to see that the variability increased when the recall period was lengthened beyond a month.

Table 15.6 ***F*** **ratios for two periods of recall, United Kingdom**

	Period of recall	
Item	*One month*	*Two months*
Number of items	0.92	1.93*
Expenditure on clothing	0.90	1.60*

* Significant at 5 percent.

15.8 JOINT USE OF INTERVIEW AND ACCOUNT-BOOK METHODS

We have outlined in the previous sections the relative merits of the two principal methods of data collection in this field, namely, the account-book method and the interview method. The reader might ask whether it is possible to use a method combining the principal advantages of both. The answer to this question is in the affirmative. In fact the problem is not whether the interview method or the account-book method should be used. The issue is to identify those items which are amenable to the interview approach and those which should be handled through the account-book method. This decision can only be made when sufficient information on consumer behavior and buying practices is known. Since such practices are likely to vary from country to country and even among the rural and urban areas in the same country, it is not possible to lay down methods which will work everywhere.

Broadly speaking, information on major items can be collected by the method of interview. These items include refrigerators, automobiles, household furniture, radios, toasters, etc., and belong to category 3 of Sec. 15.7. Such items are not forgotten even when the reference period is somewhat long, say, three months. This type of information is best collected by staggering the sample of households uniformly over time. Some households are dropped every month and others are added to take their place. The rotation system can be arranged to reduce the effect of conditioning of response. The effect of telescoping can be minimized by presenting to the sample households the list of expenditures made for the previous quarter.

Information on items belonging to category 4 can also be collected by the interview method. Payments for these items are made regularly at more-or-less fixed intervals. There is considerable merit in asking when the last payment was made and what the expenditure involved was. When about an equal number of households is interviewed every month throughout the year, seasonal variations are taken care of.

As shown in Table 15.6, it is unnecessary to use a reference period as long as three months for collecting information on items falling in category 2. The interview method may be used asking information for the month previous to the date of survey. No telescoping will be involved if the bounded method is used; that is, the data obtained for the previous month are at hand at the time of interview. The effect of conditioning can be reduced by keeping a household in the sample no longer than is really necessary.

Information on the remaining items, namely, daily purchases such as food and transportation, can perhaps be obtained by the record-book method when it is found possible to secure the cooperation of the households in the sample. Many of these items are so small that they are likely to be forgotten in an interview in which the period of recall is longer than just a few days. When the account-book method is used, these items can be recorded on a daily basis by making occasional reference to records and consulting other household members whenever necessary. The accuracy of recording can be further improved by asking interviewers to make periodic checks and review some of the completed record books from time to time. The recording period should not exceed two weeks for any household. A household is unlikely to cooperate for longer periods. The effect of telescoping can be reduced by using this method on a subsample of households from those selected for interview and by making the recording follow the interview. The risk of conditioning must be closely watched but it is unlikely to be very important. The common man does not stop purchasing bread or do more traveling by bus simply because he has to enter these expenditures in a record book. When the risk is there, we may follow the suggestion that the records be kept for 17 days to permit exclusion of the first 2 and the last days (9)

It is time to consider the question of the integration of the two samples, one for the interview method and the other (a subsample) for the record-book method. All households in the interview sample will provide information on all items except those in category 1. The subsample asked to keep records of expenditure for two weeks will record all expenditure whether major or minor. The expenditures meant to be covered by the interview method can thus be excluded from the record book at the analysis stage. This may as well be done by the interviewer who goes to pick up the account book at the end of the recording period. There is a reason that a household is required to record all expenditures and not just a part of the total budget. In one of the studies (6) it was found that the expenditure recorded was significantly higher when the record was limited to a certain category of purchases only (Sec. 15.9).

15.9 RESEARCH AND EXPERIMENTATION

Considerable research and experimentation is needed in order to evolve appropriate procedures for the collection of expenditures data. The data are subject to recall lapse and other response errors when the information is collected by interview. There may be conditioning, telescoping, and fatigue effects present in the figures recorded in the account book when this method is used. A series of pilot studies should be undertaken to determine the major sources of error in the method proposed and the modifications needed to make the data reasonably free of defects.

The methods of experimental design can be successfully employed to carry out these investigations. By subjecting different random samples of households to the various treatments to be tested and by comparing the results so obtained, it is possible to verify different hypotheses regarding conditioning, telescoping, and other effects likely to be present. An investigation on these lines has been carried out in Israel (6). A random sample of 3,000 households was divided up into three subsamples. The first subsample recorded expenditures on food for both fortnights A_1 and A_2 in a particular month. The second subsample recorded information for the first fortnight only B_1 and the third for the second fortnight only C_2. The effect of telescoping was assessed by testing the hypothesis $A_1 + A_2 = B_1 + C_2$, and that of fatigue by testing $A_2 = C_2$. The end-period effect was judged by testing the hypothesis $A_1 = B_1$. In this study expenditure on food was found to be 5 percent higher for fortnightly recording as against monthly recording. Those households that had only just begun recording expenditure on foods gave a total that was 4 percent higher than those that had been engaged in recording for the previous fortnight. Further, the households corresponding to B_1 recorded 7 percent greater expenditure on all foodstuffs than did the A_1 households. As a part of the study, all households in the sample were asked to record expenditure on nonfood items for the whole month. Thus there were two types of households in the second and third subsamples: those that recorded both types of expenditure in a fortnight and others which recorded only one type (nonfoods). A comparison of the results showed that expenditures on nonfood items were higher for households which were not keeping food records during the fortnight. It was concluded that concentration on only certain kinds of expenditure could lead to an overstatement and that it was better to record all expenditures.

15.10 THE GREEK BUDGET SURVEY OF 1958

The reader will now be introduced to some of the consumer expenditure surveys conducted in different countries. The first survey we describe comes from Greece. This survey covered all private households in towns with 1951 population of 10,000 or more. Out of 49 such towns 25 were selected in the sample: 15 with probability proportional to population and the 10 largest with certainty. A sample of dwellings was taken from each town and the households occupying the dwelling units at the time of visit by the interviewer constituted the sample of households.

At the initial interview information was collected on household composition and the type of living accommodation. The interviewer visited the household daily for a period of seven consecutive days in order to record household purchases of all types. During the next three weeks the interviewer visited the household twice a week to collect data on purchases of utensils, clothing, etc. Expenditure on furniture and major equipment was obtained by interview, the reference period being 12 months. In case of regular payments such as rent and electricity, the last payment was recorded along with the period for which it was made. Information on education, licenses, weddings, etc., related to the past 12 months.

Every spending member of the household was required to record all personal expenditure for a period of seven days. Another form used was the record of motor vehicle expenses asking for information on the annual cost of circulation tax, license fee, and insurance and expenditure on maintenance and repair for the previous month and gasoline and oil purchases for seven days. Furthermore, all goods and services received free of payment by the household during the week were recorded.

The delivery basis (that is, the point at which products are purchased or services are received) was adopted for most classes of expenditure. But certain services such as education or medical care were considered as expenditures at the time of payment. For owner-occupied units an imputed rental value was calculated to represent housing cost. Table 15.7 gives the relative sampling errors of the average purchases per household for selected groups of items (7).

Table 15.7 Relative sampling errors of average weekly purchases per household, urban Greece, 1958

Goods and services	*Reference period*	*Relative sampling error* (%)
Food	Week	1.22
Bread, flour, cereals, etc.	Week	1.36
Meat	Week	1.87
Fish	Week	2.22
Oils and fats	Week	2.37
Dairy products and eggs	Week	1.63
Pulses, vegetables, and fruits	Week	1.38
Sugar and confectionery	Week	2.54
Other food	Week	2.86
Beverages (excluding alcoholic drinks)	Week	2.55
Alcoholic drinks and tobacco	Week	2.03
Household supplies and services	Week	3.35
Medical care	Week	15.01
Recreation	Week	3.84
Transportation	Week	4.14
Clothing	Month	3.08
Furniture	Year	8.64
Education	Year	7.49

Sample: 25 towns 2,830 households

15.11 THE BRITISH FAMILY EXPENDITURE SURVEY

The primary purpose of this continuous survey is to know whether the weights used for the price index are becoming unrepresentative of current expenditure patterns and to indicate when a revision is necessary. The survey covers private households only. The register of electors is used as the frame for the selection of the sample from Great Britain while the frame in Northern Ireland is based on tax lists. A two-stage sampling procedure is employed. The primary sampling units are parishes or electoral wards which are allocated to strata on the basis of degree of urbanization, region, and a quantity correlated with the ratable value of dwellings. Selection within a stratum is made with probability proportional to the size of parliamentary electorate. In all, 128 primary sampling units are selected. Within a selected psu a systematic sample of addresses is taken.

The sample of 128 psu's is divided up into 8 systematic subsamples of 16 psu's each. During each quarter of the year two subsamples are used, one after the other. An interviewer visits six to seven households per week. If the selected address contains more than one household, a maximum of three is taken in the sample. In all about 5,000 households are visited every year for the collection of expenditures data.

A household is defined as a person living alone or a group of persons living together in the sense of sharing meals and other household expenses. Expenditures are recorded on the payment basis, that is, at the time the payments are actually made during the reference period of two weeks. An exception is made in the case of advance payments made to a store to be redeemed later in goods. In this case, only items actually delivered during the reference period are recorded.

The interviewer calls on all selected addresses to find out who is living there and to explain the purpose of the survey. At this time information on household composition, demographic characteristics, and housing facilities is

Table 15.8 Percentage standard errors of expenditure, family expenditure survey, United Kingdom, 1962

Commodity or service	*Average weekly household expenditure (shillings and pence)*	*Relative standard error (%)*
Fuel, light, and power	22 8.7	1.6
Housing	34 3.9	2.0
Food	108 2.3	0.9
Fish	2 11.3	1.8
Bread	6 1.6	1.2
Eggs	3 9.8	1.4
Sugar	2 4.0	1.3
Alcoholic drinks	13 5.2	3.1
Tobacco	21 7.7	1.7
Clothing and footwear	33 6.4	2.8
Durable household goods	23 9.4	6.0
Furniture	5 5.1	16.0
Other goods	26 7.1	2.2
Transport and vehicles	34 1.8	4.2
Services	31 11.3	3.2
Recreation	2 3.1	5.0
Medical care	0 8.0	12.0
Miscellaneous	1 7.2	7.0
Total	351 10.9	1.2

Sample: 138 areas 3,594 households

collected. Information on regular household expenses such as rent and house tax is collected on the last-payment basis. The other questions relate to items received free or at concessionary prices. The expenditure is to be recorded in account books of the semiopen type, allowing two pages for each day, with general product headings. Calls are made to ensure that the records are kept satisfactorily. The completed books are collected by the interviewer at the end of the period. Provided all the spenders in a household cooperate fully, they are each paid £1 as a reward.

There is another form to be completed by the interviewer for each spender in the household. It covers certain kinds of regular payments such as for television, radio, and driver's licenses, etc. It also lists income of the spender providing separate headings for various components such as wages and salaries, business profits, pensions, etc. Income tax and other deductions are to be entered on the same form.

The results show that the relative standard errors for durable goods items are rather high (see Table 15.8). Comparisons with independent national accounts data show that information on tobacco and alcoholic drinks is deficient (5).

15.12 THE UNITED STATES SURVEY OF CONSUMER EXPENDITURES

This is a large-scale survey conducted every 10 years. The main purpose is to develop weighting patterns and market basket items for the consumer price index. The interview method is used, the recall period being one week for detailed food expenditures and one year for the other items. Except for residents of institutions and persons living on military posts, the entire population is covered by the survey. The basic survey unit is a person living alone or a group of persons usually living together who pool their income for their major items of expense. The delivery principle is used for recording expenditure. This means that all goods purchased or services received during the year are considered, whether fully paid or not.

The sample design for the 1960–61 survey was as follows. Strata were formed by putting together contiguous counties of similar economic characteristics. Within a stratum counties were selected with probability proportional to size. This gave a sample of 126 counties from the rural sector. In order to select a sample from the urban sector, the 12 largest Standard Metropolitan Statistical Areas were taken with certainty. The remaining areas were chosen by the method of controlled selection to give a wide geographic spread to the sample. There were in all 66 urban areas in the sample. The second-stage unit was the block in urban areas and the enumeration district elsewhere. A list of all housing units was prepared within each selected second-stage unit and a sample of housing units was taken. A systematic selection of housing units spaced halfway between the selected units served as the reserve sample for substitutions where necessary. About 17,000 housing units were selected in the initial sample. The rural sample was investigated in 1961 while the 12 largest urban areas were enumerated both in 1960 and 1961. One-half of the remaining urban sample was completed each year.

The interviewer visited the housing unit in the sample and listed all occupants there. Information was collected on household characteristics and the composition of consumer units during the reference year. Then a very detailed questionnaire relating to annual income, expenditure, and savings was completed by interviewing the housewife and/or other members who appeared to be the best respondents for different sections of the questionnaire. Sometimes parts of the questionnaire were left with the household for completion. In the case of urban and rural nonfarm households a supplementary questionnaire was used to collect information on some expenditures relating to the previous week. Food, personal care, household supplies, and some other items were included in this questionnaire.

The survey describes the spending and saving of all families in the United States at the regional and national levels. The data are used for obtaining estimates of a number of components in the household sector of the national accounts. In addition standard quantity budgets for selected types of families are derived (10).

REFERENCES

1. Dominion Bureau of Statistics, Canada (1960), Family Budget Enquiries, *Proc. Fifth Conf. British Commonwealth Statist.*, Wellington.
2. Fowler, R. F., and L. Moss (1961), The Continuous Budget Survey in the United Kingdom, in "Family Living Studies," International Labor Office, Geneva.
3. Kemsley, W. F. F. (1965), Interviewer Variability in Expenditure Surveys, *J. Roy. Statist. Soc.*, (A)**128**.
4. Kemsley, W. F. F., and J. L. Nicholson (1960), Some Experiments in Methods of Conducting Family Expenditure Surveys, *J. Roy. Statist. Soc.*, (A)**123**.
5. Ministry of Labor (1963), "Family Expenditure Survey: Report for 1962," Her Majesty's Stationery Office, London.
6. Prais, S. J. (1958), Some Problems in the Measurement of Price Changes with Special Reference to the Cost of Living, *J. Roy. Statist. Soc.*, (A)**121**.
7. Reisz, A. B. (1961), A Budget Survey in the Urban Areas of Greece, in "Family Living Studies," International Labor Office, Geneva.
8. Turner, R. (1961), Inter-week Variations in Expenditure Recorded during a Two-week Survey of Family Expenditure, *Appl. Statist.*, **9**.
9. U.S. Bureau of the Census (1968), Methodology of Consumer Expenditure Surveys, *Working Paper* 27, Washington.
10. U.S. Bureau of Labor Statistics (1965), Consumer Expenditure and Income, Survey of Consumer Expenditures,1960–61, *Rept.* 237–93, Washington.
11. von Hofsten, E. (1961), A Budget Survey in Sweden, in "Family Living Studies," International Labor Office, Geneva.

16
Surveys of Health

16.1 INTRODUCTION

The state of health of a nation is one of the most important aspects in the study of a human population. The frequency and duration of sickness, its severity, and the number of deaths determine to a great extent the amount of misery in the community. In view of the social disturbance and economic loss caused by sickness and the cost of medical care, accurate information is needed to plan preventive measures on an adequate scale. The total volume of sickness and its characteristic variation by season or by age, sex, or occupation are some of the problems to be studied by the health agency in the community. Since no system has been devised for the reporting of all morbid conditions to this agency, attempts are made to interpret data of a rather selective nature from hospitals and other institutions. Alternatively, morbidity statistics are collected through household surveys by using the method of interview. In countries where such surveys are being undertaken, a number of problems have been encountered. The purpose of this chapter is to discuss some of these.

16.2 THE CONCEPT OF ILL HEALTH

It is difficult to draw a definite dividing line between illness and health. Apparently healthy individuals are known to vary widely with respect to measurements such as blood pressure, pulse rate, and blood counts. Two persons may appear to be in the same bodily condition as judged by observation and measurement and yet one is sick and the other is not. Thus the presence of sickness has to be largely determined by the person himself. If he believes he is sick, he may be legitimately regarded so. On the other hand, he may not be aware that he is sick until a routine medical checkup reveals that all is not well. We may, therefore, say that an illness has started when the person begins to be conscious of symptoms of some disability or when someone else has decided that the disease is present and cannot be ignored without risk to the patient.

As a matter of fact, the purpose of the study should determine how sickness is to be defined. Once this is done, it is possible to distinguish between new, recurrent, and continued sickness and to define grades of sickness such as major, minor, or ill-defined. Furthermore, definitions of onset, termination, and duration may be formulated and the question of the statistical units to be used may be considered. Ordinarily the statistical units will be persons, illnesses, or spells of illness. The information collected may relate to sickness beginning within the period of observation or ending during the period or current at any time or at some particular point of time during the period. Alternatively, the statistics may refer to the duration of sickness current or ending during the period.

16.3 DEFINITION OF SICKNESS

We shall use the terms *sickness*, *illness*, and *morbid condition* interchangeably to denote any departure, subjective or objective, from a state of physiological well-being resulting from disease, injury, or impairment. The operational definition of sickness can be formulated to suit the purposes of the survey. In the Japanese health survey, for example, sickness is defined (7) to include any unusual disturbance of physical and mental well-being (1) treated by a physician, a dentist, or some other therapeutist, or treated with medicines, drugs, or other medical materials; or when (1) is not applicable, (2) in which the patient is confined to bed or incapacitated for work for two days or more. However, the following cases are excluded: (1) normal pregnancy, delivery, puerperium, menstruation, (2) handicapped conditions (blindness, deafness, dumbness, impairment of trunk extremities, and disturbance of central nervous system) with symptoms stabilized and fixed, receiving no medical attention, (3) myopia, hypermetropia, and astigmatism.

In the United States National Health Survey (9) morbidity is defined as a departure from a state of physical or mental well-being, resulting from disease or injury, of which the affected individual is aware. It includes not only active or progressive disease but also impairments, that is, chronic or permanent defects that are static in nature, resulting from disease, injury, or congenital malformation. The existence of morbidity in an individual caused by a particular disease, injury, or impairment is called a *morbidity condition* or simply a *condition*.

16.4 METHODS OF MEASUREMENT

There are two methods of measurement of sickness, one based on detailed clinical and laboratory examinations and the other by interview. The interview method is usually employed for it is almost impossible to secure the services of a large number of professional physicians for making clinical and laboratory examinations. When the interview method is used, the household has been found to be a very convenient unit of enumeration. In a household survey the data on morbidity can be linked to a variety of social and economic conditions. And more important, other types of data such as the number disabled or invalid or the amount of medical care and cost of medical treatment can be collected at the same time.

16.5 THE PREVALENCE SURVEY

In a household survey of the prevalence type, the household is asked about illness existing at the time of the enumerator's visit or during a specified period. All current illnesses are recorded whether they started before or during the period. Alternatively, all persons found to be ill during the period are counted; the count is made irrespective of the fact that the person fell ill before or during the period. When the number of persons found to be sick sometime during the period is divided by the average number of persons exposed to risk during that period, the rate so obtained is called the *period prevalence rate*. If the numerator used is the number of spells of sickness, the rate obtained refers to spells of sickness. The period prevalence rate becomes the point prevalence rate if the number of sicknesses or of persons sick relates to a given point of time such as the first day or the last day of the period. As an example of period prevalence rates, consider Table 16.1, based on a Canadian survey (2). In this case females were found to be subject to higher morbidity than males.

When information on illness relates to the day the enumerator visits the household, the response obtained is expected to be reasonably complete. The

Table 16.1 Prevalence rate per thousand persons by sex and age, Canada, 1950–51

Age-group	*Male*	*Female*	*All persons*
Under 15	258	256	257
15–24	134	178	157
25–44	167	244	206
45–64	157	207	181
65 or over	175	216	195
All ages	189	229	209

reason is that in this case members of the household are not required to recall past illness. About the same situation holds when the period involved is short. The response should be particularly complete if the enumeration is restricted to cases that can be identified objectively such as those causing inability to work or pursue normal duties. However, the recorded cases will be heavily weighted by chronic disease; such cases are expected to be present when the enumerator calls but not the many minor colds and digestive disturbances which have come and gone.

The average situation brought out by Table 16.1 is not very useful in view of the considerable seasonal variation present in the records of sickness. A clearer picture can be obtained by asking about illness during a short period and by staggering the households uniformly over the calendar quarter of the year. Table 16.2 is an example of the kind of information obtained when this method is used (5).

Table 16.2 Prevalence rate per thousand persons by sex, age, and quarter, England and Wales, 1951

Age-group	*Quarter* 1	*Quarter* 2	*Quarter* 3	*Quarter* 4
		Males		
21–44	1,120	960	930	1,080
45–64	1,490	1,220	1,210	1,300
65 or over	1,970	1,760	1,800	1,880
		Females		
21–44	1,490	1,280	1,260	1,430
45–64	1,920	1,750	1,700	1,740
65 or over	2,410	2,310	2,200	2,270

16.6 THE INCIDENCE SURVEY

In the incidence survey the members of the household are asked for illnesses that occurred or commenced during a specified period, whether or not the person is sick on the day the enumerator has called. Any illness that started before the period and continued into it is excluded. An illness commencing within the period is counted. Similar information can be obtained for persons becoming ill within the period or for spells of illness beginning within that period. A number of rates can be calculated. One is the incidence rate (persons), in which the numerator is the number of persons who start at least one spell of sickness during the period, the denominator being the average number of persons exposed to risk during that period. The incidence rate (spells) is defined as the number of spells of sickness which start during the period divided by the average number of persons exposed to risk during that period. Table 16.3 presents an example of incidence rates (2).

Table 16.3 Incidence rate per thousand persons by sex and age, Canada, 1950–51

Age-group	*Male*	*Female*	*All persons*
Under 15	254	232	253
15–24	132	174	153
25–44	161	234	198
45–64	147	193	169
65 or over	154	193	173
All ages	182	220	201

As compared with the prevalence survey, the incidence survey presents greater balance between acute and chronic diseases, inasmuch as the many colds and stomach disturbances find a place in the record along with the chronic ailments that started during the period. The disadvantage is that the members of the household cannot recall all the minor illnesses if the period involved is rather long. In this case it is better to stagger the households uniformly over time and collect information on the incidence of illness for a short period from each household. When this is done the survey is called the *periodic incidence survey*.

16.7 DURATION OF SICKNESS

With respect to a given reference period all sicknesses can be divided into four classes: (1) those commencing and ending within the period, (2) those commencing within the period but continuing on the date of survey, (3) those beginning before the period but ending within it, and (4) those beginning prior to the period and continuing on the date of survey. By interviewing the members of the household it is possible to determine the entire duration of the spells of sickness in categories 1 and 3. When the total of these entire durations is divided by the number of spells involved, the average duration per spell is obtained. In order to obtain the average duration per sick person, the denominator to be used is the number of persons who experienced at least one sick spell during the period. The frequency distribution of durations by period of entire duration is another useful mode of condensation of data. Sometimes the entire duration is not considered; only that part is taken into account which is confined to the reference period. In this case the total duration (within the period) of all spells of sickness that occurred wholly or partly within that period is determined. When this is divided by the average number of persons exposed to risk, the average duration of sickness per person is obtained. Table 16.4 serves as an illustration of average duration within a period (5).

Table 16.4 Average duration (days per spell) of sickness in the reference month, England and Wales, 1951

Age-group	21–	25–	35–	45–	55–	65–	75–	*Total*
Male	1.04	0.84	0.71	0.98	1.27	0.95	0.80	0.94
Female	0.76	0.61	0.63	0.63	0.71	0.69	0.66	0.65

16.8 CHOICE OF RESPONDENT

An important question to decide is whether information on morbidity must be collected from the respondent directly concerned. Some evidence on this issue is available from the United States health survey (4). In this survey

Table 16.5 Condition prevalence rate for proxy and nonproxy type of interviews, United States health survey

Type of condition	*Proxy interview not permitted*	*Proxy interview permitted*
Chronic	11.48	10.20
Major nonchronic	1.31	1.26
Minor nonchronic	0.96	0.76
All conditions	13.74	12.23

two subsamples were used, one in which all adults were interviewed for themselves and the other in which proxy interviews were permitted. A comparison of the results was made by calculating the condition prevalence rate per 1,000 persons for the three types of conditions: chronic, major nonchronic, and minor nonchronic. The results showed (see Table 16.5) that there was an underreporting of 11 percent when proxy interviews were permitted. The minor nonchronic cases were underreported by as much as about 21 percent.

16.9 CHOICE OF PROBES

Another question to be decided is whether the respondent should be given a checklist of diseases as an aid to recall of conditions prevailing during the reference period. An alternative is to use probes having the effect of defining diseases in terms of the kinds of action people take as a result of their illness. Both methods were tried in the United States health survey (4). Actually the interviewers started with the probes and presented the checklist at the end. The results obtained on chronic conditions are given in Table 16.6. It is clear that the use of initial probes could elicit less than 50 percent of the chronic conditions and that the final checklist was found to be necessary to pick up the other half. Just one-third of the minor chronic cases were reported initially. There is no doubt that the checklist is a useful aid to memory.

Table 16.6 Percentage of chronic conditions, picked up by initial probes and final checklist, United States health survey

Condition type	*Initial probe*	*Final checklist*	*Total*
With medical care and activity restriction	68	32	100
With medical care or activity restriction	47	53	100
Without medical care or activity restriction	33	67	100
All chronic conditions	47	53	100

16.10 CHOICE OF REFERENCE PERIOD

As stated before, morbidity statistics are usually collected by the method of interview. The members of the household are required to recall the various types of sicknesses from which they suffered during a specified period. The longer the period of reference the greater is the chance that the event is not reported due to memory failure. Evidence on this point is provided by a British survey (3) in which informants were asked about illnesses in the three calendar months prior to interview. A simple cross tabulation of the data by month of interview and month of sickness experience (see Table 16.7) showed that the incidence rate for a particular month decreased considerably when the period of recall was lengthened. The conclusion drawn was that the recollection of minor complaints three months ago was not reliable. Accordingly, it was recommended to use two months as the period of recall in the survey of sickness.

Table 16.7 Monthly incidence rates per thousand persons, survey of sickness, England and Wales, 1944–49

Type of illness	*Month of interview*	*Month of sickness experience*				
		Oct.	*Nov.*	*Dec.*	*Jan.*	*Feb.*
Serious illness	Jan.	17	21	20		
	Feb.		18	20	18	
	Mar.			20	20	21
Influenza	Jan.	15	85	142		
	Feb.		52	107	50	
	Mar.			81	53	23
Colds	Jan.	49	98	193		
	Feb.		83	140	126	
	Mar.			113	137	158

With the two-month reference period it was found possible to compare the morbidity data for the "last month" with that obtained for "the month before last." The comparison was made by calculating the ratio of the number of illnesses experienced, using the "last month" as the denominator. The results obtained are given in Table 16.8 (5). It is clear that the new and less severe illnesses are underreported when the recall period is lengthened.

Table 16.8 Ratio of number of illnesses in month before last to number of illnesses last month, British survey of sickness, 1947

Type of illness	*Persons aged 16–64*				*Persons aged 65 or over*
	Serious illness	*Moderate and mild*	*Minor and ill-defined*	*All illnesses*	*All illnesses*
New	99	72	64	66	64
Recurrent	92	92	98	98	103
Continued	100	83	98	97	100

As another example, we consider the California health survey of 1954–55 (6) in which respondents were asked about illnesses in the four calendar weeks prior to interview along with the week in which the onset took place. As the sample was randomized over the year, it was possible to tabulate the incidence of illness by the period of recall. The results obtained are given in Table 16.9. In this table the week prior to interview is taken as the base. The

Table 16.9 Index of incidence rates of illness, California health survey, 1954–55

	Number of weeks prior to interview						
Ilness category	1	2		3		4	
Doctor seen	100	109	58	101	38	95	41
Disabling	100	99	72	76	52	65	52
Doctor seen and disabling	100	130	79	112	52	100	55
Doctor seen or disabling	100	96	58	78	39	70	41
Doctor not seen, not disabling	100	57	42	33	25	24	32
All illnesses	100	73	53	51	35	42	38

figures in the first half of a column relate to acute conditions while those in the second half relate to chronic conditions. It is clear that recall was better for acute conditions than for chronic conditions and for medically attended acute cases than for acute cases involving neither disability nor medical care. As a result of these findings it was decided to collect information for just two calendar weeks prior to interview in the Charlotte pretest, 1957. The index of incidence rate of illness for the second week (expressed as a ratio to the rate for the first week before interview) obtained at this pretest is presented in Table 16.10. The results show that there was no consistent drop-off for the

Table 16.10 Index of illness rate for the second week with first week as the base, Charlotte pretest, United States National Health Survey, February, 1957

	Number of weeks prior to interview	
Illness category	1	2
Doctor seen	100	111
Disabling	100	97
Doctor seen and disabling	100	110
Doctor seen or disabling	100	100
Doctor not seen, not disabling	100	87
All illnesses	100	95

second week before interview as compared with the first, with the possible exception of illnesses involving neither medical care nor disability. Thus a two-week reference period was used in the national survey for all illnesses, injuries, and medical and dental care visits.

The memory factor is considered to be so important in Japan that just one day is preferred as the reference period for the national survey of sickness. However, monthly data are collected from time to time to serve as a check. Based on the 1956 survey a comparison was made between the first day, the last day, and the monthly average prevalence rates per 100,000 persons (7). Some of the results are given in Table 16.11. It is apparent that the reporting of a disease for a one-day reference period is, in general, less complete than for a monthly reference period. But the proportion of each disease to the total does not vary appreciably with the method used. There is, however, a tendency for respondents to report more completely diseases which are serious or of the prolonged type when the reference period is just one day. Furthermore, reporting on the last day is more complete than that on the first day.

Table 16.11 Prevalence rates of sickness per 100,000 persons based on one-day and monthly data, Japan, 1956

	Percent distribution			*One-day as percent of monthly average*	
Selected diseases and injuries	*First day*	*Last day*	*Monthly average*	*First day*	*Last day*
Tuberculosis	13.1	11.9	9.8	95.8	100.4
Eye diseases	3.4	3.6	3.6	67.6	83.1
Diseases of respiratory system	11.2	11.6	15.6	50.9	61.3
Diseases of digestive system	18.5	18.0	18.9	69.6	78.4
Skin diseases	5.2	5.6	5.6	66.5	83.1
Diseases of bones, etc.	4.0	3.9	4.2	69.1	76.5
Accidents and poisoning	5.7	6.2	6.8	59.1	74.7
Other diseases	38.9	39.2	35.5	77.1	79.3
All diseases and injuries	100.0	100.0	100.0	71.3	82.4

16.11 VALIDATION OF SURVEY DATA

The reader may ask whether the data collected by lay interviewers by asking lay respondents on a subject as complex as illness are sufficiently accurate to justify the expense involved in the conduct of household surveys. A partial answer is provided by a study made in San Jose, California (1). In this study reports on hospitalization were obtained during a preceding period ranging from 7 to 11 months by interview of households. These reports were checked against hospital records in five hospitals. It was found possible to match records for 249 periods of hospitalization. Thirty reports in the survey could not be matched with hospital records while thirty-nine periods of hospitalization were not reported in the survey. Calculations were made on a number of measures of hospitalization using both the survey and the hospital records. The results obtained are given in Table 16.12. The table shows that there is no significant difference between the survey figures and the hospital reports for the items considered. This means that the reports of hospitalization obtained in household surveys can be sufficiently accurate.

Table 16.12 Comparison of measures of hospitalization as obtained from the household survey and from hospital records, San Jose, 1952

Measure	*From survey reports*	*From hospital records*	*Ratio*
Annual admissions per thousand persons	65.5	67.9	1.04
Annual days of hospitalization per person	0.609	0.655	1.08
Average length of stay per period of hospitalization in days	9.1	9.5	1.04
Percent of admissions with surgery	43.4	44.4	1.02

Another type of check of household survey data was made in Japan (7). A sample of institutions (hospitals, clinics, and dental clinics) was taken from the complete list of medical institutions. The number of inpatients and outpatients who visited physician's offices, had physician's visits at their home, or were being hospitalized on a designated day was determined. The estimates of medical consultation rates per 100,000 persons as calculated from

Table 16.13 Medical consultation rates per 100,000 persons per day by selected diseases, Japanese health survey and patient census sample, 1956

Disease	*Survey*	*Census*	*Disease*	*Survey*	*Census*
Tuberculosis	295	406	Ear	75	102
Venereal	3	23	Nose	63	64
Malignant	15	20	Genital organs	23	55
Mental	28	72	Skin	117	231
Eye	92	108	Accidents	166	179
			All diseases	2,062	3,145

this sample were compared with the corresponding estimates made from the household survey. Table 16.13 gives the results obtained. It is evident that the number of consultations reported by the household survey is less than that enumerated by the patient census sample, the former being about two-thirds of the latter. Furthermore, diseases which people tend to conceal such as venereal diseases and diseases of genital organs as well as minor diseases such as diseases of the skin are less completely reported in the survey than others.

16.12 THE QUESTIONNAIRE

Since sickness is a complex and delicate subject, the questionnaire used to elicit response should be as simple as possible. Perhaps it is best to start with the question, "Were you sick at any time during the last 14 days?" If the answer is yes, ask, "Did you talk to a doctor about your illness?" If yes, ask, "What did the doctor say?" If no, ask, "What was the matter (nature of illness)?" In order to find the duration, ask, "When did it start? Are you now free from all symptoms? When did you consider yourself free from all symptoms?" To determine the economic loss involved, ask, "Did this illness cause you to cut down on your usual activities? How many days of the last 14?" The entire sequence of questions should be repeated until all illnesses are mentioned.

To collect information on injuries, we may start by asking, "Did you have any accident or injury at any time during the last 14 days? How did it happen? What kind of injury resulted? Did this injury cause you to cut down on your usual activities? How many days of the last 14? Did you have any other accident or injury during the last two weeks?" The sequence may be repeated until all accidents or injuries have been recorded. To obtain information on health care visits, ask, "Have you been attended by a doctor during the last 14 days? How many times at each of these places (home, office, etc.)? What did the doctor do for you?"

Information on hospitalization may be obtained by asking, "Were you admitted to the hospital during the last 12 months? How many times were you admitted during that period? How many days were you in the hospital during the last 12 months?" A further question may be asked, "Do you have any impairments? What impairment?" And finally ask, "Were there any medical expenses to be met during the last 12 months?" These expenses may be listed by category such as hospital fees, cost of drugs, etc.

16.13 THE UNITED STATES NATIONAL HEALTH SURVEY

A brief description of some of the current health surveys will now be given. We first take up the United States survey, which has been designed to give data on the prevalence and incidence of disease, injuries, and impairments, the nature and duration of the resulting disability, and the amount and type of medical care received. Every week about 720 households selected from the entire civilian population are interviewed. This gives an annual sample of about 36,000 households. Interviewing is conducted in the home, whenever possible with the individual person if over 18 years of age, and otherwise with a responsible adult member of the family. The questionnaire contains 40 items for identification of households and socioeconomic description of respondents, 12 general questions on the presence or absence of illness, accidents, impairments or conditions of each member of the household, and 54 detailed questions for each person regarding details of illness, accidents, impairments, and medical, dental, and hospital care. For most questions, the recall period is the previous two weeks. But for some items of low incidence, for which memory is reliable, such as hospitalization, the recall period extends over the year previous to the interview.

In order to select the sample, the country is divided up into about 1,900 primary sampling units. A psu is a county, a group of contiguous counties, or a Standard Metropolitan Area. The psu's are allocated to strata on the basis of geographical location, density of population, rate of population growth, proportion nonwhite, type of industry in predominantly urban areas, and type of farming in rural areas. Since separate estimates are needed for each of 41 regions, called *tab areas*, this requirement has resulted in classification of the psu's into 372 strata. From each stratum one psu is selected with probability proportionate to population. Within each selected psu segments are defined geographically in such a manner that each segment contains an expected six households in the sample. A random sample of segments is taken. The segments enumerated during each calendar quarter of the year form an independent sample of the land area of the country. The samples

Table 16.14 Relative sampling errors of selected items, United States National Health Survey, July to September, 1957

Item	*Estimate (millions)*	*Relative standard error* (%)
Number of bed days for medically attended chronic conditions in last 12 months	756	1.0
Number of visits to the doctor	199	2.2
Number of acute conditions	70	3.0
Number of acute conditions, medically attended	47	4.2
Number of persons with chronic limitation of activity	17	3.2
Number of persons injured in accidents	14	5.1
Number of persons injured in motor vehicle accidents	1	17.5

are also randomized by weeks within each quarter, so that each week's interviews become a random sample of the population and the weekly samples are additive over the quarter.

The supervisory staff of the survey recontact about one-sixth of the households enumerated by interviewers and reinterview 1 predesignated member of the household. This reinterview program serves three purposes: training and control of quality of interviewers, measurement of interviewer variability, and detection of interviewer bias. In addition to this program, several internal editing and consistency checks have been introduced to evaluate and control the quality of data. The relative sampling errors of a number of items are presented in Table 16.14 (8).

16.14 THE JAPANESE NATIONAL HEALTH SURVEY

This is an annual survey carried out in October or November when conditions are most favorable for holding such an inquiry. The purpose of the survey is to keep the medical care system in the country under regular review and to obtain essential information on the volume and nature of diseases, the method of treatment, and the cost of treatment.

The entire population of Japan is the object of the survey. The frame used is the list of national census tracts covering the whole country. The communities are stratified on the basis of population and an initial 1 percent sample of areas is taken from each stratum. This sample is enumerated for obtaining an overall picture of the relation between sickness and living conditions. A subsample of 1 in 18 of these areas is used in the national survey. This produced 12,041 households and 50,030 persons for the 1956 survey.

Two kinds of schedules are used, the household schedule and the individual case schedule. The type of social insurance coverage, economic and social status, and expenditure are some of the items on the household schedule. The other schedule lists the name of the disease, dates of onset and termination, duration of sickness, length of treatment, method of treatment, and cost of treatment. The interviewer visits the household twice a week to collect information from the housewife. During a preliminary visit all household members are requested to maintain diaries to record illnesses during the week. On an average 10 to 15 households are enumerated by an interviewer.

Three types of prevalence rates are calculated. One refers to the beginning of the period, the other to the end, and the third to the average over the period. The average is generally found to be the largest, the end-of-period rate comes next, and the prevalence rate at the beginning of the period is found to be the lowest every year (7).

16.15 THE BRITISH SURVEY OF SICKNESS

This survey was designed to provide information on the volume and nature of diseases and injuries among the population. The survey continued for 10 years from 1943 to 1952. About 4,000 adult civilians were interviewed every month to obtain information on illness.

Illness was defined as a condition (included in codes 001–795 of the international classification of diseases, injuries, and causes of death) which caused some disability during the period covered by the survey inquiry. By disability was meant that the subject was suffering from it and was aware of its existence as something disturbing his state of health during the time. Three kinds of illnesses were recognized. A *new* illness was one which began at some time during the month and which was not a recurrence. A *recurrent* illness was defined as an attack of illness similar to one experienced previously, any such former attack having subsided at least one week before the present one began. By a *continued* illness was meant one present throughout or terminating during the period, having started before the beginning of the period.

Illnesses were classified by degree of severity. A *serious* illness was one which involved considerable risk of death or which produced a total incapacity to work for four weeks or more. A *moderate* illness did not generally involve the risk of death but usually produced from 7 to 27 days of incapacity. *Mild* and *minor* illnesses were recognized according as the number of days of incapacity ranged from 3 to 6 days and 0 to 2 days respectively.

The sample for the survey was selected in a number of stages. The administrative district was the primary sampling unit. The psu's were stratified by the 11 civil defense regions and within these by urban and rural districts. Another variable used for stratification was the index of industrialization, which was the ratio of industrial to total net annual value of property in the district.

In order to select the sample of rural districts within a stratum, the districts were arranged in descending order of the index of industrialization. A systematic sample of two districts was selected with probability proportional to population.

As a first step to the selection of the urban sample, all towns with 300,000 or more inhabitants were taken with certainty. The remaining urban districts were divided up into the required number of strata of equal population by listing the districts in descending order of industrialization index. Two districts were selected from a stratum by following the procedure of systematic sampling with probability proportionate to population.

Within a selected district the electoral register was used for selecting a sample of addresses. A systematic sample of addresses was selected for this purpose. Every month about 4,000 adults were taken in the sample. The total number of interviews was divided among the 11 regions in proportion to population. Within a region the number of interviews in the rural and urban sectors was based on their respective populations.

The interviewers began by explaining the purposes of the survey. The sequence of the opening questions was, "How was your general health during the last month? Did you have any illness, ailment, poisoning, or injury or trouble with long-standing complaints during the month?" The interviewers were instructed to record the name of the complaint, if known, with as full a description as possible. They were warned not to suggest an answer or to make a diagnosis from the symptoms described by the respondent. After the respondents had reported all the complaints, the interviewer read a checklist to help the informants recall anything they might have forgotten.

If the interviewer did not get a full interview on the first call, he would call again. A proxy was allowed only after three unsuccessful calls. When a satisfactory proxy was not available, a substitute was selected from the reserve sample. The substitute had to be of the same sex and age as the original person and geographically as near as possible.

Various changes were made in the layout of the schedule so as to simplify the fieldwork and the processing of the data. The order of questions, however, had to be retained as trial studies showed that a change in order would produce differences in response.

Four rates were calculated from the survey: monthly prevalence rate (persons), monthly prevalence rate (spells), average duration of incapacity per 100 persons, and consultation rate. It was found that, out of 100 persons complaining of some illness during the month, about 77 of them did not visit the doctor. This shows that only an incomplete picture of total morbidity can be obtained solely from medical records. Furthermore, at least 50 percent of the adults who were asked about their health complained of an illness of one kind or another. Although many of these complaints may be considered to be trivial, these high figures cannot be disregarded.

REFERENCES

1. Belloc, N. B. (1954), Validation of Morbidity Survey Data by Comparison with Hospital Records, *J. Am. Statist. Assoc.*, **49**.
2. Dominion Bureau of Statistics (1957), "Canadian Sickness Survey No. 11," Government of Canada.
3. Gray, P. G. (1955), The Memory Factor in Social Surveys, *J. Am. Statist. Assoc.*, **50**.
4. Linder, F. E. (1959), Comparability in Field Studies in General Health, *Proc. Work Conf. Problems Field Studies Mental Disorders*, American Psychopathological Association, New York.
5. Logan, W. P. D., and E. M. Brooke (1957), "The Survey of Sickness, 1943 to 1952," Studies on Medical and Population Subjects, no. 12, General Register Office, London.
6. Nisselson, H., and T. D. Worlsey (1959), Some Problems of the Household Interview Design for the National Health Survey, *J. Am. Statist. Assoc.*, **54**.

7. Soda, T. (1961), A Health Survey in Japan, in "Family Living Studies," International Labor Office, Geneva.
8. U.S. Public Health Service (1958), The Statistical Design of the Health Household Survey, *Publ.* 586-A2, Washington.
9. U.S. Public Health Service (1958), Concepts and Definitions in the Health Household: Interview Survey, *Publ.* 586-A3, Washington.

17
Industrial Surveys

17.1 INTRODUCTION

Today, governments are extensively concerned with the amount of industrial activity in the country, especially the output of goods and services and the employment of labor. Reliable information on industry is required for many policy decisions. For example, adequate national security cannot be provided without pertinent facts on the nation's industry and its location. As a result, governments have expanded their collection of basic statistics on industry—mining, manufacturing, construction, and the production of gas and electricity. These statistics are found to be useful in ascertaining the kinds of industrial establishments on which development efforts should be focused and some of the measures needed to promote industrial growth. The industrial census is designed to provide a bench mark at infrequent intervals for measuring the productive activity of industry in the country while annual changes in the bench-mark magnitudes are usually measured through annual industrial surveys. In addition, monthly and quarterly surveys are instituted in order to follow short-term developments and to assess the usefulness of measures taken to expand industrial activity. Useful as these data are, their collection and analysis present difficult problems of measurement and classification. The purpose of this chapter is to discuss these problems.

17.2 THE DATA TO BE COLLECTED

Industrial activity can be measured from the input or the output side. In the former case data on the amounts of labor, materials, and capital consumed are needed; in the latter case the amounts of products turned out are relevant. A brief discussion of the important items involved is given below.

EMPLOYMENT

The number of persons employed in industry gives an idea of the industrial activity in the country. Included in this group are production workers, whose number is most directly related to the current volume of industrial output. Such workers are usually wage earners. Then there are other employees such as managers, supervisors, and salesmen who are usually paid salaries. The activities of this group of employees are not tied so directly to the current level of manufacturing activity. For this reason it is customary to tabulate the production workers separately (see Table 17.1). A special problem exists in the case of large concerns whose range of activity extends beyond industry such as into trade or transportation. And then there are auxiliary units such as warehouses and repair shops whose main job is to provide services that help in keeping the industry going. The problem is to separate those engaged in industry from the others. Since some employees may be engaged in both types of activities, some arbitrary criteria will have to be used to separate the two categories. Alternatively, the total of all employees may be taken regardless of their activities.

The number of people on leave or quitting their jobs or starting them increases with the length of the period for which the count is made. Thus data on the number employed are more meaningful if the count is based on a short period, say, of one or two weeks duration. Ideally such information should be collected over the entire year but this places considerable burden on the respondents. When the seasonal fluctuations are not pronounced, quarterly figures on employment may serve most purposes.

MAN-HOURS

The number of employees gives only an approximation to the labor input in industry since the number of hours worked may vary considerably. Man-hours worked is a much more precise measure of labor input. Further, man-hours of production workers are more relevant for measuring labor input. The other employees are salaried workers and differences in their man-hour input are of less consequence. Ordinarily it should suffice to collect quarterly figures on man-hours worked for both categories of workers. When man-hours data are summarized weekly rather than monthly, it is easier to report data quarterly since 13 whole weeks represent a quarter.

WAGES AND SALARIES

It is usual to collect annual or quarterly data on wages and salaries paid to production workers and other employees respectively (Table 17.1). The average hourly earnings for production workers can be obtained by dividing the total wages received by the number of production-worker man-hours.

MACHINERY AND OTHER EQUIPMENT

Another way of measuring industrial activity is through its potential capacity. This is done by collecting information on the capital plant and equipment. The number of machines capable of performing a limited range of operations gives an idea of the capacity to achieve a specified level of production. Furthermore, the amount of power equipment available puts an upper limit on many types of manufacturing operations. The amount of electricity generated provides information on one category of this equipment, namely, electricity-generation equipment. A more direct measure of the capacity of this class of equipment is provided by data on the horsepower rating of the prime movers which drive it and other types of machinery.

GROSS OUTPUT

The output of goods and services is the most direct way of measuring the level of industrial activity. Gross output may be measured in terms of cash receipts or the total value of goods and services. The two items of data are not identical. Cash receipts measure the sales of goods and services during the year under inquiry, part of which may have been produced in other years and some of the output of goods during the year under inquiry may be held in inventories. When the value of shipments or sales is known, the value of goods produced can be found by adjusting the figures on sales by changes between the beginning and the end of the period in the value of inventories of semifinished and finished products (6).

COST OF MATERIALS

Materials form an important part of the nonlabor costs in industry. It is therefore essential to collect information on the goods and contract services consumed in production. This information may be collected by asking for the value of raw materials, containers and supplies, fuels, electricity, and contract work consumed in production. In the case of raw materials and fuel costs, an alternative method is to ask for purchases during the year under inquiry and make an adjustment for change in inventory.

It should be pointed out that there is likely to be some duplication when the costs of material for several plants within an industry or between industries are added up. This happens because the materials used by one plant may consist of the finished products of another. As a result, cost of materials figures are usually shown for individual industries when the within-industry duplication is believed to be less than 15 percent (1).

VALUE ADDED

When the value of total gross output and the cost of materials have been obtained for the year under inquiry, it is possible to calculate the value added by manufacture, i.e., the value of output net of the goods and services consumed in producing it. Since value added is a net concept and is additive without involving duplications, its total gives a good idea of the economic value of the various operations performed in the industrial sector (8).

Table 17.1 General statistics for manufacturing industries, United States, 1964 and 1965 (based on Annual Survey of Manufactures)

	All establishments		*Operating establishments*		*Relative standard error* (%)	
	1965	1964	1965	1964		
	(*A*)	(*B*)	(*C*)	(*D*)	(*A*)	(*C*)
All employees total (millions)	17.99	17.27	17.14	16.49	1	1
Payroll (million dollars)	113,962	106,048	105,573	98,685		
Production workers total (millions)	12.98	12.40	12.98	12.40		
Man-hours (millions)	26,397	25,245	26,397	25,245		
Wages (million dollars)	71,095	65,838	71,095	65,838		
Value added (million dollars)	225,365	206,193	225,365	206,193	1	1
Cost of materials (million dollars)	266,254	247,080	266,254	247,080		
Value of shipments (million dollars)	489,596	447,985	489,596	447,985		
Capital expenditures (million dollars)	16,534	13,287	16,534	13,287	1	1
End-of-year inventories (million dollars)	67.92	63.26	67.92	63.26		

17.3 THE UNIT OF ENUMERATION

To facilitate the collection of data a suitable reporting unit must be chosen. There is a choice between the establishment (the individual workshop or factory at which goods and services are produced under one ownership), and the enterprise or company (one or more establishments under common ownership). The use of the establishment as the reporting unit is quite common. The reason is that the establishment is expected to engage primarily in one type of industrial activity and is located in one area. Thus it becomes easy to distribute statistics by industry and area, which is an important requirement. The enterprise is at a disadvantage in this regard since it may consist of establishments spread over a number of areas and engaged in widely different types of work. In the second place, separate and complete records are usually available for the establishment. Enterprises are unlikely to have centralized records of employment, power equipment, inventories, and so on. Thus the establishment is a more practical unit of enumeration. At the same time it approximates the unit used in making and executing decisions on activities to be carried out and resources to be utilized.

This does not mean that there are no problems involved in the use of the establishment as the reporting unit. A number of different products may be produced in the same establishment. This distorts the tabulations by type of industrial activity. And then the same type of activities may be carried on under common control at a number of locations. Since the same staff and machinery may be employed at the different locations, separate records may not be available. In this case the enterprise is a better reporting unit. Finally, it is implicit in the definition of an establishment that auxiliary units such as warehouses and offices be combined with the productive unit. This raises problems when the auxiliary units are in an area different from the units they serve. With the establishment as the reporting unit, the usefulness of data by area is impaired. A solution is to treat the auxiliary units as establishments by themselves.

It should be pointed out that the unit of sampling may not be the same as the reporting unit. Thus the company or the enterprise may be used as the sampling unit while data are collected by establishment for each company in the sample. When an enterprise is required to report data on one establishment and not on the other under its ownership, the management often does not understand the reason for it.

17.4 CLASSIFICATION OF INDUSTRIAL ACTIVITY

The data collected must be classified by industry in order to facilitate economic analysis. Industrial classification can be used as an instrument for organizing statistical materials in such a manner that they reveal the principal structural aspects of the economy in terms of concentrations of employment, types of products, and processes employed. By its means an industrialist can estimate the performance of his plant against the totality of similar ones and the magnitudes of the various industrial markets for his products. Such classification also helps in the comparison of statistics produced by several agencies over time or statistics of different countries.

What criteria should be used for developing the classificatory system? Broadly speaking, the system should conform to the existing structure of industry in the country. The establishments should be used as the building blocks rather than companies. Each establishment should be classified according to its major activity. In order that the intra-industry relationships shown by the published statistics be statistically significant, a group of just a few establishments with few paid employees should not ordinarily be recognized as a separate industry. An industry should be defined almost entirely in terms of products, and every product should be primary to one and only one industry. Furthermore, no industry should be created unless the establishments included in it account for a large proportion of the output of the products which are primary to it. It is desirable that the alien products (i.e., products which are primary to another industry) of an industry be limited. The number of plants not classifiable to any industry should be kept down to a minimum.

A number of problems are encountered in applying a selected system of classification. It is true that a large number of establishments produce a single product and thus can be classified unambiguously in almost any type of classification system. But then there are many establishments producing products ordinarily associated with two or more different industries. This results in disagreements between product and industry statistics. Second, in many important industries, establishments within the same industry differ as to the number of processes encompassed. The classification system cannot ordinarily recognize these differences. Another consideration is confidentiality of the returns. This implies that, when a single company is responsible for the output of a particular product, the data in this regard cannot be published. The burden of this discussion is that any system of classification will have to be a compromise between what is feasible and what is ideal.

17.5 THE INTERNATIONAL SYSTEM OF CLASSIFICATION

Comparability of the industrial statistics of one country with those of another can be achieved by following the International Standard Industrial Classification (ISIC) of all economic activities (7). In this system the unit to be classified is the establishment which is defined operationally as the combination of activities and resources directed by a single owning or controlling entity toward the production of the most homogeneous group of goods or services, usually at one location but sometimes over a wider area, for which separate records are maintained that can provide the data needed. Establishments are classified according to their major kind of activity, based mainly on the principal class of goods produced or services rendered. No distinction is made between kind of ownership, type of economic organization, or mode of operation. In drawing up the classification an effort has been made to ensure that the system is compatible with the economic structure, statistical practices, and needs of most countries of the world. The entire field of economic activity is broken down into 9 major divisions (one-digit codes),

33 divisions (two-digit codes), 72 major groups (three-digit codes), and 158 groups (four-digit codes) of industries. There is also an additional major division (code 0) for nonclassifiable activities (see Appendix 1).

The United States with its highly industrialized economy uses its own system of classification, which may be called the USSIC (9). Except for government activity this system provides 8 divisions further subdivided into 74 two-digit major groups, 383 three-digit groups, and 896 detailed four-digit industries. There is an additional division for government, with 4 separate two-digit major groups, each of which is further subdivided into four-digit industries representing the scope of the two-digit major groups of the private sector. There is also a division for nonclassifiable establishments.

17.6 CLASSIFICATION OF INDUSTRIAL PRODUCTS

The number of products made on which information is to be collected may run into several thousand. It is impossible to tabulate data for each single product. And for analytical purposes only significant cross-tabulations are needed. To achieve this use can be made of the idea of class of product. The product class is intermediate between the detailed level for which data on products are collected and the group of products made by one industry. A further use of this concept is that it facilitates the collection of data on value of sales on a write-in basis in survey questionnaires since the contents of the class of product can be identified in terms of items on the survey. However, considerable effort is needed in constructing, testing, and finalizing the product-class basis on which products can be combined into classes of product. Although no firm rules can be given for making product classes, some possible guidelines can be provided. The class should be homogeneous with respect to the products it contains. All the individual products constituting the class should fall within the same industry. An individual product must be included in one and only one product class. The number of classes of product within an industry should be small, say, 10 or fewer. Wherever possible, product classes should be made subdivisions of primary categories in industrial classifications.

17.7 CLASSIFICATION OF ESTABLISHMENTS BY SIZE

The size of an establishment may be measured by the number of persons paid or engaged, by the type and amount of power equipment used, or by the value added by manufacture. These measures are correlated. The measure of size usually used is employment as it is universal and, unlike monetary units, is not subject to inflationary fluctuations. A question to consider is whether to use all persons engaged (including proprietors, unpaid family workers, and paid employees) as the basis for classification or only those on the payroll of the establishment. In a developing country where it is very common for family members to help with certain tasks, it is useful to make a separate study of those establishments which have no paid employees. The identification of this group helps to focus attention on the problem of distinguishing between the cottage industry and the handicraft establishment. In such a situation, classification of establishments by the total number of persons engaged should be avoided in order that the "unpaid family member" establishments should not be mixed with others. However, these establishments can be sorted out in the zero group if the basis of classification is the number of paid employees.

17.8 SKEWNESS OF DISTRIBUTION

When the manufacturing plants are stratified by employment, it is usually found that the distribution obtained is highly skew; that is, there are too many small plants and too few large ones. Consider, for example, Table 17.2,

Table 17.2 Size distribution of manufacturing plants, United States, 1958 (Census of Manufactures)

Number of employees	*Number of plants*	*Percent of plants*	*Value added (thousand dollars)*	*Percent of value added*	*Sampling fraction (%)*
1–4	105,641	35.428	1,831,999	1.297	1
5–9	50,660	16.990	2,543,890	1.801	2
10–19	46,820	15.702	4,837,994	3.425	5
20–49	40,307	15.530	11,089,100	7.850	10
50–99	21,764	7.299	12,023,698	8.511	20
100–249	16,132	5.410	21,162,398	14.980	50
250–499	6,240	2.092	19,291,058	13.655	100
500–999	2,757	0.925	18,103,226	12.815	100
1,000–2,499	1,363	0.457	21,448,943	15.182	100
2,500 or more	498	0.167	28,937,990	20.484	100
Total	298,182	100.000	141,270,297	100.000	

giving the distribution of the manufacturing plants in the United States for 1958. More than one-third of the plants are very small, employing 4 persons or fewer, and just 1.5 percent employ 500 or more persons. And the relative contribution of the small plants to value added is very low (1.3 percent) while the upper 1.5 percent of the plants contribute about 50 percent of the total value added. This means that a higher proportion of the larger plants will have to be selected in the sample (Example 4.5) in order to minimize the variance of the estimate for a fixed cost of the survey. The actual sampling fractions used in a particular year in the annual survey are shown in the last column of Table 17.2. It may be noted that all plants employing 250 or more persons are taken in the sample with certainty.

17.9 SAMPLE ALLOCATION

In order to provide improved estimates the census data on employment can be used for allocating the sample to strata when such records are available.

Table 17.3 Sample allocation to strata, annual industrial survey, Greece, 1958

	Number of establishments in strata			*Number of establishments in sample*		
Industry	10–19	20–49	50–	10–19	20–49	50–
20	516	266	126	12	22	126
21	78	32	17	7	15	17
22	19	51	150	9	14	76
23	363	255	192	11	30	192
24	410	142	34	14	15	34
25	208	68	15	14	14	15
26	129	37	16	12	12	16
27	45	24	16	9	11	16
28	103	59	33	9	14	33
29	65	27	10	9	13	10
30	14	10	10	7	8	10
31	681	146	43	14	10	43
32	7	2	3	2	2	3
33	326	207	41	9	17	41
34	8	6	10	1	2	10
35	190	74	35	12	12	35
36–38	318	140	92	40	51	92
39	76	25	11	14	10	11
Total	3,556	1,571	854	205	274	780

In fact one can calculate the mean and the standard deviation of employment within each stratum to use the optimum allocation formula. Such a calculation was made in Greece where it was desired to estimate the total employment for each two-digit industry group with a coefficient of variation no more than 5 percent. The establishments were assigned to three strata on the basis of the number of persons engaged. The following allocation (Table 17.3) was obtained for the annual industrial survey for the year 1958 (5). The mean and standard deviation (SD) of employment in the two noncertainty strata are presented in Table 17.4. It is clear from Table 17.4 that the coefficient of variation does not differ very widely between the two strata and for different industries within the same stratum. Thus the allocation based on total employment (Example 4.5) is equally effective. This finding is reassuring since often the optimum allocation cannot be used for want of information on strata variances.

Table 17.4 Mean and standard deviation of employment for non-certainty strata, Greece, 1958

	Stratum 10–19			*Stratum* 20–49		
Industry	*Mean*	*SD*	*CV* (%)	*Mean*	*SD*	*CV* (%)
20	13.0	2.6	20.0	30.2	8.4	27.7
21	12.8	2.3	18.0	32.6	9.4	28.8
22	14.1	3.2	22.7	31.3	8.2	26.3
23	13.7	2.9	21.2	30.8	8.5	27.5
24	12.9	2.7	20.9	30.1	7.9	26.3
25	12.9	2.9	22.5	27.1	6.5	23.8
26	13.1	2.8	21.4	28.5	7.7	27.0
27	13.4	2.5	18.7	30.8	8.0	25.8
28	13.7	2.5	18.2	28.1	8.0	28.3
29	13.0	2.6	20.0	31.2	8.7	28.0
30	12.9	3.3	25.6	31.6	8.9	28.1
31	12.8	2.5	19.5	28.7	8.1	28.2
32	12.8	1.7	13.2	27.0	5.4	19.8
33	12.8	2.4	18.8	29.7	8.1	27.3
34	12.8	1.8	14.1	36.2	10.6	29.3
35	13.2	2.7	20.4	29.5	7.7	26.0
36	13.5	2.4	17.8	29.8	9.5	32.0
37	13.1	2.8	21.3	29.6	8.2	27.8
38	12.8	2.5	19.5	30.8	9.4	30.5
39	13.8	3.3	23.9	28.4	7.7	27.1
Total	13.1	2.7	20.6	29.9	8.3	27.8

In Table 17.3 the sample sizes have been calculated on the assumption that the estimate is to be based on simple averages within strata. When bench-mark figures from a census are available, it is, however, possible to make improved estimates for intercensal years by using the difference estimate (Sec. 4.6). When this can be done, the sampling errors of estimates are considerably reduced, sometimes by 40 to 50 percent.

17.10 THE METHOD OF CUTOFF

It is not uncommon in industrial surveys to use the method of cutoff, by which data collection is confined to the larger, better-organized establishments with superior records and reporting facilities. In many surveys only those establishments have been considered which employed 10 or more persons at the time of the census. If this method is used, it must be remembered that the results apply only to the stratum of establishments in the census (say, employing 10 or more persons) from which the sample has been selected, unless some means can be found to bring into the sample plants which have at least reached the cutoff point at the time of the survey. This difficulty can be overcome by making sampling universal; that is, the sample is selected from the totality of establishments existing at the time of the survey. If the sample taken from the entire universe enumerated in the census can be supplemented by a sample of the newborn plants, this gives a complete sample from the existing universe. This can be done by treating the births as a separate stratum and preparing independent estimates of their contribution to the total. The infant mortality of such plants, the lack of firm classificatory information, the prevalence of part-year operations, and the excess of nonresponse from newly queried plants all seem to indicate the desirability of oversampling this group. Plants closing down after the census present no problem since they are automatically excluded from the sample during the course of the survey.

Sometimes the method of cutoff is taken to its logical conclusion. As a result of the highly skewed size distribution of the plants and the fact that the relatively small number of large plants contribute the major portion to the aggregates, it is decided not to include any small plants in the sample at all. The ratio of, say, the present employment of large plants in the survey and their employment in the census is calculated. This ratio, when multiplied by the total employment in the census, is taken as an estimate of employment of all plants in the universe at the time of the survey. The assumption made is that the relative positions of large and small plants remain essentially constant. When this assumption does not hold, the results may be far from the truth. And then there is no means of knowing whether the assumption involved is justified.

17.11 THE SAMPLING UNIT

There are several advantages in using the enterprise (or company) as the sampling unit rather than the establishment (or plant) in industrial surveys. In the first place, several types of data can be reported better if the company supplies it for each plant owned by it. Second, it makes poor public relations to ask for data on some plants belonging to a company and not on others. Third, when the company is used as the unit of sampling, some of the smaller and otherwise inaccessible plants automatically are included in the sample. One method of selecting a sample of companies is to stratify them by size measures as obtained from a recent census and take a simple random or systematic sample from each industry-stratum cell. But then the question arises as to how a company should be allocated to an industry when it produces goods classifiable to several industries. The solution of this problem is not difficult but it is unnecessary. One can always select a sample of companies through a sample of plants by associating with each company its largest plant. In that case a sample of plants automatically gives a sample of companies. When the company is required to report data on each of its plants separately, the reports obtained can be easily classified by area or by industry.

17.12 CLASSIFIED DIRECTORY

As information on industry is usually collected by mail in the first instance, it is essential to have a classified industrial directory showing the geographic location, address, industrial classification, ownership, and some indication of the size of the reporting units. When this information is available, it is possible to identify each reporting unit and distinguish it from the others. Such a directory can be used as a frame for the selection of the sample. Without adequate information on the size of each plant, it is impossible to make use of the technique of stratification in order to reduce sampling errors. And the classified information contained in the directory can be put to good use when it is proposed to change the sample design, say, in order to provide figures for three-digit groups of industry rather than for two-digit groups. Furthermore, the smaller plants can be sent a short form for collecting worthwhile data and the larger ones a more detailed form when information on the size of a plant is available from the directory. And finally, an up-to-date directory is by itself a source of statistical information on the number of plants and their distribution by size, by industry, and by location.

Useful as the classified directory is, it is no easy task to make it and keep it up-to-date. Perhaps the census of establishments can be used as the starting point, the directory being supplemented regularly by the births taking place since the census. In this connection all available sources of information should be tapped to ensure that the coverage is complete. Extraordinary diligence is needed to do this job. There are several difficulties. New manufacturing activity may suddenly start at a location where there was none before. There may be a change of ownership, a change in the composition of partners, or a change in the physical location of the plant. All this has to be taken into account to decide whether a birth has taken place, and it needs to be added to the directory.

17.13 SMALL-PLANT ECONOMIES

In highly industrialized countries the large plants account for all but a small fraction of production. As a result countries with large plant economies have tended to exclude the smaller plants (with a few paid employees) in their annual surveys. It has, however, become the custom everywhere, in both developed and developing countries, to use the method of cutoff and exclude the smaller plants. This is unfortunate as the small plants contribute considerably to production in the less industrialized countries (called *small-plant economies*). Consider, for example, the data presented in Table 17.5 regarding

Table 17.5 Percentage distributions of manufacturing plants in Greece (1958) and the United States (1954)

Number of employees	*Greece*		*United States*	
	Plants	*Employees*	*Plants*	*Employees*
1–4	82	37	37	2
5–9	11	16	17	2
10–19	4	10	15	4
20–49	2	10	15	9
More than 49	1	27	16	83
Total	100	100	100	100

Greece (1958) and the United States (1954). In the United States plants employing nine or fewer employees account for just 4 percent of total employment while the corresponding percentage is 53 in Greece, a small-plant economy. It is clear that in a developing country such as Greece it is unwise to restrict the survey to the larger plants only.

What is not commonly understood is that in small-plant economies the reliability of the estimates made from the survey depends heavily on the quality of the data collected from the smaller plants (2). This idea can be made clear by taking the following size distribution of plants observed in a particular small-plant economy (Table 17.6). The table gives for each size group the number of plants and the total number of persons engaged as obtained from a country-wide census. Suppose it is decided to use a sample of 1,600 plants for the survey and furthermore that plants in the highest size group are to be taken with certainty. If the allocation of sample size to other strata is made in proportion to the number of persons engaged, it is fairly simple to see that the raising factors to be used at the stage of analysis for the different strata are given by the last column of Table 17.6. The implications of these calculations are that a report from a plant in the lowest stratum is to be multiplied by 220 while the multiplying factor is just unity for a report from the certainty stratum. This points to the need for greater accuracy in the reporting by the smallest plants, which, unfortunately, have the poorest accounting records. On the other hand, the developed countries can tolerate relatively larger errors in the returns from the smaller plants as such plants contribute little to the aggregate.

Table 17.6 Distribution of plants by size

Size group	*Number of plants*	*Number of persons engaged*	*Average number of persons engaged*	*Raising factor*
1–4	64,370	96,551	1.5	219.9
5–9	8,635	56,123	6.5	50.7
10–19	3,556	46,227	13.0	25.4
20–49	1,571	47,130	30.0	11.0
More than 49	854	142,619	167.0	1.0
Total	78,986	388,650		

With regard to employment it can be said that accurate figures can be reported by even the smallest plants. It is data on items such as value of product, cost of raw materials, and value added for which reporting errors are likely to be serious. Conceptual difficulties coupled with the non-availability of records covering the whole year point to the need for considerable research and experimentation before a program for the collection of basic industrial statistics from the small plants can be devised.

17.14 HOUSEHOLD INDUSTRY

We shall now consider small-scale manufacturing. Experience shows that it is possible to collect information on small-scale manufacturing through a sample of households when a general-purpose household survey is already running in the country. Through pilot studies suitable measurement techniques can be devised to obtain acceptable material on the important aspects of small-scale manufacturing. The Indian National Sample Survey provides an example of this approach. From time to time a questionnaire has been added to the main survey asking for information on small-scale manufacture, that is, activities concerned with the transformation of material objects in which raw materials are procured and finished products are sold. Also included are all types of repair and constructional services rendered on own account by artisans. The results obtained from the survey covering the period from July, 1958, to June, 1959, are presented in Table 17.7 (4).

These results can be improved by augmenting the general purpose sample of households with a special sample taken from areas with considerable

Table 17.7 Small-scale manufacturing, India, 1958–59 (based on National Sample Survey)

Item	*Rural*	*Urban*	*Total*
Number of households engaged (millions)	11.1	2.3	13.4
Number of workers (millions)	12.9	4.4	17.3
Value of input (million rupees)	220.2	207.4	427.6
Value of output (million rupees)	384.8	451.0	835.8
Value added (million rupees)	164.6	243.6	408.2

industrial activity. These areas can be identified on the basis of a recent population census which gives for each village and town the number of workers in different economic activities. Variable sampling fractions can be used for the selection of villages in the rural areas and of towns in the urban areas.

A brief description of the sample design of the Indian National Sample Survey will now be given to enable the reader to appreciate the point made in the previous paragraph. This survey covers the entire country. The design of this continuing survey has undergone several modifications since its inception in 1950. In the year 1970 the federal sample contains about 8,400 villages and 4,600 urban blocks. The state sample is about the same size. This sample has been selected from about 622,000 villages and 2,700 cities and towns in the country. In order to select the sample, all the 330 districts in India have been allocated to regions on the basis of topography, population density, and pattern of cropping. The regions do not cut across states. There are 350 rural strata (of about equal population) formed by putting together within regions contiguous subdistricts called *tahsils*. From each stratum 4 interpenetrating subsamples of six villages each are selected systematically with probability proportional to population. The number of households to be selected from a village is determined in order to apply a uniform raising factor at the state level. In the urban areas the four big cities are strata by themselves. The other towns with a population of 50,000 or more form another stratum. Within strata, the towns are selected with probability proportional to population. The selection of blocks within towns is similar to the selection of villages in the rural areas. The time reference of the survey is one year divided up into six subrounds of two months each. One-sixth of the total sample is investigated in each subround of the survey.

17.15 THE UNITED STATES ANNUAL SURVEY OF MANUFACTURES

We shall now describe two important sample surveys being carried out in the field of industry. The most comprehensive survey on this subject is the

United States Annual Survey of Manufactures. The Census Bureau has been conducting this survey since 1949. The purpose is to bring forward from the year of the last complete quinquennial census of manufactures the principal measures of manufacturing activity in the nation. All manufacturing plants having at least one paid employee are covered by the survey. The survey relates to the calendar year.

The data collected include total employment; number of production workers; total wages and salaries; production worker wages and man-hours; cost of materials consumed; inventories; the quantity of electricity generated, purchased, and sold; capital expenditures; total value of shipments (and in detail for approximately 1,000 individual product classes); the value of products bought and resold without further processing; receipts for contract work; and miscellaneous receipts.

The frame used for selecting the sample is the census list of manufacturing plants. This is supplemented for the most part by lists of new manufacturing companies obtained from the Bureau of Old-Age and Survivors Insurance (BOASI). In 1955 when the survey was redesigned, about 50,000 plants were selected for inclusion in the survey from a population of about 300,000 plants. As a general rule, all plants with 100 or more employees were taken in the sample with certainty. Smaller plants were stratified by industry, ranked according to size, and selected systematically with probability proportional to size. The sampling fractions varied from industry to industry in accordance with the survey specifications in terms of standard errors for the primary product classes of the industry, size distribution of the industry, and the degree of concentration of the product classes there. Although the initial selection was that of plants, companies rather than their individual plants were used as the ultimate sampling units. This was done by retaining in the the sample only those companies whose largest plant (with the highest probability of selection) had been selected in the initial selection. In such a case the entire company was designated as being in the sample and all its plants were assigned the probability of the largest unit.

The method of difference estimation is used to obtain most of the totals. With the census figures as the base, the sample is used to estimate the change between the current year's data and the census year's data. The estimate of change when added to the census total produces an improved estimate of the current year's total for the characteristic under consideration.

The data are collected by mail. The report forms are mailed to the plant at the end of the year or early in the following year. The initial mail canvass is followed by successive waves of reminders to the nonrespondents. More than 98 percent of the reports are usually received in time for tabulation. The individual figures reported are tested on the computer for plausibility and consistency by comparison with data previously reported by the same plant and by comparisons with industry-wide averages. All suspected reports are screened by clerks for errors. The advice of professional staff is sought in difficult cases. Related figures from the survey and data from independent sources are used to check the tables made. From time to time area sampling is used to check the coverage of the survey. It has been found that the frame covers about 90 percent of the plants and about 99 percent of employment and wages.

17.16 THE INDIAN ANNUAL SURVEY OF INDUSTRIES

This is a continuing survey conducted to compile industrial statistics of registered factories in the nation. The registered factories are all those which use power and employ 10 or more workers or which do not use power but employ 20 or more workers. A complete census is taken of all factories which use power and employ 50 or more workers or which do not use power but employ 100 or more workers. The remaining factories are covered on a sample basis. Defense installations, units engaged in storage and distribution, and all units registered as hotels or cafés are excluded from the survey.

Information referring to the calendar year is collected on almost all aspects of manufacture, production, and sale. The items include labor employed, their emoluments, working hours, shifts and working days, inventory and transactions relating to capital structures, raw materials, products, expenses, repair and maintenance, services received and rendered, taxes, transport, trade, capital expenditure, and installed capacity.

In order to select the sample of factories, the factories within a state are grouped by industry. A sample of factories is selected from each industry in the state. The allocation of the sample size to the strata is made to ensure a coefficient of variation of about 5 percent for the important characteristics in each industry for the whole nation and a coefficient of variation of about 10 percent for the two biggest industries in each state. This was achieved in the following manner in 1961 (3). A sample of about 6,000 factories was first allocated to the different industries in proportion to their total all-India employment. When the allocation for an industry was considered excessive, it was reduced to provide a larger allocation to some major industries within states. Furthermore, all factories were taken into the sample when the number in the stratum was found to be very small. This resulted in a total sample of 6,139 factories from the sample sector; the total number of factories investigated was about 16,000. The sample in a stratum was selected in the form of two interpenetrating subsamples of the same size. Systematic sampling with a random start was the method actually used to select the factories.

All factories under investigation were informed of the purpose of the survey. The investigator left the schedule with the manager of the factory for completion after explaining the nature of the information needed. He revisited the factory to clarify difficult points and to collect the completed schedule. About 3 percent of the factories failed to provide the information needed. The total cost of the survey was about 1.1 million rupees.

REFERENCES

1. Hanna, F. A. (1959), "The Compilation of Manufacturing Statistics," U.S. Department of Commerce, Washington.
2. Hanna, F. A. (1966), Measuring the Activity of Small Manufacturers, *Appl. Statist.*, **15**.
3. National Sample Survey (1965), Annual Survey of Industries, 1961, Sample Sector, *Govt. of India Rept.* 139, New Delhi.
4. National Sample Survey (1962), Small Scale Manufacture: Rural and Urban, *Govt. of India Rept.* 94, New Delhi.
5. National Statistical Service (1960), "The 1958 Census of Industrial and Commercial Establishments," Government of Greece, Athens.
6. United Nations Statistical Office (1953), "Industrial Censuses and Related Enquiries," Studies in Methods, ser. F, vol. 1, no. 4, New York.
7. United Nations Statistical Office (1968), "International Standard Industrial Classification of All Economic Activities," ser. M, no. 4, rev. 2, New York.
8. United Nations Statistical Office (1960), "International Recommendations on the 1963 World Programme of Basic Industrial Statistics," ser. M, no. 17, New York.
9. U.S. Bureau of the Budget (1957), "Standard Industrial Classification Manual," Washington.

18
Surveys of Distributive Trade

18.1 INTRODUCTION

Distributive trade and related services are an important part of the economic activity in a nation. With an increasing proportion of the labor force engaged in them, the importance of collection of information on this sector of the economy can hardly be underestimated. It is widely realized that an up-to-date and sound perspective on distribution is a prerequisite for successful balancing of supply and demand for goods and services and for smooth economic growth. The census of distribution provides bench-mark data once in 5 or 10 years at considerable cost. Between censuses, regular information is needed about the main elements of the distributive system and the direction of changes at the national level. Here sampling methods can play an important role by providing information more economically and in a more useful form for both government and business. The data to be collected, the methods of measurement and classification, the sample to be used, and the errors associated with inquiries of this type form the subject of this chapter.

18.2 THE DATA NEEDED

In order to define the coverage of the distribution inquiries, information is needed on kind of activity, that is, whether it is a wholesale, retail, or other type of establishment. Wholesale establishments are further subdivided as those buying and selling on their own account, those who sell for a manufacturing enterprise of which they are a part, and those who buy and sell on the accounts of others (4). All wholesale and retail stores are classified by major kinds of commodities sold. A distinction is made between single-establishment and multiestablishment enterprises. The total number of establishments, existing, say, on a particular day, gives a count of the population involved. Their distribution by area can be obtained by collecting information on the physical location of each establishment.

Another characteristic of great interest is the number of persons engaged. This number gives information on the employment provided by the distributive sector. In order to facilitate reporting, the number of persons engaged usually refers to a week or month or the payroll period. For socioeconomic studies it is useful to give a breakdown of the number engaged as working proprietors, unpaid family workers, and employees. Working proprietors are usually restricted to owners actively engaged in the work of the establishment. Furthermore, only those unpaid family workers are considered who worked for a specified minimum period of time during the reference period.

Payroll records can be used to determine the wages and salaries paid to employees in connection with their work. Included in this category are all payments, whether in cash or in kind, for fringe benefits, overtime, or bonuses. Payments made by the employer for social security and similar obligations are reported separately.

Information is wanted on the value of goods intended for sale at the beginning and at the end of the reporting period. The goods must be owned by the enterprise but may be located anywhere. Included are goods ready for sale, goods processed before sale, and materials consumed in rendering services. Excluded are articles such as packing supplies, fuels, and repair, maintenance, and office supplies. As far as practicable, stocks are to be valued at replacement cost. It may, however, be more convenient for stores to report values from their accounting records. In any case the basis of valuation is kept uniform from period to period.

The value of sales is a key figure. It is the value of all goods whose ownership is transferred to others, and of all services rendered during the reporting period. Included are time-payment or installment contracts, goods sold by an establishment on its own account and that of others. Goods shipped for sale on consignment for display are excluded. The sales value of services includes commission and fees received by the establishment for the sales and purchases made by it on the account of others. The time at which the transactions are recorded by the establishment ordinarily determines the sales to be reported for the period of inquiry. Sales price is net of discounts, rebates, and similar concessions granted. Indirect taxes and duties collected are to be reported separately.

Another useful item is the accounts receivable. This is the credit outstanding at the end of the inquiry year that was advanced in selling goods on own account and in rendering services. This item of data helps measure the volume of credit in the economy at a particular time. Another item of interest is the value of purchases of goods intended for sale. When the respondents suspect that data on the value of purchases may be used to reveal gross margins, they are reluctant to provide information on this item.

18.3 THE DISTRIBUTION OF STORES BY SIZE

The size of a store may be measured by the number of persons engaged or the value of sales. When information on the value of sales is available, say, from a census of distribution, this measure of size should be preferred for purposes of stratification. Quite often, the number of persons engaged is the only relevant piece of information collected during the census of nonagricultural establishments. In that case there is no real choice to be exercised.

A particularly important characteristic of the distribution of sales of stores is that the distribution is highly skewed. A relatively few large stores, constituting a small proportion of all stores, account for a substantial part of the total sales. At the same time, the very large number of remaining stores also account, in the aggregate, for a considerable proportion of total sales. Consider, for example, the distribution of independent retail stores in the United States (Table 18.1) obtained from the census of 1939 (10). The stores with sales of $300,000 or more were just 0.5 percent of the total but their sales were as high as 21.2 percent of the total. This means that a higher

Table 18.1 Independent retail stores by sales size, United States, 1939

Sales size group (thousand dollars)	*Percent of stores*	*Percent of sales*
Under 50	93.36	48.7
50–99	4.11	14.0
100–299	2.03	16.1
300–499	0.28	4.3
500–999	0.16	4.6
1,000–4,999	0.05	6.4
5,000 or more	0.01	5.9
Total	100.00	100.0

proportion of the larger stores will have to be taken in order to achieve an efficient sample. The exact proportion can, however, be calculated only when approximate measures of size are available before the survey is made, at least for the larger stores.

18.4 THE FRAME

The census of distribution is the appropriate frame from which to select the sample. But such a frame serves a limited purpose. Experience shows that there is considerable turnover in the distributive sector. Within a few months of the census it may be found that many stores have closed down, some have changed their character, some others have changed hands, and many new ones have come up. Thus it is nearly impossible to keep the list up-to-date. The strategy then is to take a sample from the available list after suitable stratification and supplement it by an area sample to take in stores not on the list. It is fairly obvious that the stores on the list will be sampled more intensively than those not on it.

The United States monthly survey of retail stores presents a good example of this strategy. The primary frame is a list of retail employers who pay social security taxes for their employees. This list is constantly changing as new employers are added and deaths are dropped. However, it is possible to keep this list reasonably up-to-date on the basis of information received from the Internal Revenue Service and the Social Security Administration. The deaths are deleted from the sample and a selection of reported births is added. The secondary frame is based on area segments spread all over the country. A number of area segments are selected and canvassed for retail stores. All stores found in this manner are asked to give their employer identification number used in the Social Security system if they are employers. Those found listed in the primary frame are disregarded. Stores not given a chance of selection in the primary frame become part of the area sample. This method has been followed since August, 1968. Previously the primary frame was restricted to the largest firms taken with certainty. The secondary frame based on area segments accounted for all other establishments. Thus the old area sample was about five times as large as the present area sample. This meant about 55 percent of the total retail sales compared with just 7 percent of total sales in the new sample (9).

A useful method of preparing the frame and selecting the sample was employed in Great Britain for the 1957 survey of retail trade. The 1950 census of distribution provided a list of the large stores which it was found easy to bring up-to-date. The remainder of the sample was taken on a geographical basis through a census. All retail stores were enumerated and a 20 percent sample was taken in new towns, central London, and a few special areas where it was believed that great changes had occurred since the census. In Greater

London a sample of electoral wards stratified by size was taken, distinguishing between the shopping and residential areas, and all stores in the selected wards were included in the survey. From the large towns a sample of streets stratified by the number of stores in 1950 was taken. In other towns a cross section of the local authority areas was taken after stratifying the areas by the total value of sales in 1950. The rural districts were stratified by population density or rate of population growth and a sample taken within each region.

The sampling methods used in the British survey apply, with some modifications, to similar surveys in many developing countries. Administrative records and the census of distribution, if any, can be used to prepare a reasonably accurate list of the large stores which usually keep some kind of accounts. In order to select a sample of the other stores in the urban areas the cities can be stratified on the basis of number of stores or number of persons and a sample of cities selected from each stratum. Within selected cities maps showing block boundaries can be used to select a sample of blocks and all stores in these blocks can be canvassed. In the rural areas the villages may be stratified by population or the number of stores and a sample of villages selected to canvass the stores contained in them. If the boundaries of enumeration areas used in the census of population are well described so that these areas can be identified on the ground, the area sample may as well be based on enumeration areas. When the results expected from the survey do not appear to be commensurate with the cost of canvassing all stores, the survey can be made part of an inquiry of all kinds of establishments—industrial, distributive, and others. In that case a distinction will have to be made at the listing stage between manufacturing, wholesale, retail, and other stores found in the sample areas.

18.5 USE OF POPULATION DATA

In order to select the sample for the retail trade survey the cities in the urban areas and the villages in the rural areas may be selected with probability proportionate to population. The reason is that retail trade exists to serve the population and therefore it is expected to be distributed between cities and villages in much the same way as population. The use of population data for selecting the sample is likely to be fairly efficient. Evidence on this point is provided by the survey of retail stores in the United States (1) in which the sample used was the same as that in the labor force survey (Sec. 14.12). The results presented in Table 18.2 show that the national estimates of retail sales were close to the census for many kinds of business.

We do not mean to imply that sales data, when available from a previous census, should not be utilized for purposes of stratification. Their use should prove very beneficial to the survey. But when such data are not available, population figures may be used with confidence.

Table 18.2 Estimates of retail sales for all retail trade and for selected kinds of business, United States, 1939

	Known 1939 census	*Estimate from sample*	*Percent difference*	*Relative sampling error of estimate (%)*
General stores (with food)	810,342	791,095	−2.4	12.6
General merchandise	5,665,007	5,678,947	+0.2	2.5
Drugs	1,562,502	1,533,235	−1.9	2.0
Apparel	3,258,772	3,187,193	−2.2	2.0
Lumber-hardware	2,734,914	2,563,790	−6.3	3.9
Furniture	1,733,257	1,657,949	−4.3	3.3
Total—all types	42,041,790	42,170,007	+0.3	1.4

18.6 NONRESPONSE

In a survey of this type asking for information on matters as touchy as value of sales, inventories, and purchases, the initial rate of nonresponse is expected to be higher than that encountered in household surveys. Accordingly it is essential to incorporate in the survey procedures by which the information needed can be obtained from the majority of the nonrespondents. Thus field visits may be found to be necessary when there is no response to mail requests. In the end some imputed value will have to be used for the stores that have refused to cooperate. The imputed value may be based on previous figures, if available, or on the size of the store reported in the list used for sample selection.

In the United States survey of retail stores, nonresponses accounted for about 14 percent of sales volume for the year 1948. Of this, about 6 percent came from the large store list, about 2 percent from the area sample, and the balance was attributed to factors such as illness or filing late. Within a period of 20 years it has been possible to reduce the nonresponse on sales to about 9 percent. This shows how difficult it is to tackle the problem of nonresponse in this field.

18.7 ERRORS OF LISTING AND CLASSIFICATION

When the enumerator goes to the sample area to make a list of the stores located in the area, some stores may be missed and some others may be included in error. For the stores on the list, further information is needed on the nature of activity of the establishment, retail sales, wholesale sales, and receipts for personal services. This information is used to decide whether the establishment is within the scope of the survey. A further classification is needed to determine the business group to which the establishment should belong. At the same time information is collected on items such as the value of sales and employment. All these operations are potential sources of error. It is therefore a matter of considerable importance that adequate steps be taken to determine the magnitudes of these errors and keep them down to a minimum.

Table 18.3 Percent change in sales by cause for different groups of business, based on list sample

Business group	*Change of scope*	*Change of group*	*Response difference*	*Net change*
Food	0.0	5.6	0.2	5.8
Eating and drinking	0.0	−6.6	1.9	−4.7
General stores with food	0.0	−27.6	0.0	−27.6
General merchandise	0.0	12.7	0.0	12.7
Apparel	−9.7	−1.5	4.5	−6.7
Furniture	0.0	30.7	−10.2	20.5
Lumber	−6.1	5.2	1.4	0.5
Automotive	−2.6	−2.8	0.6	−4.8
Gasoline stations	0.0	6.0	6.5	12.5
Drugstores	0.0	0.0	−1.3	−1.3
Liquor stores	0.0	100.5	0.0	100.5
Other retail	0.0	−35.4	1.4	−34.0
All retail	−2.5	...	0.3	−2.2

Table 18.4 Percent change in sales by cause for different groups of business, based on area sample

Business group	*Change of scope*	*Change of group*	*Listing error*	*Response difference*	*Net change*
Food	6.4	1.5	17.9	4.2	30.0
Eating and drinking	8.7	−1.8	1.1	−2.1	5.9
General merchandise	0.0	−41.8	0.0	−3.4	45.2
Apparel	0.0	18.7	0.0	1.3	20.0
Furniture	−0.4	1.2	0.0	−9.3	−8.5
Lumber	−13.6	0.0	0.0	−0.6	−14.2
Automotive	4.0	−14.7	6.9	−0.7	−4.5
Gasoline stations	−9.5	−2.6	0.0	2.0	−10.1
Drugstores	0.0	−2.9	0.0	0.9	−2.0
Liquor stores	0.0	0.0	0.0	0.0	0.0
Other retail	1.4	10.7	10.6	−4.0	18.7
All retail	0.6	0.0	5.1	−1.1	4.6

One of the methods of ascertaining the magnitude of these errors consists in checking a sample of the material through reinterviews. This technique has been used in the United States survey of retail stores. In 1948 a sample of 212 cases was taken from the large-store list to study the differences involved in the recheck. The effect of the different errors on the total value of sales is brought out in Table 18.3 (6).

It is clear that the differences caused by allocating an establishment to the wrong business group are quite important for individual groups. But the response differences are not large. A similar study of the effects of errors in the area sample gave a net change of 4.6 percent in the total values of sales originally reported in the sample. The detailed results of this investigation are given in Table 18.4 (6).

18.8 SAMPLE ROTATION

In continuing surveys it is customary to rotate the sample in order to reduce the burden of response on respondents. By a suitable choice of the scheme of rotation it is also possible to reduce the sampling errors of the estimates made from the survey. The sample is usually replaced partially over a period of time in household surveys. In surveys of establishments there is an opportunity to use a more efficient type of rotation when it is known that establishments can provide data for the current period as well as for the previous period with reasonable accuracy (3). In a monthly survey, for example, the total sample can be divided into 12 subsamples and a different subsample used each month of the year. In a given month, say i, the establishments give data y_i for the current month and x_{i-1} for the previous month. Using the data y_i, a simple unbiased estimate U_i can be made for the current month. At the same time a simple unbiased estimate V_{i-1} can be made for the previous month. The ratio U_i/V_{i-1} when multiplied by the estimate W_{i-1} for the previous month [based on figures collected at the $(i-1)$st month and the previous months] provides another estimate for the ith month. A linear combination of this estimate and U_i using weights W and $1-W$ respectively gives an improved preliminary estimate Y_i for the current month. This estimate can be revised next month when data on month i are available from the subsample enumerated then. This is done by forming a linear combination of the preliminary estimate Y_i and the simple unbiased estimate V_i, the weights used being K and $1-K$. The estimate so obtained may be called the *final composite estimate* (11).

There are several advantages in using the plan in which a subsample of stores is enumerated every month. In a survey taken at the end of a year the stores that go out of business during the year make no contribution to the annual estimate. With the scheme suggested, only those deaths are missed which occur during a one-month period. Second, it is possible to estimate trend from one month to the same month a year later by using an identical sample at the two time periods. And then the quality of the data collected improves when the burden of enumeration is spread over the whole year instead of just at the end of the year. And finally, an establishment is required to give data no more than just once during the entire year. Viewed as a sampling problem, the main advantage is that a new subsample is available in successive months and that the entire subsample can be used for identical links (Sec. 6.2).

The United States survey of retail stores provides considerable information on the efficiency of this rotation system. In this survey the weighting coefficients W and K are taken as 0.8 and 0.83 respectively. These are not necessarily the best values of the coefficients to use as the optimum values depend on the month-to-month correlation coefficient which is found to be

Table 18.5 Standard error of monthly sales based on a rotation sample as a fraction of standard error based on nonrotating sample, United States monthly survey of retail trade

Correlation coefficient	*Optimum constants*		*With optimum weights*		*With given weights*	
	W	K	*Preliminary*	*Final*	*Preliminary*	*Final*
0.99	0.868	0.876	0.37	0.35	0.39	0.37
0.98	0.817	0.834	0.45	0.40	0.45	0.41
0.95	0.724	0.762	0.56	0.49	0.58	0.50

0.98 for all kinds of business combined. A comparison can be made among the situations by (1) use of optimum weights, (2) use of actual weights, and (3) use of a nonrotating sample. Such a comparison is presented in Table 18.5 for three values of the correlation coefficient (11). In all cases the estimate involved is the value of sales for a single month. It is seen that the weights actually used in the survey do not lower the precision to any great extent. Furthermore, the use of the rotation scheme reduces the standard error by about 50 percent as compared with the situation in which no rotation is used.

When information is collected for the current as well as for the previous month from a sample of stores, it is usually found that there is a downward bias associated with figures relating to the previous month. The reason is that there is a tendency on the part of enumerators to disregard stores not in business during the current month although they were in business during the previous month. The bias is reduced when the enumerators are specially instructed not to omit such stores.

18.9 HOUSEHOLD TRADE

In the developing countries a part of the trade is carried on by enterprises which are physically indistinguishable from the household or whose transactions cannot be easily separated from the domestic transactions of the household. This part of the distributive trade can best be measured through household surveys. Such an attempt has been made in India using the National Sample Survey as the vehicle. In a number of rounds of the survey information has been collected on purchases and sales of merchandise, trading costs and other expenses, man-days utilized, and wages paid to hired laborers. The reference period used is the month prior to the date of survey. Some of the important findings of the 1956 survey are given in Table 18.6. The results are presented by subsample in order to throw light on the internal consistency of the data collected (2).

Table 18.6 Household retail trade, National Sample Survey of India, December, 1955, to May, 1956

	Rural		*Urban*	
	Sample 1	*Sample* 2	*Sample* 1	*Sample* 2
Percentage of households engaged in retail trade	3.7	3.8	9.9	10.2
Number of persons engaged per household per day	1.3	1.3	1.6	1.6
Number of man-days per household per day	1.2	1.2	1.4	1.4
Trading costs per month per household (rupees)	4	5	16	14
Monthly payments to hired laborers per household (rupees)	1	1	12	13
Value of purchases per month per household (rupees)	223	297	997	1,100
Value of sales per month per household (rupees)	222	287	1,040	1,155

18.10 THE UNITED STATES SURVEY OF RETAIL STORES

We shall now describe a few important surveys being conducted in the field of distributive trade (5). Perhaps the best-known survey in this area is the United States monthly survey of retail stores. The purpose of this survey is to provide estimates of sales of retail stores by kind of business and geographic area and to provide national estimates of accounts receivable balances of retail stores by kind of business. All establishments engaged primarily in selling at retail are covered by the survey. By and large information from establishments is collected by mail; only 7 percent of the information is collected by personal interview. The reference period is the month in which the retail sales take place.

Many improvements in sample design have been made since the survey started in 1951. Since August, 1968, two frames have been used for the selection of the sample. The primary frame is based on the 1963 census of business supplemented by information received from the Social Security Administration (Sec. 18.4). The secondary frame comprises land areas spread over the length and breadth of the country. The sampling unit in the primary frame is the employer identification number. In the original draw of the sample from the census of business these sampling units were grouped by size in each of the 40 kinds of business, giving about 400 strata. From the strata the sample is selected using variable sampling fractions so as to minimize the sampling error of the estimate of retail trade. About 50,000 identification numbers are selected in the sample. The largest establishments are selected with certainty. The other establishments in the sample are subdivided into four equal-sized panels which are canvassed each month in rotation, each providing information for three months. This sample is supplemented each month by a sample drawn from new employer identification numbers which are issued. A large sample of these births is taken and more reliable information on their size and kind of business is collected. When this information has been obtained the establishments are stratified by size and by kind of group. A subsample of establishments is then selected, taking care that the total probability of each birth within a stratum is equal to that of the original sample (9).

The sample from the secondary frame is selected in this manner. Following the procedures used in the Current Population Survey (Sec. 14.12), the first-stage units, which are counties or groups of counties, are allocated to 58 strata. One first-stage unit is selected from each stratum with probability proportionate to size. The selected first-stage unit is divided up into segments containing about four retail stores on the average, from which a random sample of segments is selected. The sample of segments is divided into 12 equal subsamples and each subsample is used just for one month during the year. About 400 segments are selected each month, the overall sampling rate being 1 in 1,000.

During a particular month data are collected for that month as well as for the previous month. This makes it possible to use a composite estimate procedure (Sec. 18.8), which reduces the sampling error considerably. The estimates of sales by kind of business and the median value of the relative standard error for the month of April, 1969, are given in Table 18.7 (7). Similar information on end-of-month accounts receivable is provided in Table 18.8. It is clear from these tables that as judged by the sampling errors, the reliability of the data is high, especially for month-to-month changes. Actually, it has been found possible to publish monthly sales data by geographic areas, too.

Table 18.7 Estimated monthly retail sales and their relative sampling errors, United States, April, 1969

Kind of business	*Estimate (million dollars)*	*Median relative error (%)*	*Median relative error of month-to-month change (%)*
Food	6,017	1.2	0.3
Eating and drinking	2,073	2.4	0.8
General merchandise	4,500	0.8	0.2
Apparel	1,642	2.3	0.7
Furniture	1,281	2.8	1.4
Lumber	1,757	3.0	1.1
Automotive	5,924	1.7	0.6
Gasoline stations	2,070	2.1	0.8
Drugstores	931	2.5	0.8
Other retail	2,619		
Total	28,814	0.6	0.2

Table 18.8 Estimated end-of-month accounts receivable and their relative sampling errors, United States, April, 1969

Kind of business	*Estimate (million dollars)*	*Median relative error (%)*	*Median relative error of month-to-month change (%)*
Food	290	6.0	0.4
Eating and drinking	167	5.5	1.4
General merchandise	8,831	0.5	0.1
Apparel	1,276	5.0	0.4
Furniture	2,350	4.2	0.5
Lumber	1,863	6.6	0.8
Automotive	1,974	2.9	0.6
Gasoline stations	358	5.4	1.0
Other retail	2,318		
Total	19,427	0.8	0.4

18.11 THE UNITED STATES SURVEY OF WHOLESALE TRADE

The purpose of this monthly survey is to provide estimates of sales of wholesalers and the trend of sales over time. The survey is limited to merchant wholesalers, who constitute the major portion of the broad field of wholesale trade. It excludes the other types of wholesale trade such as manufacturers' sales branches and sales offices, petroleum bulk stations and agents, and brokers and assemblers of farm products. All kinds of business in which merchant wholesalers operate are represented in the survey (8). Enterprises which, in addition to merchant wholesale establishments, operate other types of establishments report in this survey only for their merchant wholesale establishments.

Beginning in 1966, the sample of firms has been selected from two frames. The principal frame is the 1963 census of business lists representing all wholesalers with paid employees. The secondary frame is the Social Security Administration lists of wholesalers with paid employees entering business since 1963. The sample is supplemented monthly for new firms entering the secondary frame. About 1,000 very large firms, selected in the sample with certainty, are required to report every month. A sample of about 16,000 firms is selected from the remaining firms after allocating them to strata on the basis of kind of business and the value of sales. The number of firms selected for each kind of business varies from 50 to 650, depending upon the total number of firms, their distribution by sales size, and whether or not data are to be provided by geographic division. This sample is divided up

Table 18.9 Estimates of value of sales of durable goods and their relative standard errors, United States, March, 1969

Kind of business	*Estimate (million dollars)*	*Median relative error (%)*	*Median relative error of month-to-month change (%)*
Automotive	1,476	4	2
Electrical	1,254	3	1
Furniture	443	5	3
Hardware	846	2	1
Lumber	1,010	4	2
Machinery	2,311	3	2
Metals	955	4	2
Scrap	421	3	2
Other durable	157		
Durable goods total	8,873	1.2	0.7

into four rotating subsamples each containing about 4,000 firms which report every fourth month during the year. These firms provide data for the current month and for the previous month.

As a result of the sample rotation employed, it is possible to provide provisional and final composite estimates (Sec. 18.8). The weighting factors W and K are 0.7 and 0.72 respectively. The final composite estimate (available a month later) is generally more reliable than the preliminary estimate. The relative sampling errors of the estimates of value of sales of durable and nondurable goods for the month of March, 1969, are given in Tables 18.9 and 18.10 respectively.

Table 18.10 Estimates of value of sales of nondurable goods and their relative standard errors, United States, March, 1969

Kind of business	*Estimate (million dollars)*	*Median relative error (%)*	*Median relative error of month-to-month change (%)*
Grocery	3,863	2	1
Beer, wine	889	2	1
Drugs	765	3	1
Tobacco	448	2	1
Dry goods	881	4	2
Paper	586	2	1
Farm products	1,130	4	3
Other nondurable	1,718	4	2
Nondurable total	10,280	0.9	0.7

In this survey imputed values are used for firms that fail to respond. About 12 percent of the total sales and 27 percent of total inventories are imputed for nonresponse. The data provided by firms on inventories are believed to be less reliable than the data on sales because inventory records are not kept by firms on a monthly basis to the same extent as for sales.

REFERENCES

1. Hansen, M. H., W. N. Hurwitz, and M. Gurney (1946), Problems and Methods of the Sample Survey of Business, *J. Am. Statist. Assoc.*, **41**.
2. National Sample Survey (1961), Household Retail Trade, *Govt. of India Rept.* 41, New Delhi.
3. Raj, D. (1968), "Sampling Theory," McGraw-Hill Book Company, New York.
4. United Nations Statistical Office (1958), "International Recommendations in Statistics of Distribution," ser. M, no. 26, New York.

5. United Nations Statistical Office (1965), "Bibliography of Industrial and Distributive-trade Statistics," ser. M, no. 36, New York.
6. U.S. Bureau of the Census (1953), The Sample Survey of Retail Stores: A Report on Methodology, *Tech. Paper* 1, Washington.
7. U.S. Bureau of the Census (1969), *Monthly Retail Trade Report* BR-69-5-17, July, 1969.
8. U.S. Bureau of the Census (1969), *Monthly Wholesale Trade Report* BW-69-4-17, June, 1969.
9. U.S. Bureau of the Census (1969), "Revised Sample for Estimating Monthly Receipts of Retail Establishments in the United States," (unpublished memorandum).
10. U.S. Bureau of the Census (1939), "Census of Business: Retail Trade, 1939," vol. 1, pt. II, Washington.
11. Woodruff, R. S. (1963), The Use of Rotating Samples in Census Bureau's Monthly Surveys, *J. Am. Statist. Assoc.*, **58**.

19
Sampling as an Aid to Censuses

19.1 INTRODUCTION

It has been stressed in the previous chapters that a census is an invaluable aid to a sample survey. It provides a frame from which the sample can be selected and furnishes bench-mark data which can be used for increasing the precision of the sample estimates. The truth, however, is that both the census and the sampling method are interdependent. Sampling methods, too, can play an important part in the conduct of the census. In the first place, the scope of the census can be broadened by asking certain questions of only a sample of the population. This reduces the burden of response and the time required for carrying out the work. Second, the important results of the census can be tabulated speedily on the basis of a sample of the returns. When this is done, the census authorities can afford to bring out the detailed results from the complete census in due course. In the third place, the census can be taken on only a sample of areas or other units in order to save costs, to save time, or to provide information under conditions that render a complete enumeration impossible. And then, the enumeration in the census can be checked for quality by collecting the same information from a sample using more skilled staff and better procedures. Finally, the errors in processing the census returns can be kept within reasonable bounds by examining a sample of the returns at various stages of the processing. The use of the methods of sampling inspection helps to control and to improve the quality of the finished product. A detailed discussion of some of these points is presented in the sections that follow.

19.2 ADVANCE TABULATION OF CENSUS

Sampling can be used effectively for obtaining advance estimates of some of the important census results for the country as a whole and large subdivisions thereof. This is done by taking a sample of returns from the census record and processing them in advance before the remainder of the returns are handled. An important consideration is that sample tabulation should not interfere with the regular processing of the census. Thus the units of sampling should ordinarily be the fundamental building blocks which are processed as one lot in the census. Quite often such lots are the enumeration areas (EAs) or enumeration districts into which the country is divided for the purpose of taking the census.

A very convenient sampling plan is to select a sample of enumeration areas and process all units contained in them. This method does not interfere with the regular processing of the census. Actually the sample returns may be conveniently added to the rest when the complete census is to be tabulated. The only disadvantage is that the plan may not be efficient from the point of view of sampling errors when the size of the cluster is large. An alternative method is two-stage sampling. In this method a further sample of subunits is taken from each of the enumeration areas selected. For example, in a population census the subunits may be households if the data have been collected by household. If the sizes of the enumeration areas are known to differ considerably, the EAs may be selected with probability proportionate to size. In that case the sampling fraction within an EA can be adjusted to make the sample self-weighting.

A variation of the plan is to tabulate the results by subsamples. Suppose a 10 percent sample of enumeration areas is to be used for advance tabulation. Then a 1 percent sample of EAs is taken and tabulated quickly. This is followed by the other nine 1 percent samples in rapid succession. The ten subsample estimates for a cell in the table can then be combined to get an improved estimate for the cell. At the same time the subsamples provide 9 degrees of freedom for the estimate of error. Some of the estimates which are found to be weak, that is, subject to large sampling errors, can be withheld from publication. The main advantage of this plan is that some of the results can be obtained very quickly, say within a few months of the census. If it is found that the first few subsamples are sufficient for most purposes, the other subsamples may not be processed as a part of this program.

The sample gets a wider geographical spread if the enumeration areas are listed in a serpentine fashion and the sample of households is selected directly from the totality of households in the population. Systematic sampling is the obvious method of selection to use in this case. Provided this plan does not interfere with the normal census tabulation, this is the most efficient procedure to use. Assuming systematic sampling equivalent to random sampling, it is fairly simple to determine the size of the sample in order to achieve stipulated precision for the important estimates to be made.

19.3 USE OF SYSTEMATIC SAMPLING

Great care should be exercised in the use of the method of systematic sampling for selecting the sample for advance tabulation. The method could lead to serious over- or underestimates if used uncritically. Take, for example, the 10 percent sample of the 1951 population census of India (4) taken to build up certain important tables by state in the country. In the census there was

Table 19.1 Comparison of sample and census for proportion male, census of India, 1951

	Number of districts		
State	*Sample exceeds census*	*Census exceeds sample*	*Total*
Uttar Pradesh	47	4	51
Bombay	20	7	27
Madras	21	5	26
Madhya Pradesh	15	7	22
Madhya Bharat	15	3	18
Bengal	12	5	17
Assam	10	7	17
Hyderabad	16	0	16
Orissa	9	4	13
Mysore	7	2	9
Vindya Pradesh	7	1	8
Pepsu	5	3	8
Saurashtra	4	2	6
Himachal	3	3	6
Travancore	2	3	5
Punjab	5	1	6
Total	219	62	281

an enumeration slip for each person enumerated. A pad of slips was provided by each enumerator for the persons enumerated in the area under his jurisdiction. These pads were cut, as in a game of bridge, before distributing the slips into 10 pigeonholes, and the sixth pigeonhole always constituted the sample. When the sample was tabulated it was discovered that in almost all the states the sample overestimated the proportion of males (Table 19.1). In as many as 80 percent of the districts in India the proportion male as given by the sample exceeded the proportion male as subsequently calculated from the complete census. It is clear that the vagaries of systematic sampling were at full play in the selection process. Systematic sampling was not found to be equivalent to simple random sampling. Further evidence on this point was obtained by finding the difference between the sample estimate of the number of males in the district and the corresponding number in the census.

Table 19.2 Distribution of the ratio *d*, census of India, 1951

d	*Number of districts*	*d*	*Number of districts*
Less than −4	10	3 to 4	25
−4 to −3	4	4 to 5	18
−3 to −2	5	5 to 6	7
−2 to −1	14	6 to 7	4
−1 to 0	29	7 to 8	4
0 to 1	55	8 to 9	1
1 to 2	46	9 to 10	2
2 to 3	35	10 or more	6

When divided by the standard error of the estimate assuming simple random sampling, the resultant ratio d should behave like a standardized variable with mean zero and variance unity, if systematic sampling is equivalent to simple random sampling. Actually, the following distribution was obtained (Table 19.2) for all the states excluding Hyderabad. The mean and the standard deviation of the distribution of d are 1.6 and 3.0 respectively. These figures are very different from the values 0 and 1 expected on the hypothesis that systematic sampling is equivalent to simple random sampling.

19.4 SUGGESTED METHOD OF SAMPLING

Experience shows that certain starting points in systematic sampling have a persistent tendency to overestimate or underestimate certain characteristics of the population. A safe procedure, therefore, is to use all possible starting points in the selection process. This can be done by randomizing the starting points and using one each in the very many groups into which the population can be conveniently divided. This method was followed in the advance tabulation of the 1961 population census of Greece and no unpleasant features of the systematic sampling were observed. In this census an ordinary enumeration district (ED) contained about 40 households and about 10 neighboring EDs formed a chief enumerators' section. In order to select the 2 percent sample, the EDs in the section are listed along with the number of private households enumerated in the census. Then cumulative totals of the number of private households in each section are obtained (see Table 19.3 as an illustrative example). At the same time the first 50 natural numbers are

Table 19.3 Sample selection, population census of Greece, 1961

Serial number of ED	*Number of households*	*Cumulated total*	*Random number*	*Serial number of selected household within ED*	*Number of persons in selected household*
1	42	42	44		
2	40	82	94	2	4
3	45	127	144	12	5
4	38	165	194	17	3
5	40	205	244	29	1
6	50	255	294	39	3
7	40	295	344	39	4
8	30	325	394		
9	35	360	444	19	3
10	45	405	494		

arranged at random and divided up into 5 groups along with their complementary numbers (Table 19.4). The random number for selecting the first sample household is to be taken from the column A_1 of Table 19.4. Other households in this section are selected at an interval of 50. The random start for the first selection in the second section is to be taken from column A_2 of Table 19.4. Columns A_1 and A_2 are to be used alternately. When the random numbers in these two columns are used up, the sampler proceeds to the next set of columns B_1, B_2, and so on (6).

The use of complementary random starts brings about greater accuracy in the estimate of the total number of households. If the random number is too low, the sample size is likely to increase by 1 when the number of households in the ED is not divisible by the sampling interval. The use of the complementary number corrects this tendency.

Table 19.4 Random sets of numbers for 2 percent sample

A_1	A_2	B_1	B_2	C_1	C_2	D_1	D_2	E_1	E_2
44	7	30	21	9	42	8	43	27	24
2	49	16	35	6	45	32	19	13	38
23	28	3	48	35	16	26	25	1	50
34	17	28	23	5	46	19	32	20	31
36	15	47	4	22	29	42	9	33	18
18	33	11	40	14	37	38	13	50	1
24	27	41	10	48	3	29	22	40	11
46	5	37	14	10	41	43	8	7	44
4	47	17	34	49	2	12	39	31	20
25	26	39	12	15	36	45	6	21	30

19.5 ADJUSTMENT FOR DEFECTS

When the number of households and the number of persons in the sample selected are found, it is simple to estimate the total number of households and the total population of the country. These estimates are likely to differ somewhat from the corresponding figures obtained from the census. At this stage it is wise to make some slight adjustments to the sample selected in order to bring the sample estimates (of the numbers of households and persons) in line with the full counts which are already known to the public. The procedure consists in removing or adding at random some households from or to the sample selected. In the Greek census no more than about 10 to 20 households were involved in the adjustment procedure for a department, there being 52 departments in the country. No adjustment was needed in the case of the institutional households, from which a systematic sample of persons was selected with a new random start in each section.

Similar problems were encountered in the sample tabulation of the 1951 population census of Great Britain (3). A 1 percent sample was taken by

Table 19.5 Comparison of sample with census, population census of Great Britain, 1951 (in thousands)

	Number of private households		*Number of persons*	
	Sample	*Census*	*Sample*	*Census*
England and Wales	13,197	13,200	43,716	43,745
Scotland	1,466	1,454	5,124	5,095
Great Britain	14,663	14,654	48,840	48,841

asking the enumerator to make copies of the schedules of all households bearing numbers ending in 25 if his ED number was odd, and ending in 76 if it was even. Table 19.5 gives a comparison of the results obtained with the census totals. The agreement on numbers of households is satisfactory for England and Wales, but there is some excess in the sample for Scotland.

19.6 POSTENUMERATION SURVEYS

Census taking is a huge undertaking requiring the use of a very large number of enumerators and other persons at various levels. Due to the sheer size of the several operations involved, errors of different types are likely to creep in. The fact that a small well-designed and well-administered sample survey can produce more accurate measurements than is feasible in the much-larger-scale operations required for a census suggests the use of the former as a vehicle for checking on the latter. Such a survey conducted after the census is called a *postenumeration survey* (PES). The purpose is to provide a measure of accuracy of the census results and to ascertain the sources of error. When the sources of error are located, steps can be taken to reduce their impact in the next census. Since the results from the PES are meant to serve as a standard against which to judge the figures obtained in the census, it is very important to take all possible steps to ensure that the survey is reasonably free of errors. This is usually done by employing in the survey the interviewers whose performance in the census has been judged as excellent. They are given special training, far more intensive than is possible in the census itself. The data are collected from the best respondent for each question, as distinguished from the situation in the census where the head of the household provides information for the entire household. The survey is conducted as closely as possible to the date of the census so that the two sets of data collected are comparable.

19.7 LIST SAMPLE

To illustrate how a postenumeration survey works, we shall take the case of a census of population. Here there are two kinds of errors to be checked: coverage errors and content errors. The total number of persons may be under- or overcounted. Or, there may be errors in the reports on age, occupation, industry, and so on. In order to check on coverage errors, two different samples are usually selected. One is the list sample, that is, a sample taken from the census lists. In this sample a check is made on the households enumerated in the census to find out whether any persons have been missed or counted in error. As an example, a sample of households was taken from the records of the 1961 population census of Greece and reenumerated within two days of the census to check on coverage. Table 19.6 gives the frequency

Table 19.6 Distribution of dwelling units by discrepancy in number of persons, Greek census of population, 1961 (list sample)

Discrepancy	*Greater Athens*	*Other urban areas*	*Rural areas*
−4	...	...	2
−3	...	...	4
−2	...	2	4
−1	4	11	26
0	259	526	1,251
+1	13	13	10
+2	...	3	1
+5	...	...	1
Total	276	555	1,299

distribution of errors of omission or erroneous inclusion (5). The net effect of these errors on the total count of the population can be assessed from Table 19.7. The estimates shown in this table have been obtained by following the sample design used.

Table 19.7 Sample estimates of population, census of Greece, 1961 (list sample)

	Estimated persons actually enumerated		*Estimated persons who should have been enumerated*	
	Number	*Sampling error*	*Number*	*Sampling error*
Greater Athens	1,576,500	94,200	1,586,500	94,600
Other urban areas	1,979,800	60,100	1,980,200	60,300
Rural areas	4,705,800	102,400	4,675,500	101,100
Total	8,262,100	151,600	8,242,200	151,000

A number of precautions were taken to ensure that the survey procedures were really superior to those of the census. The enumerator was required to locate the listed address and enumerate all those who usually lived at the address, including any visitors or any other persons who spent the night of the census there. A large number of questions had been designed to remind the household of the different categories of people, such as babies, relatives, and servants, who might have been associated with the household. For all persons listed in this manner an inquiry was made as to where the person had spent the night of the census. If it was found that the present occupants were not staying at the listed address at the time of the census, the whereabouts of the previous occupants were asked for. In case the listed address was found to be vacant, neighbors were asked whether any persons occupied the premises at the time of the census.

19.8 AREA SAMPLE

The list sample provides information on errors of inclusion or exclusion made in dwelling units listed in the census. This must be supplemented by an area sample in order to find the dwelling units, if any, that were entirely omitted in the census. The persons associated with these dwelling units are to be

Table 19.8 Distribution of households omitted in the area sample

	Number of households omitted		
Number of persons in household	*Greater Athens*	*Other urban areas*	*Rural areas*
1	18	10	12
2	5	2	0
3	4	1	1
4	2	2	1
5	2	0	2
6	0	1	0
Total	31	16	16

enumerated as of the date of the census. This procedure was followed in the sample check of the 1961 Greek census of population. A sample of 141 segments from Greater Athens, 232 from the other urban areas, and 544 from the rest of the country was selected for this purpose. The supervisory staff of the census were required to make fresh lists of all places of abode in the selected segments. A comparison was then made with the census lists to determine whether or not the dwelling units in the new lists were included in the census. All dwelling units missed in the census were enumerated. Table 19.8 provides information on the households missed in the area sample (5). This shows that it was usually the single person living on his own who was missed in this way. Using proper weighting factors, it was found that about 12,000 households were missed in their entirety from the census. Adding over the list and area samples, the following picture emerged (Table 19.9).

Table 19.9 Sample estimates of population of Greece in private households

	Estimated persons actually enumerated		*Estimated persons who should have been enumerated*	
	Number	*Sampling error*	*Number*	*Sampling error*
Greater Athens	1,586,300	95,200	1,602,100	95,600
Other towns	1,984,200	60,700	1,987,800	60,900
Rural areas	4,706,800	103,500	4,680,300	102,100
Total	8,277,300	153,100	8,270,200	152,500

It is now possible to adjust the sample estimates by making use of the known census counts in the different strata of the survey. This procedure is expected to give an improved estimate of the total number of persons. The estimate made is shown in Table 19.10. It can be said that there was an underenumeration of about 10 per thousand in Greater Athens and an overenumeration of about 6 per thousand in the rural areas. For the country as a whole there was no significant difference between the PES estimate and the census count. As judged by the PES, the job of census taking was well done.

Table 19.10 Adjusted population of Greece in private households

	Census population	*Adjusted population estimate*	*Standard error of estimate*
Greater Athens	1,775,815	1,793,500	9,200
Other towns	1,894,396	1,897,800	5,000
Rural areas	4,434,169	4,409,300	10,700
Total	8,104,380	8,100,600	15,000

19.9 OTHER COVERAGE CHECKS

Experience shows that the PES procedures cannot cover adequately those groups of the population for which the risk of underenumeration is highest. Persons without a close attachment to a dwelling unit or a household are most likely to be omitted both in the census and in the PES. This calls for new methods to check coverage. The coverage check of the United States population census of 1950 is a good illustration of this point. The best available evidence indicated that the total count may have been deficient by about 2.4 percent, but the PES could account for no more than half of the under-enumeration (9). Thus alternative methods were tried in order to obtain information on coverage in the 1960 census. Essentially the method consisted in constructing an independent sample of the population as of the census date and in determining how many persons appearing in the sample were not enumerated in the census. For purposes of sample selection the population of the United States as of the 1960 census date was assumed to be divided into four classes: persons enumerated in the 1950 census, births occurring in the population since the 1950 census, persons missed in the 1950 census, and aliens living in the country just before the 1960 census date. A probability sample was selected from each of these classes. Use was made of the Current Population Survey (CPS) to select a sample of those who were enumerated in the 1950 census. A sample of EDs was taken from the primary sampling units of the CPS and a systematic sample of about 2,600 persons was selected (10) from the selected EDs. In order to select the sample of births, about 4,500 birth registrations were taken systematically from the files of births in the primary sampling units of the CPS. The PES of the population census of 1950 provided a convenient frame for the selection of the sample of those missed in the 1950 census. In all 273 persons were selected from those found in the PES but missed in the census. A systematic sample of 209 aliens was selected systematically from the lists of registered aliens available from the relevant department of the government.

Having taken the sample from the population, an attempt was made to locate current addresses and thus determine whether the persons in the sample were enumerated in the 1960 census. The difficulties encountered in locating and identifying persons were tremendous. Table 19.11 shows the way in which sample persons for each of the record checks were accounted

Table 19.11 Results of location operations for sample persons, record check studies, United States population census, 1960

Record source	*Number in sample*			*Percent of sample located*
	Total	*Located*	*Not located*	
1950 census	2,605	2,371	234	91.0
Births	4,525	3,873	652	85.6
1950 PES	273	227	46	83.2
Aliens	209	209	0	100.0

for. Various assumptions were made regarding the persons who could not be located. This gave rise to several minimum and maximum rates of net undercount. The various estimates were found to lie in the range of 1.3 percent to 3.4 percent. Subsequently, when all evidence from different sources was assembled, a minimum reasonable estimate of the undercount of the population in 1960 was found to lie in the range of 1.7 to 2.0 percent. It may be noted that the corresponding estimate for the 1950 census was 2.4 percent (9).

19.10 CONTENT CHECK

The quality of the responses obtained in the census on items such as age, school enrollment, and income can be measured by reinterviewing a sample of the persons enumerated in the census. In the reinterview a more intensive approach to questioning is adopted than can be done in the census. Instead of one question, a battery of probing questions designed to uncover the truth can be used in the check. One can then compare responses in the census with responses in reinterviews and study the differences involved. Take, for example, the class of persons in the age-group 0–4 years. As judged by the PES, some persons might have been included improperly in this age-group and some others excluded erroneously. The total number of errors of classification affecting this group may be called the *gross number of errors.* However, the net error is the difference between the number improperly included and the number erroneously excluded. Under certain conditions (2) the square root of the gross number of errors can be used as the standard deviation of the net error. Table 19.12, based on the PES of the 1950 population census of the United States, is a good illustration of this procedure. It is clear from Table 19.12 that the net error in the age-group 0–4 exceeds four times its standard deviation, the corresponding figure being 2.5 for the class 15–19. This suggests a systematic bias in census reports of age in these groups.

Table 19.12 Gross and net errors in classification by age

Age-group	*Number of persons in census*	*Number included in error*	*Number excluded in error*	*Gross number of errors*	*Net error in census class total*	
					Number	*Standard deviation*
0–4	2,897	40	88	128	−48	11
5–9	2,377	71	50	121	21	11
10–14	2,023	65	62	127	3	11
15–19	1,907	86	55	141	31	12

This type of evaluation was used extensively in the PES of the 1950 population census of the United States. A count of the errors of misclassification was made in each of the several age-groups, occupation and industry groups, and income groups. Some of the results obtained (2) are presented in Table 19.13. The conclusion drawn is that the PES procedures did not yield substantially different results from the census procedures for items such as age, occupation, and industry. On income the PES results were different from those of the census and presumably more accurate.

Table 19.13 Gross and net errors of classification, 1950 population census of the United States

				Average class		
					Net error in class total	
Classification	*Number of classes*	*Number of cases*	*Percent misclassified*	*Number in class*	*Amount (%)*	*Standard error (%)*
Age	16	26,980	6	1,686	0.8	0.8
Occupation	10	9,502	15	950	1.8	1.8
Industry	14	9,464	11	676	1.9	1.8
Income	15	9,012	31	601	10.8	3.2
Number of rooms	9	19,300	18	2,144	2.5	1.3

19.11 SAMPLE AS PART OF CENSUS

The main purpose of a census is to produce data on basic items for each of the small administrative divisions of the country. Actually it is a legal requirement in many countries to count the number of persons by locality. It is also considered important to have local data on the distribution of persons by age, sex, and marital status. However, there are many other items, such as fertility, for which estimates for the country as a whole or fairly large regions are usually adequate. Such information may be collected from a sample of the population. In this manner the program of the census can be broadened by asking supplementary questions of a sample of persons only.

Since the sample is to be taken in conjunction with the census, it is extremely important that the method of selecting the sample be simple and automatic in its operation. Nothing should be left to the discretion of the enumerator. At the same time the enumerator should not be burdened with complex rules. Furthermore, the method should require little if any change in well-established census procedures.

When there is danger that the task of sample selection would add materially to the burden of the enumerators' work, the sample can be selected centrally in the national statistical office. The simplest plan is to select every kth enumeration district from the list prepared for purposes of the census. The first selection is made at random and the other EDs are selected systematically. When information from a previous census is available, the EDs may be stratified suitably and a systematic sample selected from each stratum. In the selected EDs information is collected both on the basic census questions and the questions on the supplementary list. Only the basic census questions are asked in the other EDs. It is clear that no biases of sample selection are involved when this method is used. Furthermore, there is no need to train all enumerators in the art of asking questions on the supplementary list. Only those who are to work in the sample EDs need this training. The main disadvantage of this procedure is that the sample becomes clustered and therefore becomes inefficient for those items which are positively correlated within clusters.

The sample can be given better geographic spread by making a selection from each ED in the population. In this connection, it is natural to use the frame created by the census-taking process. Thus, for example, every kth household or person may be asked supplementary questions by the census enumerator. But then who should select the sample? If the enumerators are asked to make the selection, they may introduce biases by deliberately or unconsciously departing from the correct procedure. If every kth person is to be selected, the order of enumeration may be changed in order to bring a designated person in the sample.

19.12 SAMPLE OF HOUSEHOLDS

The census of population provides data both on persons and households. When household data on the supplementary items are needed, the sampling unit must be the household. In that case a convenient sampling method is the selection of every kth household, say, every tenth. Since the list of EDs is available in advance, a random number between 1 and 10 can be allocated to each ED. This number is to be used as the starting point for the selection of the systematic sample in the ED. So that no single starting point can predominate, the frequencies of the 10 starting points can be equalized. For each ED the sample and the nonsample questionnaires are then interleaved in the required order. The enumerator delivers them by taking them one by one off the top of his pack. The enumerator will not know until the moment of delivery whether a particular household is to receive the sample questionnaire. The bias of selection is therefore reduced. This method was used in the 1961 population census of Great Britain (1).

19.13 SAMPLE OF PERSONS

Sometimes, when the number of items on the census is small, each person can be enumerated on one line of the schedule, there being 30 to 40 lines to a page. The lines are numbered serially from 1 downward on the page. If a 10 percent sample is to be used for the supplementary data, every tenth line can be marked on the schedule as the sample line. When the enumerator reaches that line, information on both the census items and the supplementary items is collected. There is a danger involved in the use of this method. The enumerator has to list the households in the ED in the order in which he finds them and within a household the members are to be listed systematically—head, wife, children in order of age, and so on. This produces a pattern of waves on the schedules. As a result of the periodic nature of the data, certain starting points will have the persistent tendency to overstate some of the population characteristics while others may tend to understate them. The bias arising from this source can be eliminated by marking all possible random starts on the schedules with the same frequency. But this involves the printing of several sets of schedules and is therefore to be avoided. There are other sources of bias, too. Some enumerators may leave the last few lines on the page blank in order to use a new schedule for the next household. Some others may use some of the lines for writing notes.

A very practical method of deciding upon the sample lines was used in the 1940 population census of the United States (8). Each schedule contained 80 lines, there being 40 lines to a page. A 5 percent sample was to be used for the collection of additional information in conjunction with the census. Instead of using all the 20 styles of schedules, just 5 were used. The material collected in the pretest showed that the initial starting points (lines 1–6), the lines at the beginning of the back page (41–46), and the lines at the end (75–80) were most subject to bias. Therefore most of these lines were included under styles II–V (Table 19.14). The other 64 lines were represented by the lines included in style I. The material available showed that line 14 gave a good idea of the population average of units listed beyond the first six lines. There-

Table 19.14 Sample lines, United States census of population, 1940

Style	*Relative frequency*	*Line numbers*			
I	16	14	29	55	68
II	1	1	5	41	75
III	1	2	6	42	77
IV	1	3	39	44	79
V	1	4	40	46	80

fore, this line was selected. The other three lines in this style were chosen by assuming a cubic regression of the characteristic y on line number. In order to avoid bias, style I was used in 16 EDs out of 20.

About the same procedure was used in the 1950 census (7). This time there were 30 lines to a page and the sampling interval was 5. This gave 6 sample lines on each page of the schedule. The last selection on each page formed a subsample on which only those items were enumerated which were to be tabulated for very large areas. Care was taken to ensure that each line on the schedule was part of the sample on one and only one of the five printings. Within each enumeration district the schedules were divided approximately equally among the five styles. There was, however, a definite change of procedure at the 1960 census. A sample of housing units, as opposed to persons, was taken for collecting additional information. The sampling fraction was increased to 25 percent to make an allowance for the clustering of the sample. The housing units listed within an ED were assigned the letters A, B, C, and D in that order, the first letter being determined by the last two digits of the serial number of the ED involved (11). The units designated A were all in the sample. In spite of the rigid rules of sample selection, it was feared that the sample would not always be selected in an unbiased manner. Nevertheless, it was decided to take the risks of such biases rather than increase the cost of the investigation by selecting the sample after the initial census canvass was completed.

19.14 CONTROL OF DATA PROCESSING

The data collected from a census must go through many stages of processing before the final tables can be produced. Thus the returns are to be edited, coded, and punched into cards for sorting and tabulating by machine. All these operations are potential sources of error. It is therefore important to know whether the work of the operators at different stages of the processing meets the specifications. This can be done by inspecting a sample of the work as it is done. It is neither necessary nor desirable to introduce 100 percent inspection. Some errors are bound to remain. Provided the errors are inconsequential, it is not worth the effort needed to track down each one of them. However, it is essential to prevent gross errors of the type that will produce material errors in the results. Thus a single misclassified card in a count will usually produce an insignificant error in the result, but the mispunching of a number such as 661 for 001 may produce a serious error in the resulting total of the class to which the unit belongs.

It should be stressed that the aim of quality control methods based on sampling is not 100 percent correction of substandard work. The aim of sample verification is to ensure that the work is satisfactory and to do so at minimum cost. This is done by suitably adjusting the process to ensure such low error rates from each operator that correction is not necessary. A minimum standard is laid down for all workers and those not attaining this standard are transferred to other work. As a practical step, the proportion of errors made by each worker during a prescribed period of time is determined, and remedial action is taken when necessary.

An example is provided by the quality control program for the coding of data collected in the 1960 population census of the United States (11). In the processing of the earlier 1950 census the work of a coder was reviewed by a verifier who determined whether or not the correct codes had been assigned to the coded items. The effectiveness of this system was evaluated by planting deliberate errors in the work and checking whether the verifier had been able to identify those errors. It was discovered that a verifier might fail to find as many as half of the errors in the work he verified. Therefore a new procedure was devised to check the quality of coding in the 1960 census. A sample of households was selected from each enumeration book, the sampling rate being 1 in 80. For a given type of coding three different persons independently coded the data. The first and second coders entered the codes on specially prepared cards called *pench cards*. The third, the regular census coder, entered the codes on all the schedules for the ED including those not in the sample. Then a matcher compared the two pench cards with each other and with the coded schedules. Where all three agreed, the code used was considered to be correct. Where two agreed, this was considered to be correct and the third in error. Cases in which all the three disagreed were referred to a specialist. This system provided a means of measuring the quality of the census coder and of the two pench coders. The quality of the work of the matcher was determined by planting deliberate errors and counting those that the matcher had not been able to detect. (See Sec. 10.16.)

REFERENCES

1. Benjamin, B. (1960), Statistical Problems Connected with the 1961 Population Census, *J. Roy. Statist. Soc.*, (A)**123**.
2. Eckler, A. R., and W. N. Hurwitz (1958), Response Variance and Biases in Censuses and Surveys, *Bull. Intern. Statist. Inst.*, **36**.
3. General Register Office (1952), "Census 1951, Great Britain, One Per Cent Sample Tables," pt. I, London.
4. Lahiri, D. B. (1956), "On the Question of Bias in Systematic Sampling in Population Censuses" (unpublished report).
5. National Statistical Service (1962), "The Coverage Check of the 1961 Population Census of Greece," Government of Greece, Athens.
6. National Statistical Service (1962), "Sample Elaboration of the 1961 Population Census," Government of Greece, Athens.
7. Steinberg, J., and J. Waksberg (1956), Sampling in the 1950 Census of Population and Housing, *U.S. Bur. of the Census Working Paper* 5, Washington.
8. Stephan, F. F., et al. (1940), The Sampling Procedure of the 1940 Population Census, *J. Am. Statist. Assoc.*, **35**.

9. U.S. Bureau of the Census (1960), The Post-enumeration Survey: 1950, *Tech. Paper* 4, Washington.
10. U.S. Bureau of the Census (1964), "Evaluation and Research Program of the U.S. Censuses of Population and Housing, 1960: Record Check Studies of Population Coverage," ser. ER 60, no. 2, Washington.
11. U.S. Bureau of the Census (1966), "1960 Censuses of Population and Housing: Procedural History," Washington.

20
Surveys in Other Fields

20.1 INTRODUCTION

We shall now discuss data collection problems in a few other important fields which could not be taken up in the previous chapters. The topics considered include surveys of road traffic, public opinion surveys, marketing research, postal traffic, payroll studies, surveys of food consumption, and sociological research.

20.2 SURVEYS OF ROAD TRAFFIC

Transportation being an essential service for economic activity, data are needed on the adequacy and use of transportation facilities. There are several forms of transport: railways, road transport, inland waterway transport, coast-to-coast shipping, international seaborne shipping, and air transport. In this book we shall consider road transport, a field in which sampling methods are being increasingly used. Two types of surveys may be distinguished: one in which a sample of vehicles is selected to collect data such as the type of vehicle, kilometers run, and freight carried, and one in which a sample of points on the road network is selected to determine the distribution of traffic.

SAMPLE OF VEHICLES

The sample of vehicles can be best selected from the list of vehicle registrations. As information is usually available on the nature of each vehicle's operations, the sample can be selected from within strata. For example, vehicles may be classified by capacity and by type (whether for hire or on own account) and variable sampling fractions used within the substrata. The main piece of information to collect from each vehicle is the total gross weight of the goods loaded into the vehicle during the period in question. Each vehicle may be asked to keep a record showing the distance run, the gross weight of each consignment loaded, and its origin and destination. This record is to be kept for just a few time periods during the period in question. However, these time periods should vary from vehicle to vehicle so that the entire period is adequately covered. A problem arises when it is found that a truck does not have to make out a bill on a weight basis. In this case the weight will have to be estimated. Another difficulty is that vehicle owners are reluctant to show overloads. In this situation the rated capacity of the vehicle should not be asked, especially when it can be obtained from other sources (26).

It is generally possible to collect the bulk of the information by mail. The nonrespondents of the survey can be stratified geographically and a sample of them taken in order to collect information by personal interview. When all the returns are in, it is possible to make estimates of the total gross weight of commercial goods loaded, the average length of haul of a ton of goods, vehicle kilometers, capacity in ton-kilometers, and freight ton-kilometers.

A suggestion is sometimes made that the sampling unit should be the carrier and not the individual vehicle. The danger in collecting information from the carrier on its whole fleet is that large carriers may not be willing to provide the information, both because of the work involved and because they would not desire to reveal the workings of their enterprise as a whole. Therefore, the individual vehicle is the more appropriate unit of sampling.

When the object is to provide estimates for the entire calendar year, the inquiry must be spread over the year. In that case it becomes important to select a random sample n of vehicles out of the totality of N and a random sample t of weeks out of the total number $T = 52$. The design can be so arranged that each vehicle reports for a specified number s of weeks and that every selected week the same number m of vehicles report the data. The precision of the estimate will, of course, depend on the variances associated with vehicles, weeks, their interaction, and the numbers n, t, m, and s where $ns = mt$. A number of designs can be tried to find the best one. Consider,

Table 20.1 Two sample designs used in survey of road traffic

	The Netherlands		*South Holland*	
	Design 1	*Design* 2	*Design* 1	*Design* 2
N	10,920	10,920	2,133	2,133
T	52	52	52	52
n	10,920	10,920	2,133	2,133
s	4	2	4	2
m	3,640	420	711	82
t	12	52	12	52

for example, the two designs shown in Table 20.1. The purpose of the survey was to estimate the volume of goods carried by road in the Netherlands (17). In the first design each vehicle reports for 4 weeks out of the 12 selected at random. In the second design the number of reports is restricted to 2 weeks only, but every week in the year is represented in the sample. The number of questionnaires used is twice as large in the first design but the standard errors of the estimates are much higher than in the second design (Table 20.2).

Table 20.2 Standard errors of estimate of annual average, survey of road traffic

	The Netherlands		*South Holland*	
Characteristics	*Design* 1	*Design* 2	*Design* 1	*Design* 2
Tons	4.5	1.6	5.0	3.6
Ton-kilometers	565	49	571	111

In a repetitive survey it is possible to make estimates of variances associated with vehicles, weeks, and their interaction and make use of this information for improving future surveys. An example of this comes from the Netherlands (17) where the 92 border customs offices were asked to count the number of Dutch cars entering the country. The optimum design was found as one in which all offices are asked to count for 14 randomly selected days with about 3.5 offices reporting on each selected day. A comparison was made with two alternative procedures: (1) one in which a random sample of 14 days is selected for each office without any other controls, and (2) one in which all the offices are asked to report on 14 fixed days of the year selected at random. The coefficients of variation of the estimates of the population total are given in Table 20.3. It is clear that it is unwise to use the same 14 days for enumeration at each post.

Table 20.3 Survey of counts of cars at 92 posts, Netherlands, 1961

Method	*Coefficient of variation of estimate (%)*
Optimum restricted selection of 14 days at a post	1.56
Unrestricted selection of 14 days at each post	3.64
The same 14 days at each post	10.97

SAMPLE OF ROAD NETWORK

Another method is to select a sample of the road system. When a complete list of the roads and their lengths is available, it is possible to imagine all the roads laid out from end to end. A random or a systematic sample of points or stretches can then be selected. At each sample a count can be made of the number of vehicles passing through it. By stopping vehicles selected at random the weight carried can be asked. This gives the average flow per sample point. When multiplied by the total length of the road system, estimates of population total can be made.

An example of this type of survey comes from the United Kingdom (24) in which a study of road traffic was made in four selected areas. Within an area the roads were stratified by class and by location (whether in built-up areas or not). A number of points were selected systematically with a random start from each stratum, the total number of points being 50 in each area. At each of these points the flow of traffic was estimated by counting for two 1-hour periods or for two 2-hour periods during one week, which was different in different areas. The method of calculation is simple. If the sampling interval in a stratum is k (one point every k miles) and the sample total of the number of vehicles passing the points in a given period is n, an unbiased estimate of the total vehicle miles in the period is kn. Table 20.4 gives the average flows in each stratum and the vehicle mileage.

Table 20.4 Average weekly flow per weekday and vehicle mileage in four selected areas

	Number of vehicles per weekday		*Percentage total vehicle mileage*	
Class of road	*Built-up areas*	*Other areas*	*Built-up areas*	*Other areas*
Trunk	6,100	6,100	7	16
Class I	4,200	2,400	16	12
Class II	2,200	900	6	4
Class III	1,500	400	6	8
Unclassified	900	200	20	5

When a large-scale survey of road traffic has taken place in the country, it is possible to estimate trends in the volume of traffic by instituting counts on a sample of points selected at random. Wherever practicable, automatic counting machines may be used. The automatic counters do not have to be very numerous, as they may be used on a rotating basis. Short counts at a large number of points are preferable to long counts at a small number of points. Automatic counts may be supplemented by manual counts wherever necessary. The method of sampling in space and time (27) can be put to good use in inquiries of this type. In this method a random sample of days over the year is taken just as a random sample of points is taken from the entire road network. Counting takes place at the sample points following the preferred design. A popular practice is to count for 14 days at each point. The sets of 14 days differ from point to point. If counting is manual, the intensity of counting may be reduced at night when the traffic is expected to be small.

20.3 PUBLIC OPINION SURVEYS

Public opinion surveys are one of the greatest contributions to democracy in the present century. By asking a sample of the population to express their views, wishes, and interests, these unofficial surveys provide very useful information on what people are thinking about current issues. The most dramatic side of these surveys is their ability to predict official elections. Besides predicting election results, opinion surveys can throw light on the influences which determine the direction of the popular vote during the course of a campaign. The manner in which opinion crystallizes during the campaign can be studied. A number of traditional theories can be verified on the basis of the facts collected. When the voters can be classified by age, sex, economic status, and other factors, polls can be used to identify different opinion alignments by group. In this manner polls can add materially to our store of political knowledge.

DETERMINANTS OF PUBLIC OPINION

From the standpoint of sampling it should be emphasized that the political public is not a homogeneous population but is divided up into groups between which there are significant differences of outlook and interest. Thus there are men and women living in different geographical areas and earning their living as merchants, farmers, physicians, miners, teachers, and so on. Different occupations produce different levels of income and different living standards. The play of these forces divides the public into groups and stimulates common attitudes within the groups. People who live differently think differently. Age, sex, religion, language, occupation, political affiliation, and general cultural background are the basic determinants of their experiences and opinions. This fact is illustrated in Table 20.5, which is based on the replies of a sample of the population who were asked whether the government should have the power to tell each citizen what to do as his part in the war effort (3).

Table 20.5 Percentage distribution of opinions classified by age, sex, economic status, and political affiliation, United States, 1942

Classification	*For*	*Against*	*Undecided*
Sex			
Men	66	29	5
Women	55	36	9
Age			
21–29	56	38	6
30–49	61	31	8
50 or over	62	30	8
Politics			
Democrats	67	26	7
Republicans	57	37	6
Economic status			
Upper income group	62	34	4
Middle income group	61	33	6
Lower income group	60	30	10
Total	61	32	7

This situation can be exploited for sampling purposes by stratifying the population on the basis of characteristics which are known to be the basic determinants of public opinion. What is crucial is that all strata are sampled in order to obtain a fair picture of the opinion of the population on the issue in question. Failure to sample from all the strata may prove to be ruinous. A classical example is provided by the *Literacy Digest* poll (8) conducted to predict the United States presidential election of 1936. One of the reasons that it pointed to the wrong candidate was that the voters selected in the sample were, for the most part, telephone subscribers and owners of automobiles. The poorer sections of society were seriously underrepresented and this proved disastrous to the poll although more than 2 million ballots were taken.

The most common method of selecting the sample is to divide the country into geographical sections and the sections into strata according to degree of urbanization. Within the selected places an area sample or a quota sample is taken and information collected from the respondents by asking questions. The size of the sample in national polls is rarely large; it ranges from 1,500 to 50,000 persons. What is important is the character of the sample and the care with which the survey is executed. An example is provided by the New York gubernatorial election of 1942 (3) in which Dewey received 53 percent of the votes. In the week before the election the Office of Public Opinion Research used a carefully stratified sample of 200 registered voters to predict that Dewey would receive 58 percent of the votes. The *New York Daily News* came out with a figure of 57 percent on the basis of a sample of 48,000 voters. The Gallup poll took a quota sample of 2,800 persons and its prediction tallied exactly with the actual returns.

SETTING THE ISSUES

It is the responsibility of the poll taker to slice the issues involved meaningfully and present them to people in such a way that answers can be reliably interpreted. The issues should be sufficiently circumscribed and the questions used should not be vague or obscure in meaning. As an example, consider the following question asked in a poll: "After the war is over, do you think people will have to work harder, about the same, or not so hard as before?" In one of the studies (3), this question was put to a sample of 40 persons. The follow-up research on the question showed that "people," "harder," and "as before" conveyed different meanings to the respondents. To a little more than half of the group the word "people" meant everybody, to a third of the group it meant a particular class, and others did not know what was meant by this word. The word "harder" turned out to mean quality to some, more competition to others, and longer hours to the rest. The phrase "as before" meant before the war started to half the group and after the war started to the rest of the group. This analysis shows that it becomes difficult to interpret the results obtained from such questions in any reliable fashion.

There are four† types of questions which are usually found in public opinion polls. One type is the free-answer question where the respondent is asked to express his opinion on an issue. No special limitation is placed on the respondent's answer. This type of question is useful in ascertaining how well informed the respondent is on the issue, what his reactions are, what factors may be influencing his judgment, and how important he thinks the issue is. When the questions concern an issue about which people have thought little or when a new and somewhat complicated problem is to be posed, the free-answer type of question is appropriate.

The second type of question requires one of three kinds of answers, namely, affirmative, negative, and undecided. This type of question is asked when it is known that opinion on the issue has become crystallized so that it can be expressed by a yes or no answer. The disadvantage with this approach is that there are many questions which cannot be accurately answered

† See Sec. 7.14 for examples.

by a yes or no without further qualifications being made. Furthermore, the intensity of opinion cannot be ascertained by this type of question.

The third type is the question that contains a pair of statements. The respondent is asked to choose the one with which he agrees or the one that is right. By implication the other statement is false or disapproved. The main difficulty with this procedure is that many human reactions are not reducible to a yes or no proposition. Thus the tabulations made on the basis of answers to such questions do not necessarily report public opinion accurately.

The fourth type is the multiple-choice type. The respondent is asked either to select one of the several answers as his choice or to rank the answers in order of preference. The multiple-choice question is appropriate when an issue does contain several genuine alternatives. In that case there is good reason to present all the alternatives rather than force them into the framework of a yes or no form. What is more, such questions may permit a partial expression of the intensity of response to an issue. The disadvantage is that the alternatives are not always mutually exclusive and there is no certainty that they are mutually exhaustive.

The split-ballot technique is very useful for deciding upon the best wording of a question or for comparing different types of questions. The technique involves the use of an *A* form and a *B* form of a ballot. Each contains a different form of the question to be tested. The two forms are then submitted to comparable groups, whose replies are checked against each other's. Some questions may be the same on each form to serve as controls. The position of some of the alternative questions may be interchanged on the two forms in order to assess the effect of the order in which questions are asked.

QUOTA SAMPLING

Since speed is an important consideration in surveys of public opinion, it has been common to use the method of quota sampling for the selection of the sample. The quota sampling procedure starts with the premise that a sample should be well spread geographically over the population and that it should contain the same fraction of individuals having certain characteristics as does the population. The characteristics usually taken into account are sex, age, occupation, economic levels, and size of place, in addition to geographical control (20). Data taken from a recent census are used as the standard. In a sense the population is divided up into a number of strata whose weights are obtained from the census or a large-scale survey. Interviewers are then assigned quotas for the number of interviews to be taken from each stratum. Thus an interviewer will be asked to select so many males and so many females, so many young and so many old persons, and similarly for the other characteristics used as controls. The interviewer is free to choose his sample as he likes provided the quota requirements are fulfilled. The essential difference between quota sampling and stratified simple random sampling is that in the former case the selection of the sample within strata is not strictly random. The interviewer may decide to omit certain parts of the area entirely if it suits his convenience. He may not like to approach certain kinds of people and so on.

COMPARISON OF QUOTA SAMPLING WITH PROBABILITY SAMPLING

From purely theoretical considerations it can be said that quota sampling lacks scientific validity as there is no element of randomization in the selection of the sample. Thus one cannot use the results obtained from such a sample without the fear that some extraneous factors might have introduced serious bias in them. However, quota samples are very popular with market and opinion researchers. There is, therefore, some point in finding out how quota sampling compares with probability sampling in practice. Considerable research has been done in this field. We shall present a few selected results.

Just before the 1948 presidential poll in the United States the Washington Public Opinion Laboratory conducted two polls in the state: one using the area sampling method and the other the quota method. The state was divided into 28 strata by size of community and within these strata 53 rural precincts and incorporated places were selected. A simple random sample of the smaller places was taken while sampling was done with probabilities proportionate to size in the case of the larger places. Within the selected areas households were selected from lists already available or specially prepared for the purpose. Persons within households were selected at random. This was the design of the area sample. The same interviewers were used for the quota sample, which was taken concurrently with the area sample. The controls used for the quota sample were sex, age, and economic level. The methods to be used for taking the quota sample were explained to the interviewers. Table 20.6 presents some of the results based on registered voters in the sample (16). There are substantial differences between the two samples. The area sample is closer to the actual vote and may, therefore,

Table 20.6 Comparison of quota and area samples, Washington State poll, 1948 (percent voting)

Candidate	*Area sample*	*Quota sample*	*Actual vote*
Truman	50.5	45.3	52.6
Dewey	46.0	52.0	42.7
Wallace	2.9	2.5	3.5
Others	0.6	0.2	1.2
Total	100.0	100.0	100.0

be said to be superior to the quota sample. However, taking into account the sampling errors associated with the area sample, the differences observed are not found to be significant.

In the summer of 1951 a socioeconomic survey was carried out in Great Britain using the quota method. The controls used were age, sex, and social class. About 8,000 interviews were taken. It was found possible to compare the estimates made from this survey with the figures obtained from the regular Government Social Survey based on probability sampling. This survey had produced about 3,800 interviews for the same period (15). On most of the questions the results from the two surveys were very close. In particular,

Table 20.7 Comparison of quota and random samples, industry/occupation distribution, Great Britain, 1951

	Men		*Women*	
	Quota	*Random*	*Quota*	*Random*
Manufacturing	6.5	24.9	4.3	7.2
Clerical	3.8	5.0	4.2	4.6
Distributive	15.8	5.9	9.0	3.0
Transportation, etc.	18.3	7.6	1.3	0.0
Professional, etc.	18.1	20.0	5.4	3.2
Mining and quarrying	1.4	4.6	0.0	0.0
Building, etc.	14.3	6.3	0.2	0.0
Agriculture	2.4	2.8	0.3	0.0
Other industries	15.5	8.1	10.3	5.5
Housewives	0.0	2.0	64.8	69.2
Retired, unemployed	4.0	12.8	0.2	6.6
Total	100.0	100.0	100.0	100.0

there was good agreement on household composition, type of dwelling, number of rooms, and the proportion of women who were housewives. The two surveys, however, differed considerably with respect to items closely related to social or economic status. Consider, for example, Table 20.7, which gives data on industry/occupation. The quota sample is very deficient in persons employed in manufacturing and has a large excess of persons in the distributive trades and transportation and building occupations. It may be noted that industry/occupation was not used as a control in the quota sample.

In view of the results of Table 20.7, an experimental study of quota sampling was made in which industry/occupation was used as one of the controls in addition to age, sex, and social class (15). Four organizations took part in the survey, each taking twelve independent quota samples, four of them in each of the three towns selected for the experiment. In half of the samples the only controls used were age, sex, and social class. In the other half, these plus one of industry/occupation and geographic controls were used. About 4,320 interviews took place. In order to make a comparison with random sampling, about 360 interviews were attempted in each town with persons selected at random. The results showed that there were relatively few differences between the quota and random samples. In the matter of education, however, the quota samples seemed to get too educated a cross section of the population. This is borne out by Table 20.8, which gives the distribution of age at which full-time education ceased.

Table 20.8 Percent distribution by age at which full-time education of males ceased

		Random samples		*Quota samples*			
Age	*Census*	1	2	1*a*	1*b*	2*a*	2*b*
0–14	81.1	78.1	76.3	69.2	68.2	74.9	70.6
15	6.0	11.3	12.0	10.7	11.5	7.9	9.8
16	7.2	4.9	6.5	9.5	10.1	9.8	8.8
17–19	3.5	4.7	3.9	6.3	5.3	4.7	6.2
20 or over	2.2	1.0	1.3	4.4	4.9	2.7	4.5
Total	100.0	100.0	100.0	100.0	100.0	100.0	100.0

The quota samples appear to underrepresent those who finished their education before the age of 15. Sample 2*a*, in which the industrial control was used, goes some way to correct this tendency. The authors of the report suggest the use of education as another control for selecting quota samples in surveys related to education.

The burden of this discussion is that sometimes the quota sample works well and sometimes it does not. The use of quota sampling can produce good agreement on some variables and poor agreement on others. There can be situations in which serious bias is present, while instances of close agreement with probability sampling are abundant. In view of these findings the method of quota sampling appears to be unsuitable for surveys in which the results are to form the basis of some administrative action.

SELECTION OF ADULTS

When there is no up-to-date list of adults in a population and a probability sample is needed for conducting the survey of opinions or attitudes, it is usual to use the area sample approach to select the sample of households. For a number of reasons it is desirable to interview a single adult member in the household. First, multiple interviews are expected to be inefficient because of the positive correlation of attitudes within the household. Second, the interview has to be obtained before the respondent has the opportunity to discuss the questions with other members of the household. Thus it becomes important to select the informant at random from the adults in the household. The selection can be made by using random numbers. The response obtained is multiplied by the number of adults in the household in order to get an unbiased estimate.

The following procedure can be used when the interviewer is not required to consult random numbers for the selection of the sample. He makes a list of household members and assigns a serial number to each adult. The males are numbered in order of decreasing age, followed by the females in the same order. Then the interviewer consults the table of selection assigned to the dwelling unit in order to determine the adult to be interviewed. This table is one of a set of tables prepared beforehand in order to ensure that the probability is $1/k$ that an adult is selected from a household with k adults. An example is given in Table 20.9 (12) in which the chances of selection are exact for all adults in households with 1 to 4 and 6 adults.

Table 20.9 Serial number of adult to be selected in the sample

Table number	*Relative frequency*	*Number of adults in the household* 1	2	3	4	5	6 *or more*
A	1/6	1	1	1	1	1	1
B_1	1/12	1	1	1	1	2	2
B_2	1/12	1	1	1	2	2	2
C	1/6	1	1	2	2	3	3
D	1/6	1	2	2	3	4	4
E_1	1/12	1	2	3	3	3	5
E_2	1/12	1	2	3	4	5	5
F	1/6	1	2	3	4	5	6

COMPARISON OF SAMPLE WITH CENSUS

In many fields such as agriculture and demography, it is possible to compare the estimates obtained from a sample with the census count taken at about the same time. This happens because the changes taking place in the population are slow. The same cannot be said of the opinions and tastes of people. Attitudes and opinions of people change very rapidly with the passage of time and as a result of the occurrence of certain events. A general election is the closest approximation to a census in the field of public opinions. It is thus natural to compare the results of a sample survey of opinions with the election returns in order to see whether the sampling method works in practice. Consider, for example, the data presented in Table 20.10, based on two surveys containing a question on voting behavior in the 1940 United States presidential election (22). There are considerable discrepancies between the responses to survey questions and the actual vote cast. Many people said they had voted while actually they did not. The proportion claiming to have voted for the winning candidate is much higher than the actual vote. Many other studies on this subject have come to about the same conclusions. In one of the studies (19) a sample of individuals was asked whether they had voted in the 1944 presidential election. Their responses were compared with the election register. It was found that about 23 percent of the 920 respondents had said that they had voted while actually they had not. All this shows that the survey results are not comparable with the election results. Whether or not

Table 20.10 Comparison of survey results with election returns

	April, 1944, *survey* (%)	*June*, 1944, *survey* (%)	*General election* (%)
Did not vote	23.7	21.6	40.8
Voted	76.3	78.4	59.2
Of those voting			
For Roosevelt	62.8	59.8	54.7
For Dewey	36.3	39.5	44.8
For others	0.9	0.7	0.5

the sampling method works in this field cannot always be judged by comparing the sample survey with the census.

A method generally used to check on the opinion sample is to collect information on auxiliary characteristics such as age, sex, education, and income in the sample survey and compare the estimates made with the known census data. If the sampling procedure is satisfactory, the sample estimates should agree, apart from sampling fluctuations, with the population values. The assumption made is that a sample which is good for these variables should also be good for attitudes and opinions because of the high correlations involved. This argument is not correct. It is well known that agreement between sample and population on one variate does not necessarily imply good agreement on other variates. There is the classical example of the sample selected from the Italian census data of 1921 (18). The sample was meant to serve as a miniature of the census. It consisted of all the returns of 29 out of the 214 areas into which the country was divided. It was selected in such a manner that seven important characteristics of the sample agreed with the population characteristics. Subsequently it was discovered that this arrangement could not ensure agreement on many other characteristics.

INTERPRETATION AND LIMITATION OF POLLS

There is considerable controversy surrounding the use of public opinion polls during an election campaign. It is felt in some quarters that polls influence and even create public sentiment. Many others hold that polls exert no such influence over the outcome of the election but merely serve as a scoreboard. Without entering into this controversy, it can be said that the results of polls should not be taken as predictions but merely as the popular sentiment at the time of the interview. The reason is that voting preferences can change drastically in the later weeks of a campaign, making it extremely difficult to make any predictions. Broadly speaking, there are three kinds of people eligible to vote. Some are uninterested, others are on the fence, and then there are the hard core. Many of the uninterested and some of those on the fence are unlikely to vote. Therefore, to get a fair picture, some means must be found to eliminate from the poll sample those who are unlikely to vote as well as those who are not qualified to vote. The use of improved questioning procedures can go a long way in separating the likely voters from the rest. Also, suitable formulas will have to be devised to allocate the undecideds to one category or the other. And then there is the problem of the nonreachables, who have to be brought into the sample in order to get unbiased results.

The answer to the question "Can polls go wrong?" must be in the affirmative. A carefully executed poll will be right most of the time and will go wrong in a small percentage of the cases. As is well known, the accuracy of a statistical procedure can only be judged in the long run. When opinion is evenly divided, it becomes difficult to predict the winner. But in this situation the polls do indicate that the contest is close, as is evident from Table 20.11 regarding the 1968 United States presidential election (4). It will be

Table 20.11 Percentage distribution of national popular votes and election results, United States presidential election, 1968

	Nixon	*Humphrey*	*Wallace*	*Undecided*
Election result	43.5	42.9	13.6	
Gallup poll, Nov. 3	42	40	14	4
Gallup poll, Nov. 3*	43	42	15	
Harris poll, Nov. 3	42	40	12	6
Harris poll, Nov. 4	40	43	13	4
Harris poll, Nov. 4*	41	45	14	
Sindlinger poll, Nov. 2	37	40	12	11

* Undecideds allocated.

seen that the Gallup poll came very close to the final election count but the Harris poll picked up the wrong candidate.

On the other hand, it should be pointed out that many factors may distort the official election as a test of poll accuracy, thereby making the election present a biased picture of popular sentiment. For example, the weather on election day may draw voters in unequal proportions from rural and urban areas. Or, political corruption may operate to vitiate the findings of the public opinion polls.

20.4 MARKETING RESEARCH

In the present-day world of mass production of most goods, the distance between the producer and the final consumer of goods is increasing more and more. With rising distribution costs it has become very important for the manufacturer to know the buying habits of the people and to use the means spent on distribution most effectively. Rather than base judgments on hunches and go out of business, it has become necessary for those engaged in various forms of economic activity to seek out scientific techniques to make a survey of the market. These techniques go by the name of *marketing research* —the collection, recording, analysis, and reporting of all facts relating to the transfer and sale of goods and services from producer to consumer. Some of the techniques which employ the sampling method are considered here (1).

PRODUCT TESTING

When a new product or a new brand of a product is contemplated by the company, it is important to investigate its acceptance by the public before production begins. Information is needed on the size and the composition of the market and the price of the product which will bring the maximum profit. Product testing is usually done in two ways. The product, such as soap, is neutrally packed and presented to a sample of likely prospective buyers, e.g., housewives. A competitive product, similarly packed, is added for comparison. After a period of time the members of the sample are visited again to note their reactions. In the other method, one or more test towns are selected in which a cross section of retailers is requested to sell the product. In one of the towns an advertising campaign is launched to measure its effect on sales. By this method it is possible to estimate the size of the market for the proposed product. The disadvantage of the method is that it can take several months before the results of the test can be known.

PACKAGE TESTING

In modern society the intrinsic worth of a product is not enough inducement to sell it to consumers. The design of the package has considerable influence on sales. Before adopting it finally, it is important to test the design in order to determine the best form, size, color and instructions for use and the advertising message on the package. Package testing can be done in a number of ways. A sample of the prospective buyers can be shown the product in several different packages and their reactions noted. Alternatively, the product in various packages is sold by retailers in a selected sample of towns. By ascertaining the numbers of each package sold in a given period, it is possible to find the best seller.

SALES RESEARCH

The manufacturer of a product would like to have information on the distributive side of his product. The number of retailers stocking his product, the number not stocking it, and the area in which it is strong or weak are some of the items in which he is vitally interested. He would also like to know the number of units sold during a given time period, the reaction of retailers to the product, and the important outlets for its distribution. Some information can be collected by taking a sample of retailers and asking their opinion of the product. A more intensive approach is through what is called *retail audit research*.† In this method a sample of retailers is taken and requested to allow specially trained investigators to make an inventory of the stock at the beginning and at the end of the survey period. When the amount of goods delivered to the retailer during the period is known, it is possible to estimate the sales of various products. Normally only those products are considered which can be identified fairly easily. This type of research is too expensive to be carried out by an individual company. Usually market research organizations undertake this research on behalf of their clients. These organizations employ qualified investigators who can differentiate one product from another with accuracy. Since there can be considerable upheaval in the selected shop as a result of inventory taking, it becomes difficult to persuade the shopkeepers to remain in the sample for long. The method, when successfully applied, can provide sufficiently accurate and useful information, even when a very small panel of 100 to 200 shops is employed.

CONSUMER RESEARCH

The manufacturer is vitally interested in knowing who are the final consumers of his product, their characteristics by age-group, social class, size of town where they live, and so on. He wants to know how frequently consumers buy his product and what use is made of it. The characteristics of those who do not buy his product and the brands preferred by them are also of interest to him. This type of information is best collected on a continuing basis by the panel method. A random sample of consumers is selected from the population and the desired information is sought from them at regular

† See Sec. 7.6. The method is also called *store audit*. The period of time for which sales are determined is usually two months.

intervals. In this manner it becomes possible to study trends of behavior of the consumers and possibly the reasons that lead them to change. By retaining the same sample over time, the sampling errors of estimates of change can be reduced considerably. But the main difficulty with panels is that it is impossible to win their cooperation over an extended period. And those who remain on the panel are known to become different from the rest by virtue of exposure to the survey for a long time.

COPY TESTING

This type of research is carried out to determine the effectiveness of the advertising message. The appeal of an advertisement depends on its size, copy, illustrations used, colors, and the position it occupies on the page. The job of the market researcher is to present the copy in such a manner that it attracts the widest attention. There are several methods of testing the effectiveness of the copy. A random sample of the readers of the newspaper or magazine in which the advertisement appeared is asked which advertisements they have seen and whether they can describe them. They are then asked whether a particular advertisement was noticed and, if so, what part or parts are remembered. In this way the attention value of a large number of advertisements can be compared. In another method, the same advertisement is placed in various positions in successive issues of the same newspaper or magazine. By asking a sample of readers the attention value of the various positions can be determined. A variation of the method is to make a hidden offer within the copy to supply a free sample or booklet and see how many people take action on it.

PUBLIC RELATIONS RESEARCH

The standing of the company with the public is an important determinant of its sales. Market research can be used to find what people think of the company, how many people are aware of the company and its products, and what methods should be used to influence people in its favor. The usual method used for this purpose is the public opinion poll (Sec. 20.3). A sample of the population is taken and their views ascertained by asking suitable questions. The method differs from those discussed before in that it aims to discover what people think about problems of public interest. It is not concerned with the behavior of people with respect to goods and services.

READERSHIP RESEARCH

Advertisers spend considerable sums of money in keeping the market informed of the existence of a product and of its attributes. It is very important for them to find the best form of an advertisement and where advertising space is to be bought. The aim of readership research is to indicate to the advertiser how many people of the type likely to be interested in his product may see his advertisement. This is done by finding from a sample of the population (1) the number of people who will be likely to see a single issue of every available publication, (2) the characteristics of readers of publications with which the advertiser may be concerned, (3) the number of people who will see any one or more of any group of publications, and (4) the number of people who will have seen any one issue of any single publication over a period of time.

An important question to decide is the definition of readership. A reader may be defined as one who reads or looks through a newspaper on the day prior to the interview. Or, reading may be defined as the conscious looking through or reading of more than one page of the specified issue of a paper. The latter definition is designed to exclude the casual readers of headlines and pictures. Alternatively, the respondent may be asked to give evidence that he has seen a particular issue of the paper and read it.† Different definitions may produce different results. An example is provided by two British surveys of readership (21). In the IPA‡ survey of 1954 a probability sample of 16,594 adults was asked a set of questions about each of some 80 publications including daily newspapers and weekly and monthly magazines. The opening questions were "Do you happen to have looked at a copy of the (name of publication) any time over the past six months? When did you look at its last?" Memory aids were used for all the publications to obtain a complete response. In the Hulton survey of 1955 a quota sample of 11,592 adults was asked questions of the type: "Which daily morning newspaper did you read yesterday?" Memory aids were used for weekly and monthly publications only. Some of the results obtained are given in Table 20.12. It will be seen from the table that there is broad agreement on the proportion of adults claiming to have read at least one of the publications in the three groups considered. But there are considerable differences between the average number of daily and Sunday newspapers read by the respondents. The IPA estimates are consistently higher than the Hulton estimates for these two groups. The nonuse of memory aids and the different definitions of readership employed in the two surveys may have caused these differences. A further contributory factor might have been the intensive questioning used in the IPA survey.

That the use of the checklist and the open-response system produce different results is shown from a British inquiry (2). A sample of persons was given a questionnaire asking whether they happened to look at any of

† For weekly magazines the standard practice used in the United States is known as the *recognition technique*. The respondent in an interview is presented with a five-week-old issue of the magazine under study. He is shown practically all of the editorial matter and then asked if he read that particular issue. The reason for using a five-week-old issue is that for all practical purposes an issue will generate its total audience before it is five weeks old.

‡ Institute of Practitioners in Advertising.

Table 20.12 Comparison of Hulton and IPA surveys of British readership, 1954–55

	Readers per hundred adult population				*Average number of publications read per reader*			
	Hulton		*IPA*		*Hulton*		*IPA*	
	Males	*Females*	*Males*	*Females*	*Males*	*Females*	*Males*	*Females*
Daily newspapers	82.5	76.4	84.0	78.0	1.25	1.20	1.68	1.43
Sunday newspapers	92.6	89.6	92.0	89.0	2.05	1.99	2.40	2.23
Weekly magazines	76.4	69.9	74.0	71.0	2.09	1.87	2.00	1.83

the listed newspapers or magazines yesterday. Another sample was asked whether they happened to look at any newspapers or magazines yesterday and were required to write down the name of each that they definitely looked at. The experiment was tried on four separate samples, each sample containing 200 to 300 persons. Table 20.13 gives the average number of publications mentioned in the checklist that were looked at by the respondents.

Table 20.13 Average number of publications looked at "yesterday" based on the checklist method and the open-response system

Type of publication	*Checklist*	*Open response*	*Recall ratio*
Daily newspapers	1.52	1.12	0.73
Sunday newspapers	1.72	1.23	0.71
Weekly and monthly magazines	0.88	0.22	0.25

It is clear from the table that the items specified in the checklist got much higher mention when listed than when it was left to the respondent to volunteer them. Furthermore, the recall ratio was much lower with respect to weekly and monthly magazines than for daily and Sunday newspapers.

National readership surveys are usually based on multistage sampling. A good example of the sample design of such a survey is the 1952 readership survey carried out by the British Market Research Bureau (14). In this survey the population covered was all civilians aged 16 or over. Apart from the large towns which were taken into the sample with certainty, a sample of 90 administrative districts was selected from about 1,500 administrative districts in the country. Before sample selection the administrative districts were stratified by geographical region. Within a region the districts were classified as rural or urban and were arranged in descending order of industrialization index—the ratio of industrial to total rateable value in the district. A systematic sample of districts was selected from each stratum with probability proportional to 1951 population. From a selected district a sample of four polling stations was selected with probability proportional to electoral population. A systematic sample of names was selected from the electoral register of each polling station in the sample. This gave a sample of households selected with probability proportional to the number of electors in it listed on the register. The number of households to be selected from a polling district was so determined that a uniform sampling rate applied to each stratum. Within a selected household an adult was taken at random for interview. In all about 3,600 interviews were attempted.

These surveys raise a number of questions. Is the response affected by the order in which questions on daily, weekly, and monthly publications are asked? Does the insertion of an advertisement in subsequent issues of a newspaper increase the potential audience for it? How does the position of the advertisement affect the audience? What is the effect of the design and layout of the advertisement on its ability to attract attention and stimulate action? The methods of sampling and experimental design can be used effectively to answer these questions.

RADIO AND TELEVISION RESEARCH

In view of the great impact the mass media of radio and television make on the people, advertisers have shown considerable interest in this type of research. The purpose is to estimate the size of the audience, their preferences for various types of programs, and the impact of the advertising messages broadcast on the audiences. A great deal of research has been done in the United States and the United Kingdom on television audience. There are four basic forms of research which are described below.

Electronic devices Electronic devices (audimeters or tammeters) are fitted to a sample of television receivers in the area under study. These devices record whether the set is switched on and to which channel or station it is tuned for each separate minute of transmission time. From this it becomes possible to compute the percentage of homes tuned to a given program minute by minute and the total number of sets which have been tuned in. It is also possible to determine the extent to which programs of different agencies are viewed in relation to one another. These devices, however, simply record the fact that the set is operating and make no allowance for the possibility that the room in which the set is installed is empty or its occupants are engaged in an activity other than viewing television. In order to get information of the latter type, further devices (called *recordimeters*) are used which remind the sample households to record the characteristics of the individuals viewing the transmission at the time of the reminder. These records are to be made on specially designed forms left with the household.

Record panels In this method record books are left with the sample households to record from time to time throughout a given period the program being viewed and the characteristics of the audience viewing it. The method is cheap but suffers from the same defects as the method of recording household expenditure (Sec. 15.5). For the first few days records may be kept accurately in considerable detail but the quality declines thereafter as a result of fatigue. As a check, *recordimeters* may be used which make possible daily recording of the total duration for which the set is switched on at all. This allows some quality control of record keeping and acts as a reminder to households in their record-keeping practice.

Interviews involving recall A sample of households is selected and interviewed by investigators in order to discover which programs were viewed on the day before the interview or for some other time period. A checklist may be used as an aid to memory. The investigation can be combined with questions as to the buying habits of viewers and nonviewers and their brand preferences. In this manner the impact of television advertising can be assessed.

Coincidental interviews A number of investigators call on a predetermined sample of households all at the same time. They carry out a personal check on the station to which the set is tuned and note the characteristics of the viewers in the room at the time. A few other questions can also be asked regarding the preceding program. The method has several limitations. The size of the sample is limited by the number of trained interviewers available. Considerable attrition of the sample is involved since many households may have no television set or may not be watching television when the investigator calls or may be away from home. While the method should provide accurate program ratings, the nature of the call often renders other information unreliable due to the reluctance of the person to be parted from the set for long. However, the method has the advantage of speed in that program ratings can be made available fairly quickly. In the United States a great deal of this type of research is carried out by telephone.†

COMPARISON OF METHODS OF MEASURING TELEVISION AUDIENCE

Several combinations of the devices mentioned in the previous section can be used. For example, meter-controlled record panels may be employed on a continuous basis providing 15-minute ratings (percentage of sets on or of people viewing). This is the method principally used in the United States. In this method members of the household and their guests record in the weekly record their presence in the room by 15-minute time periods; the meter records automatically minute by minute whether the television set is switched

† See Sec. 7.8.

on and to which channel it is tuned. Other techniques of interest are non-meter one-week record panels, coincidental interviewing, and interviewing by recall. The main question of practical importance is to what extent these different techniques compare with the meter-controlled record-panel method—a method currently used as a matter of routine. It can be said that the precise timing of the interview is crucial in the case of coincidental interviewing and so is reporting by the person directly concerned. For the success of the method of interviewing involving recall, the standard of interviewing and the memory of the respondent are crucial factors. With record panels the response rate may be no higher than 40 percent and there is the danger of conditioning. Also, daily records may not be completed at the time of viewing or by every household member individually.

A number of studies have been made in the United States and Great Britain to compare audience ratings produced by these methods. In one of the studies (5), meter-controlled record panels were compared with recordimeter records and with coincidental interviewing. The differences observed are given in Table 20.14. The table shows that there are no great differences between the three methods. It is true that coincidental ratings tend to be a little lower than meter-record ratings in the latter period when the rating levels are high. But this may be due to the fact that some people do not like to answer the door when they are watching television.

Table 20.14 Comparison of meter records, recordimeter records, and coincidental interviews, adult ratings, TAM* 1961

	Meter records minus recordimeter records		*Meter records minus coincidental interviews*	
	Two hours starting		*Two hours starting*	
	5:30 P.M.	7:30 P.M.	5:30 P.M.	7:30 P.M.
Tuesday	−1	2	0	6
Thursday	0	0	−1	4
Sunday	0	1	−1	3
Average	0	1	−1	4

* Television Audience Measurement, Ltd.

In another study (2), percentage distributions of persons seeing different numbers of television programs were compared for two methods of interview—one using a checklist of programs and the other based on the open-response system. Four subsamples were used in the investigation. The results given in Table 20.15 show that the aided-recall method produces a higher average number than the open-response system.

Table 20.15 Percentage distribution of television programs seen "yesterday" based on the checklist method and the open-response system

Number of programs seen	*Group 1*		*Group 2*		*Group 3*		*Group 4*	
	Check-list	*Open response*	*Check-list*	*Open response*	*Check-list*	*Open response*	*Check-list*	*Open response*
0	42	59	44	52	39	63	43	68
1	17	23	9	14	12	12	13	18
2	15	13	11	20	7	16	14	10
3+	26	5	36	14	42	9	30	4
Total	100	100	100	100	100	100	100	100
Sample size	99	110	148	140	142	135	158	140
Sample mean	1.37	0.63	1.97	1.05	2.08	0.71	1.60	0.49

Another finding worth reporting is that one-week self-completion records placed by telephone and mail, as used by the American Research Bureau, have given consistent results compared with meter-panel ratings (13). In Great Britain seven-day aided-recall interviews have produced results not very different from meter-record ratings (25). Furthermore, the length of the recall period appears to exert no material influence on the results (5). The burden of this research is that the different methods of measuring television audience appear to give comparable results.

20.5 PAYROLL SURVEYS

The payroll is the bookkeeping record that establishments maintain to list each worker and show the number of hours he worked and the total pay he earned. The statistics compiled from payroll records can be used to assess the progress made in each industry by way of the amount of labor used and the average earnings of the workers. The emphasis is on trends; absolute figures are of secondary interest. For this reason, published figures are usually in the form of index numbers. Table 20.16 presents an example (11).

While abstracting data from payrolls, a distinction is made between wage earners and salaried employees. The idea is to differentiate between employees who are directly associated with the productive activities of the establishment and those who are not. For each category of workers, information is collected on the number of hours that are paid for. Data on the number of hours actually worked is obtained by excluding from the number

Table 20.16 Index numbers of wage rates in Canada (for the month of October each year)

Year	*Mining*	*Manufacturing*	*Construction*	*Total*
1939	100.0	100.0	100.0	100.0
1940	102.5	104.3	104.5	103.9
1941	111.2	115.2	111.6	113.1
1942	116.6	125.5	118.6	122.5
1943	123.7	136.8	127.7	133.7
1944	134.8	141.4	129.6	137.9
1945	136.5	146.5	131.1	141.8
1946	140.6	161.5	143.9	155.2
1947	161.7	183.3	155.0	173.7
1948	182.1	206.4	176.3	196.3

of hours paid for the time spent on paid vacation, sick leave, holidays, and so on. Wherever bookkeeping practices permit, earnings data are separated to show that part of the wages and salaries that is deducted by the employer.

Several useful estimates can be made from the payroll figures. By dividing the number of man-hours by the number of wage earners, the average weekly hours can be obtained. Payrolls when divided by man-hours give average hourly earnings, from which average weekly earnings can be computed.

Although a payroll survey provides information on employment, hours, and earnings at one particular point of time, the major use of such surveys consists in providing trends of these items over time. Therefore, payroll surveys are usually conducted on a continuing basis. As a minimum, a payroll survey should be quarterly. Annual surveys cannot provide information on seasonal changes. The reference period in a quarterly survey may be a period of 13 weeks, 1 month, 4 weeks, or 1 week. The 13-week or 4-week period is preferred to the quarter or the month as these periods eliminate the effect of changes in the number of working days. The record-keeping practices of establishments are studied in order to determine the reference period to be used. The reason is that an establishment can easily supply information for the entire payroll period but not for an arbitrary period.

Another useful item of data that can be collected along with payroll studies is labor turnover, that is, the gross movement of workers into and out of employment. While figures on employment show the net changes in employment from period to period, labor turnover shows the gross volume of changes and hence serves as an indicator of economic health. Two broad categories are distinguished in collecting statistics on labor turnover. One category is accessions, that is, all permanent or temporary additions to the employment roll. The other is separations, that is, terminations of employment of workers who have left or been taken off the rolls for reasons such as retirement, death, and physical disability. With the number of people in pay status during the payroll period as the base, turnover rates such as the number of separations per 100 employees or the number of accessions per 100 employees can be calculated.

The selection of a sample of establishments from a classified directory which is kept up-to-date presents no new problems. These topics have already been discussed in considerable detail in Secs. 17.8 to 17.12.

20.6 POSTAL TRAFFIC SURVEYS

Another field in which sampling methods can play an important part is postal traffic. Surveys of postal traffic permit the administration to have fuller information on which to base their operations. The purpose is to estimate the total volume of postal traffic and the different categories such as first-class mail and book mail parcels handled by the post offices in the country. At the present moment there is no uniformity of practice in the matter of obtaining postal statistics. While some countries count postal articles for one week in the whole year, there are others where this is done once every week. Furthermore, the offices at which counting takes place are not selected following probability sampling and the same applies to the periods for which the data are collected.

The practice of selecting a period believed to be representative is not without its hazards. The method is subject to unknown biases. Experience at different places has shown that there is no evidence of pronounced periodicity of traffic within the week. Therefore, it appears safe to take a systematic sample of days spread over the whole year for the count of postal articles. When data are needed by quarter of the year, the sampling interval of 13 days appears to be a good choice. This means that counting of postal articles takes place at selected post offices every thirteenth day. The days, however, should differ for the different groups of post offices selected in the survey. At the very large post offices it may not be found practicable to count all articles coming in on the selected day. In that case the articles may be weighed by category or every kth bag can be enumerated completely. Norms can be set up through properly planned investigations as to the average weight of articles in the various categories. This information can be used for determining the number of articles.

A list of post offices all over the country is always available. Sometimes the number of persons employed at each post office is also known. When this information is available, the post offices can be stratified by size and a random or systematic sample selected from each stratum. If a countrywide count of postal articles has taken place in the past, this information can be used for finding the strata means and standard deviations and for determining the sample allocation to strata. Consider, for example, Table 20.17, which gives details of a sample selected in Greece. The total sample size is determined in order to achieve a coefficient of variation of 2 percent for the estimate of the total number of letters handled per quarter.

Table 20.17 Sample selection, postal traffic survey, Greece

Stratum based on number of employees	*Number of post offices*		*Average number of letters per post office*	*Standard deviation of number of letters*
	Stratum	*Sample*		
1–2	127	7	414	396
3	190	11	627	450
4–5	181	18	1,047	726
6–9	72	13	2,188	1,370
10–13	23	4	4,095	1,426
14–22	36	17	8,665	3,543
More than 22	32	32	60,914	
Total	661	102	4,296	

20.7 FOOD CONSUMPTION SURVEYS

In surveys of consumer expenditure discussed in Chap. 15, the emphasis is on the monetary value of household consumption. In food consumption surveys information is collected on the quantity and nutritional value of food consumed by households. Analysis of data from such surveys indicates the nutrient content of the food consumed for comparison with given nutritional standards. In order to collect usable information, the food intended for consumption is actually weighed for a specified period. Therefore such surveys are conducted on a small sample of households. When the survey is continuous it provides information on changes in the dietary habits of people and the factors influencing them.

The housewife, who is the caterer to the household, is asked to make a record of the foods entering the kitchen. The energy value and nutrient content of the recorded quantities are evaluated by means of appropriate tables of food consumption. Usually these tables are designed for direct application to the weights of food as they enter the home and make an allowance for the inedible part of the food such as bones, and skins of fruits and vegetables. From this the distribution of the proportion of calories derived from different types of food can be obtained. Table 20.18 is an example of this distribution (10). The table shows that the British people derived most of their calories from bread, flour, meats, and fats.

Table 20.18 Energy value of food obtained for domestic consumption, United Kingdom, 1958

Food	*Calories per head per day*	*Percentage of total*
Milk and cheese	315	12
Meat	381	15
Fish	24	1
Eggs	50	2
Fats	386	15
Sugar	322	12
Vegetables	184	7
Fruit	49	2
Bread and flour	587	23
Other cereals	268	10
Other foods	29	1
Total	2,595	100

Data on the calorie content of the food consumed by households can also be used for obtaining the distribution of households by calorie intake. An example of this distribution is given in Table 20.19. In this survey 369 Burmese households were visited twice daily at the time of preparation for cooking of the principal meals. An allowance was made for meals taken by guests and for members absent from the household (7). If the calorie requirement per household in Burma is assumed to be 2,700, it is evident that about two-thirds of the households fell short of the requirement.

Table 20.19 Distribution of households by calorie intake, Burma, 1948

Calories	*Percent frequency*	*Calories*	*Percent frequency*
Under 1,300	0.3	2,900–	10.4
1,300–	5.2	3,300–	5.7
1,700–	20.3	3,700–	3.3
2,100–	29.4	4,100–	1.0
2,500–	23.9	4,500–	0.5

Many of the difficulties encountered in consumer expenditure surveys apply to food consumption surveys as well. A question of particular relevance to these surveys is that of wastage. It is possible to estimate the quantity of food prepared but not the amount wasted. Probably the smaller families waste more than the larger families. But accurate measurement is difficult. Any attempt to record wastage means drawing the attention of the household to it. Once wastage has been brought to the notice of the housewife, she is likely to reduce it. In some countries a reduction of 10 percent is made for wastage irrespective of the size or type of household. Sometimes a graduated scale of reduction is used, the reduction being as high as 20 percent for the smaller families. Whether or not the estimate of wastage is appropriate is usually judged by a comparison of the energy value of consumption with estimated requirements.

Ascertainment of energy requirements is a complex problem. The amount of energy needed by a person depends on several factors such as age, sex, weight, occupation, and climate. It is impossible to take all these factors into consideration for all persons in the survey in a manner that will be universally acceptable. Usually the scales of nutritional requirements are devised to fit adults of both sexes of average weight and various age-groups. The FAO scale, for example, is based on an approach involving the use of a defined "reference" man and woman aged between 20 and 30 years, living in a climate with mean annual temperature of 10°C, with a weight of 65 kg for the reference man and 55 for the reference woman (6). Both man and woman are supposed to live a healthy and active life. The average number of calories needed per day by the reference man is 3,200 and by the reference woman 2,300. The reference requirement scale is given in Table 20.20. The application of this table in practice is limited by the insufficiency of information on the different determining factors. But, with the knowledge available at present, this is probably the best method (23) that can be used.

Table 20.20 FAO reference requirement scale (number of calories per caput per day)

Age-group	*Male*	*Female*	*Age-group*	*Male*	*Female*
Under 1	1,120	1,120	20–29	3,200	2,300
1–3	1,300	1,300	30–39	3,104	2,211
4–6	1,700	1,700	40–49	3,008	2,162
7–9	2,100	2,100	50–59	2,768	1,990
10–12	2,500	2,400	60–69	2,528	1,817
13–15	3,100	2,600	Over 69	2,208	1,587
16–19	3,600	2,400			

20.8 SOCIOLOGICAL RESEARCH

Surveys encountered in fields such as industry, trade, and agriculture are mostly descriptive in nature. In sociological research, interest centers both on description and explanation (Sec. 1.7). The questions asked vary with respect to the point in time to which they refer. Some relate to the present (see column 1 of Table 20.21), some about the immediate past (columns 2 and 3), and others about the distant past (column 4). Given this body of data, a number of operations can be performed. The distribution of the answers to each individual question can be tabulated. This operation gives what are called *marginal distributions.* Again, the distributions of answers to questions located at one point of time (e.g., A_1, A_2, or C_2, C_3) may be tabulated to study *time-bound association,* that is, whether or not the responses tend to covary. Furthermore, the distributions of answers to questions in different time dimensions (e.g., A_1, B_2, or B_2, C_4) may be related to study the causal relationship—*time-ordered association* (9).

Table 20.21 Responses to questions referring to different points in time

	Time dimension		
1	2	3	4
A	B	C	D
A_1	B_1	C_1	D_1
A_2	B_2	C_2	D_2
—	—	—	—

The goal is not only to discover the strength of association but to learn what brought it about. The manner in which sociologists try to determine whether the relationship between two variables, say, A and B, is causal is the following. If A and B are found to be descriptively related, a search is made for variables antecedent to both A and B (e.g., C, D), which might explain away the original relationship. The result of this procedure may be explanation or

Table 20.22 Percentage distributions exhibiting explanation of relation between A and B by C

	Original relation		*Partial relations based on* C		$\bar{C}$	
	B	$\bar{B}$	B	$\bar{B}$	B	$\bar{B}$
A	70	50	75	75	35	35
Ā	30	50	25	25	65	65
Total	100	100	100	100	100	100

replication. Explanation occurs when an antecedent third variable explains away the original relation. Replication occurs when the original relationship is repeated after the third variable is taken into account. The original relation between A and B is explained away by C, if the relation disappears in the two partial tables—one in which C is present and the other in which it is not ($\bar{C}$). Table 20.22 gives an example of this phenomenon.

The example shows that the original relation between A and B is spurious and that the joint distribution of A and B is produced by the antecedent variable C. We shall say that the variables A and B are descriptively related but the relationship is not causal. The hypothesis that A and B are causally related is said to have been confirmed when replication is the result of introducing a third variable, say D. An example of this is provided by Table 20.23.

Table 20.23 Percentage distributions exhibiting replication of relation between A and B

	Original relation		*Partial relations based on* D		$\bar{D}$	
	B	$\bar{B}$	B	$\bar{B}$	B	$\bar{B}$
A	70	50	65	55	60	45
Ā	30	50	35	45	40	55
Total	100	100	100	100	100	100

Here the original relation is more or less repeated in the partial tables. The variable D does not explain away the relation between A and B. If the same result is obtained by introducing all other variables which are believed to prove the relation to be spurious, it might be said that the relation between A and B is found to be causal.

CROSS-SECTIONAL SURVEY

The most commonly used design in sociological work is the cross-sectional survey. Here standardized information is collected from a sample by interview or by mail or otherwise. The data which are produced by the survey comprise the answers to the questions which the subjects of the survey have been asked. The logic of survey analysis, however, has been developed on the assumption that the time order of data can be established (see Table 20.21). This is not always so. Consider, for example, the data collected by an investigator on his subjects' attitudes toward a tribe and the amount of their social contact with it. Suppose the investigator wants to assess whether or not attitudes are improved with greater social contact. Such an assessment may not be possible because of the ambiguity of the time order of the two measurements. It is difficult to say which came first—attitudes or contact. The extent of the covariation of the two measures (i.e., time-bound association) can be studied but it cannot be established that one factor led to the other. There is, however, no point in establishing causal relationship if it is found that there is no covariation between the two factors.

CONTEXTUAL DESIGNS

Sociologists are considerably interested in studying social contexts and their influence on individual behavior. A probability sample of persons selected in a cross-sectional survey may not be adequate for this purpose. The reason is that individuals cannot give complete information about the social structure of the context in which they live or work. A contextual design is more appropriate for such studies. This design is usually based on a multistage sample in which the context is one of the stages. As an example, a contextual sample for the study of the effect of social class composition of schools on the aspirations of students can be selected by stratifying the schools, selecting a probability sample of schools within each stratum, and a sample of students from each selected school. By taking an adequate number of students, it is possible to obtain a statistical description of the context and to construct contextual variables. Since this design provides data on both the individual and his context, it is possible to study the joint influence of the individual's characteristics and of the contextual characteristics on the individual's behavior. Again, this kind of design can also be used for analysis at the level of the context since it provides a probability sample of the contexts.

SOCIOMETRIC DESIGNS

Another problem in sociology is the study of networks of interpersonal relations and their influence on behavior. For example, it is of interest to study the manner in which members of a group interact to produce the social structure of the group. The cross-sectional survey is not well-suited to such problems as the possibility is remote that an unrestricted random sample of persons will know each other and be interacting with each other. A more appropriate vehicle is the contextual design in which information is collected from all persons in a given context in order to be able to describe the network of interpersonal relations and study their influence on individual behavior. Such designs are called sociometric designs.

To illustrate a sociometric design, suppose it is desired to study the processes by which physicians come to adopt innovations in medical practice. A number of factors are believed to influence the physician's adoption practices. His background and training, nature of practice, exposure to information about the innovation, and his place in the social structure of the medical community are some of the variables to be considered. In order to take into account the status of physicians in the medical community, it is important to make use of a design in which all physicians serving patients in selected communities are interviewed. Information on the relative standing of the physicians in the medical community is collected by asking each physician sociometric questions about his interaction with other physicians.

PANEL STUDIES

The basic problem in sociology is the study of social change and the social processes which underlie it. This can best be done with the help of panel studies in which the same sample is investigated over time. In this manner it is possible to establish the time order of events with considerable precision and thus undertake explanatory analysis. Another use of panels is the possibility of detection of gross as well as net changes in the population. This point can be made clear with the help of an artificial example presented in Table 20.24. A sample of individuals is interviewed both in July and August to estimate the proportion supporting the two candidates *A* and *B*. The marginal totals show a net shift of four percentage points. The entries in the body of the table, however, tell a more detailed story. Only 80 percent of the individuals were consistent in their support of the candidates and 20 percent shifted their support.

Table 20.24 Percentage of voters favoring candidates *A* and *B*

		July		
		A	*B*	*Total*
August	*A*	40	12	52
	B	8	40	48
	Total	48	52	100

If independent samples had been taken in the two months, the net change could have been estimated but not the gross change.

By making use of differentiated description, panel studies can be used for inquiring into a number of questions concerning the dynamics of change. To pursue the example in the previous paragraph, one may determine the kinds of people most likely to change and the direction of change, the conditions under which changes come about, the relationship between present attitudes and future behavior and the variables that are likely to change simultaneously over time. But only short-term changes can be studied for it is difficult to persuade people to remain on the panel over an extended period of time (Sec. 6.2) and to locate them at successive interview points.

REFERENCES

1. Adler, M. K. (1959), "Modern Market Research," Crosby Lockwood & Son, Ltd., London.
2. Belson, W., and J. A. Duncan (1962), A Comparison of the Check-list and the Open Response Questioning Systems, *Appl. Statist.*, **11**.
3. Cantril, H. (1947), "Gauging Public Opinion," Princeton University Press, Princeton, N.J.
4. Crossley, A. M., and H. M. Crossley (1969), Polling in 1968, *Pub. Opinion Quart.*, **33**.
5. Ehrenberg, A. S. C., and W. A. Twyman (1967), On Measuring Television Audiences, *J. Roy. Statist. Soc.*, (A)**130**.
6. Food and Agriculture Organization (1957), "Calorie Requirements," Rome.
7. Gale, U. M. (1948), "Report on the Dietary and Nutritional Surveys Conducted in Certain Areas of Burma," Government of Burma, Rangoon.
8. Gallup, G., and S. F. Rae (1968), "The Pulse of Democracy," Greenwood Press, New York.
9. Glock, C. Y. (1967), "Survey Research in the Social Sciences," Russell Sage Foundation, New York.
10. Hollingsworth, D. F., et al. (1961), A Survey of Food Consumption in Great Britain, in "Family Living Studies," International Labor Office, Geneva.
11. International Labor Office (1949), "Wages and Payroll Statistics," New Ser., no 16, Geneva.
12. Kish, L. (1949), A Procedure for Objective Respondent Selection within the Household, *J. Am. Statist. Assoc.*, **44**.
13. Mayer, M. (1966), "How Good Are Television Ratings?" Television Information Office, New York.
14. Moser, C. A. (1958), "Survey Methods in Social Investigation," Heinemann, London.
15. Moser, C. A., and A. Stuart (1953), An Experimental Study of Quota Sampling, *J. Roy. Statist. Soc.*, (A)**116**.
16. Mosteller, F., et al. (1949), "The Pre-election Polls of 1948," *Soc. Sci. Res. Council Bull.* 60, New York.
17. Netherlands Central Bureau of Statistics (1963), "Statistical and Econometric Studies: 3rd and 4th Quarters of 1962," W. Dc. Haan, Ltd., Zcist, Netherlands.
18. Neyman, J. (1934), On the Two Different Aspects of the Representative Method: The Method of Stratified Sampling and the Method of Purposive Selection, *J. Roy. Statist. Soc.*, **97**,
19. Parry, H. J., and H. M. Crossley (1950), Validity of Responses to Survey Questions, *Pub. Opinion Quart.*, **14**.
20. Roper, E. (1940), Sampling Public Opinion, *J. Am. Statist. Assoc.*, **35**.
21. Shankleman, E. (1955), Measuring the Readership of Newspapers and Magazines, *Appl. Statist.*, **4**.
22. Stephan, F. F., and P. J. McCarthy (1958), "Sampling Opinions," John Wiley & Sons, Inc., New York.
23. Sukhatme, P. V. (1961), The World's Hunger and Future Needs in Food Supplies, *J. Roy. Statist. Soc.*, (A)**124**.
24. Tanner, J. C. (1957), The Sampling of Road Traffic, *Appl. Statist.*, **6**.

25. Television Audience Measurement (1961), "Comparison Survey of Audience Composition Techniques," Ltd., London.
26. United Nations Economic and Social Council (1950), "Report to the Statistical Commission on the 4th Session of the Subcommission on Statistical Sampling," New York.
27. Vos, J. W. E. (1964), Sampling in Space and Time, *Rev. Intern. Statist. Inst.*, **32**.

appendix 1

International Standard Industrial Classification of All Economic Activities

List of Major Divisions, Divisions, and Major Groups

Major division	*Division*	*Major group*	
1			AGRICULTURE, HUNTING, FORESTRY, AND FISHING
	11		Agriculture and Hunting
		111	Agriculture and livestock production
		112	Agricultural services
		113	Hunting, trapping, and game propagation
	12		Forestry and Logging
		121	Forestry
		122	Logging
	13	130	Fishing

Major division	*Division*	*Major group*	
2			MINING AND QUARRYING
	21	210	Coal Mining
	22	220	Crude Petroleum and Natural Gas Production
	23	230	Metal Ore Mining
	29	290	Other Mining
3			MANUFACTURING
	31		Manufacture of Food, Beverages, and Tobacco
		311–312	Food manufacturing
		313	Beverage industries
		314	Tobacco manufacture
	32		Textile, Wearing Apparel, and Leather Industries
		321	Manufacture of textiles
		322	Manufacture of wearing apparel, except footwear
		323	Manufacture of leather and products of leather, leather substitutes and fur, except footwear and wearing apparel
		324	Manufacture of footwear, except vulcanized or molded rubber or plastic footwear
	33		Manufacture of Wood and Wood Products, Including Furniture
		331	Manufacture of wood and wood and cork products, except furniture
		332	Manufacture of furniture and fixtures, except primarily of metal
	34		Manufacture of Paper and Paper Products, Printing, and Publishing
		341	Manufacture of paper and paper products
		342	Printing, publishing, and allied industries
	35		Manufacture of Chemicals and Chemical, Petroleum, Coal, Rubber, and Plastic Products
		351	Manufacture of industrial chemicals
		352	Manufacture of other chemical products
		353	Petroleum refineries
		354	Manufacture of miscellaneous products of petroleum and coal
		355	Manufacture of rubber products
		356	Manufacture of plastic products not elsewhere classified
	36		Manufacture of Nonmetallic Mineral Products, except Products of Petroleum and Coal
		361	Manufacture of pottery, china, and earthenware
		362	Manufacture of glass and glass products
		369	Manufacture of other nonmetallic mineral products
	37		Basic Metal Industries
		371	Iron and steel basic industries
		372	Nonferrous metal basic industries

Major division	*Division*	*Major group*	
	38		Manufacture of Fabricated Metal Products, Machinery, and Equipment
		381	Manufacture of fabricated metal products, except machinery and equipment
		382	Manufacture of machinery except electrical
		383	Manufacture of electrical machinery apparatus, appliances, and supplies
		384	Manufacture of transportation equipment
		385	Manufacture of professional and scientific and measuring and controlling equipment not elsewhere classified, and of photographic and optical goods
	39	390	Other Manufacturing Industries
4			ELECTRICITY, GAS, AND WATER
	41	410	Electricity, Gas, and Steam
	42	420	Water Works and Supply
5			CONSTRUCTION
	50	500	Construction
6			WHOLESALE AND RETAIL TRADE AND RESTAURANTS AND HOTELS
	61	610	Wholesale Trade
	62	620	Retail Trade
	63		Restaurants and Hotels
		631	Restaurants, cafés, and other eating and drinking places
		632	Hotels, rooming houses, camps, and other lodging places
7			TRANSPORTATION, STORAGE, AND COMMUNICATION
	71		Transportation and Storage
		711	Land transportation
		712	Water transportation
		713	Air transportation
		719	Services allied to transportation
	72	720	Communication
8			FINANCING, INSURANCE, REAL ESTATE, AND BUSINESS SERVICES
	81	810	Financial Institutions
	82	820	Insurance
	83		Real Estate and Business Services
		831	Real estate
		832	Business services except machinery and equipment rental and leasing
		833	Machinery and equipment rental and leasing

Major division	*Division*	*Major group*	
9			COMMUNITY, SOCIAL, AND PERSONAL SERVICES
	91	910	Public Administration and Defense
	92	920	Sanitary and Similar Services
	93		Social and Related Community Services
		931	Education services
		932	Research and scientific institutes
		933	Medical, dental, and other health and veterinary services
		934	Welfare institutions
		935	Business, professional, and labor associations
		939	Other social and related community services
	94		Recreational and Cultural Services
		941	Motion picture and other entertainment services
		942	Libraries, museums, botanical and zoological gardens, and other cultural services not elsewhere classified
		949	Amusement and recreational services not elsewhere classified
	95		Personal and Household Services
		951	Repair services not elsewhere classified
		952	Laundries, laundry services, and cleaning and dyeing plants
		953	Domestic services
		959	Miscellaneous personal services
	96	960	International and Other Extraterritorial Bodies
0			ACTIVITIES NOT ADEQUATELY DEFINED
	0	000	Activities not adequately defined

appendix 2

Random Numbers

01	02	03	04	05	06	07	08	09	10	11	12	13	14	15
13	70	43	69	38	81	87	42	12	20	41	15	76	96	85
26	99	82	78	99	05	22	99	52	32	80	91	38	51	09
72	53	95	81	07	98	14	74	52	58	73	10	40	91	90
22	08	08	68	37	16	36	62	20	02	35	98	44	53	23
21	61	90	53	85	72	86	94	87	18	50	11	31	25	22
47	38	55	66	50	96	96	78	34	45	52	78	34	35	20
96	68	13	07	31	29	70	09	16	66	81	09	36	12	17
45	92	93	44	87	72	26	75	82	31	72	69	25	51	40
51	99	50	88	62	54	90	51	01	39	18	70	17	20	75
32	31	32	26	03	55	74	15	28	81	04	55	20	72	79
67	62	30	02	88	17	37	25	42	86	00	32	75	57	37
03	08	89	77	12	41	15	25	52	30	93	11	12	47	35
45	10	04	66	94	70	33	74	97	23	40	97	73	67	55
62	48	46	97	04	36	31	27	29	84	85	35	16	02	29
59	59	33	63	53	43	60	30	15	81	67	59	48	98	13

16	17	18	19	20	21	22	23	24	25	26	27	28	29	30
72	63	67	17	24	55	68	32	24	80	13	92	73	65	42
46	28	15	70	28	98	53	36	03	89	83	74	22	96	06
21	03	09	16	31	48	05	10	98	62	14	15	57	26	11
84	82	53	39	92	14	07	84	04	01	66	17	47	76	60
75	68	40	90	39	95	46	10	94	68	39	10	31	80	30
42	77	29	80	73	38	92	11	81	72	50	88	91	55	48
63	55	09	84	66	56	92	13	97	14	87	27	83	70	10
54	29	70	14	85	95	79	72	77	48	57	92	28	35	53
42	97	50	61	19	55	38	55	85	57	85	08	86	91	62
52	30	47	73	26	54	18	05	75	92	95	08	24	86	86
27	81	21	75	39	43	77	80	81	72	55	33	32	54	38
17	41	85	13	20	66	59	22	20	93	15	11	02	14	89
51	74	23	54	88	84	12	16	77	01	89	83	44	23	49
87	91	53	86	97	42	80	83	37	31	97	12	11	84	69
30	16	17	32	34	00	07	25	52	79	77	77	69	76	38
92	81	12	15	28	42	98	67	52	38	30	12	85	98	68
03	83	93	48	64	50	32	57	94	64	87	55	68	72	06
74	85	16	86	09	22	62	06	38	16	74	71	27	69	83
97	36	58	90	91	23	91	19	04	16	31	25	96	65	32
03	85	53	06	41	29	78	51	15	49	01	26	88	45	76
77	67	60	70	44	56	91	03	19	66	19	69	66	27	28
37	15	17	96	24	95	08	39	55	15	33	19	50	98	26
64	16	38	58	74	29	71	49	62	13	29	90	80	93	66
14	16	78	44	49	34	05	46	96	88	74	51	03	39	64
29	19	71	98	71	19	51	86	82	95	83	84	13	02	62
09	39	92	56	68	36	54	55	46	13	58	83	61	66	77
41	55	75	08	62	55	19	15	75	77	74	65	03	42	78
28	98	16	85	39	67	49	02	30	47	55	67	10	59	34
92	22	79	70	66	78	13	97	42	81	54	10	57	42	17
86	08	54	39	88	38	46	74	21	13	74	36	85	52	19
36	26	40	17	70	39	94	05	76	12	98	65	97	74	18
91	20	64	12	33	15	59	43	28	75	88	60	64	80	35
14	30	57	07	34	09	56	26	81	41	14	99	96	72	06
94	83	96	96	17	02	10	89	71	76	53	37	80	03	58
52	67	59	63	22	28	76	43	45	97	87	11	68	57	74
86	17	98	62	44	62	67	18	02	15	79	36	90	21	60
34	17	10	17	43	68	47	09	66	06	96	96	18	97	02
23	83	25	22	31	25	09	32	57	52	88	05	95	82	56
64	42	07	46	19	56	27	48	22	87	41	90	47	81	74
28	01	52	52	24	90	69	59	70	66	73	13	25	50	70

31	32	33	34	35	36	37	38	39	40	41	42	43	44	45
73	27	49	30	71	93	45	23	86	40	53	13	33	09	79
68	96	82	12	70	61	57	03	27	55	72	07	87	82	47
49	27	30	17	92	45	96	75	06	25	10	97	52	41	85
81	91	89	43	17	60	76	59	96	38	96	50	35	50	90
70	11	90	78	54	31	75	14	38	49	91	76	14	39	00
76	60	35	52	47	78	49	74	59	95	14	00	98	70	66
32	04	40	57	05	72	16	19	54	78	61	03	98	31	08
28	04	32	97	32	25	47	51	21	49	54	26	17	95	15
87	23	34	82	02	05	65	10	67	08	62	72	33	16	66
16	41	58	73	76	51	37	53	22	55	41	89	56	21	52
01	13	30	72	03	34	30	08	29	67	58	00	65	94	00
04	66	49	07	80	83	53	28	77	26	75	61	19	20	88
63	72	23	44	91	97	54	05	88	16	45	44	11	17	87
05	81	81	56	04	93	91	92	94	29	49	60	13	73	14
18	80	74	51	53	88	53	90	86	97	55	59	58	01	75
22	64	46	10	22	05	37	98	77	78	83	69	31	75	90
52	48	31	16	69	39	76	94	21	69	90	25	96	40	50
58	85	06	34	69	82	18	68	65	29	90	78	29	64	83
87	04	04	58	00	78	09	82	57	19	43	05	45	99	21
85	96	61	47	26	03	43	29	11	72	73	48	08	29	41
25	19	17	50	50	46	26	92	62	41	27	66	85	60	70
54	61	41	41	91	88	83	30	32	75	59	03	58	58	83
97	50	71	35	65	67	15	45	73	09	17	60	68	38	05
96	17	27	35	82	80	77	28	97	11	26	72	02	88	96
21	48	84	49	72	93	48	66	75	82	36	33	77	97	35
85	12	09	36	72	81	06	73	04	02	03	10	81	34	44
49	57	40	54	64	88	97	69	03	12	94	45	86	74	66
07	43	79	37	60	96	75	39	46	33	42	41	29	83	73
80	07	51	15	59	55	24	80	49	12	61	68	00	44	58
40	71	81	93	03	03	60	02	42	53	38	35	05	67	73
50	24	44	84	14	02	13	95	71	17	46	16	45	72	36
51	36	08	02	99	65	46	51	84	51	20	85	22	94	38
62	81	28	56	90	81	19	95	58	41	50	80	91	11	62
83	33	85	65	91	68	33	17	85	77	15	53	18	87	75
24	05	75	46	93	05	64	39	09	20	73	52	84	82	88
28	40	31	45	53	96	36	84	57	60	99	82	84	93	66
21	23	47	38	68	53	19	50	06	54	28	00	56	78	63
00	78	78	51	53	72	74	90	79	03	63	27	02	60	44
66	96	71	70	61	05	98	64	67	41	35	00	84	20	51
46	24	17	92	11	04	92	17	17	89	52	52	65	59	36

46	47	48	49	50	51	52	53	54	55	56	57	58	59	60
55	69	47	19	10	36	47	63	23	35	15	03	79	56	48
75	17	81	21	31	84	98	99	77	96	71	72	67	99	24
35	04	66	64	83	34	75	18	40	58	65	35	98	48	02
05	83	68	55	63	72	35	53	51	58	26	41	11	16	45
45	48	17	48	46	21	44	18	99	41	51	94	64	83	03
23	99	75	91	68	71	77	35	56	29	30	26	83	97	11
24	50	63	43	58	96	59	56	76	39	86	04	11	71	07
63	99	50	54	63	11	67	06	79	37	52	17	84	84	58
67	42	45	26	06	30	13	96	47	25	54	31	33	77	79
23	73	62	45	06	38	83	49	66	35	06	02	45	45	02
09	98	28	71	65	35	74	94	87	13	79	77	27	40	88
77	97	74	00	04	18	42	69	79	74	66	90	72	83	13
81	33	58	28	11	28	89	53	29	23	21	91	15	16	83
90	47	83	90	03	18	14	38	28	21	72	09	99	32	35
59	01	09	70	86	58	53	77	74	15	15	95	71	50	87
56	96	60	28	80	20	35	34	07	89	78	85	71	49	57
69	65	85	38	62	53	81	97	00	30	43	05	36	04	39
28	23	54	79	10	52	77	02	87	74	17	42	96	00	80
40	71	41	62	18	93	71	73	71	36	10	92	91	99	73
88	83	39	47	97	05	17	97	88	09	38	28	22	06	26
04	96	83	09	23	49	00	32	23	47	09	51	83	36	02
14	72	41	35	69	48	93	00	72	10	63	46	98	63	46
94	02	46	56	58	48	99	48	02	27	38	71	95	05	17
56	04	35	54	37	10	49	21	77	05	99	06	65	11	63
05	22	71	74	19	45	22	30	57	72	47	46	77	14	01
59	61	33	54	59	83	87	10	50	31	66	32	27	95	11
66	87	57	15	87	25	06	59	67	13	45	95	50	73	90
34	42	45	30	64	25	16	53	13	25	88	11	13	23	54
85	41	23	96	62	77	66	47	21	52	93	02	61	15	30
92	08	10	47	20	83	55	04	30	53	42	16	71	13	32
50	01	66	62	10	38	47	86	17	59	64	26	02	36	17
13	35	98	13	29	61	37	85	44	14	96	63	98	71	28
68	37	23	74	77	92	37	14	25	88	78	96	90	90	00
40	43	78	99	64	47	23	89	80	49	91	95	59	60	06
90	37	63	74	14	30	64	66	72	38	19	28	01	63	44
04	84	87	41	64	03	89	57	82	34	07	71	33	49	80
43	95	90	88	46	27	34	43	61	52	24	53	09	84	27
94	20	01	52	38	82	74	59	52	76	29	85	59	84	16
64	04	67	90	38	25	44	69	32	35	04	27	03	98	84
91	89	73	11	07	29	69	79	89	36	79	99	56	05	63

61	62	63	64	65	66	67	68	69	70	71	72	73	74	75
24	43	43	01	91	48	33	23	60	63	87	15	15	27	59
77	67	34	95	86	99	27	54	40	61	32	54	74	63	89
09	91	95	96	96	59	13	33	76	69	65	15	88	82	08
36	59	12	33	44	28	85	77	72	84	23	05	57	14	43
67	03	48	83	77	15	39	38	60	87	93	20	89	37	55
87	07	87	94	15	70	33	87	92	20	44	52	85	28	63
70	83	47	08	44	92	03	01	69	36	54	02	85	92	92
35	61	24	35	08	63	55	43	88	72	23	80	06	83	24
33	90	47	53	07	64	57	02	75	91	23	41	95	06	18
10	86	00	20	21	25	38	66	72	50	88	21	00	24	82
77	56	37	85	54	76	52	11	75	37	11	21	26	61	05
12	87	87	16	41	19	24	45	50	63	61	48	53	29	34
95	03	95	57	45	95	95	29	81	58	75	81	44	15	27
41	99	20	67	81	61	53	17	68	79	40	87	93	89	23
01	80	04	99	83	84	81	35	81	62	49	64	35	74	06
44	08	81	86	13	79	07	55	16	61	27	03	43	78	93
02	96	85	29	01	71	25	75	16	12	96	88	01	45	02
57	32	28	08	02	76	67	87	64	68	02	96	09	50	78
85	58	76	47	47	46	64	91	31	90	58	42	77	01	02
93	26	47	78	64	08	47	06	70	10	57	11	66	84	10
00	02	43	05	54	97	13	33	25	19	54	57	17	23	06
82	74	28	07	31	44	49	00	20	61	44	67	71	86	23
97	46	06	85	53	95	62	24	18	87	88	37	68	12	48
27	95	04	75	66	23	46	61	58	85	84	68	51	43	60
21	55	66	63	67	54	98	39	58	34	31	53	16	63	44
30	29	75	20	04	09	21	69	24	49	43	40	97	36	36
74	52	39	96	51	43	56	08	83	97	62	18	12	93	87
51	47	38	83	16	00	02	97	68	22	58	98	54	82	50
20	45	89	39	70	21	98	91	18	94	52	94	62	30	28
64	35	70	08	57	48	86	77	42	41	25	40	72	51	35
90	52	67	66	55	33	43	44	28	84	95	45	42	09	67
87	61	28	11	65	42	70	83	95	49	82	21	10	06	23
68	97	71	36	56	22	04	71	48	17	20	02	56	25	47
74	29	89	73	99	66	99	43	46	53	89	28	42	36	55
43	75	00	82	43	53	56	70	35	64	16	69	05	11	49
54	62	73	36	69	81	11	26	79	41	02	69	54	26	61
69	38	82	26	28	87	13	61	19	47	08	09	68	85	60
47	41	61	98	00	29	74	10	37	39	14	38	62	28	71
93	45	12	64	87	97	11	76	33	05	38	29	15	92	98
01	36	46	57	62	78	74	99	00	99	05	41	48	21	67

appendix 3

Notations Used

b	sample regression coefficient
CV	coefficient of variation
E	expected value
$\doteq$	approximately equal to
f	sampling fraction
h	stratum h
λ	proportion of the sample common to two occasions
MSE	mean-square error
μ_k	kth moment about the mean
N	number of units in the population
N_g	number of units in the subgroup
n	number of units in the sample
P	population proportion
p	sample proportion
pps	probability proportional to size
Pr	probability

psu	primary sampling unit
p_i'	probability that the unit U_i is selected in the sample
$R = Y/X$	population ratio of y to x
ρ	correlation coefficient in the population
r	sample correlation coefficient
S	summation over all units in the sample
S_y^2	$\sum (Y_i - \bar{Y})^2/(N-1)$
s_y^2	$S(y_i - \bar{y})^2/(n-1)$
σ_y^2	$\sum (Y_i - \bar{Y})^2/N$
$\sigma(\bar{y})$	standard deviation of $\bar{y}$
$\sum$	summation over all units in the population
V	variance
v	estimate of variance
W_h	weight of stratum h
$\bar{x}'$	mean of preliminary sample of size n'
X	population total for x
Y	population total for the character y
$\bar{Y}, M, \mu$	population mean for y
$\bar{y}$	sample mean for y
$\hat{Y}$	an estimator of Y
$\|Z\|$	positive value of Z

appendix 4

Answers to Problems

Chapter 3

3.1. The average number of cattle per farm is 12.50, the relative error being 1.93 percent. The total number of cattle is 258,810. At least 336 farms should be selected in a future survey.

3.2. The average size is 4.67 and its standard error is 0.572. The total number of persons is 1,190 and its standard error is 239.

3.3. (*a*) Proportion is 0.234, standard error 0.013. (*b*) Proportion is 0.298, standard error 0.0182. (*c*) Average size is 3.188. (*d*) Number of persons is 3,920.

3.4. (*a*) Standard deviation is 1.5 percent for $P = 10$ percent, and 2.5 percent for $P = 60$ percent. (*b*) Sample size to be used is 588. (*c*) Sample size is 826. (*d*) Estimate is within 20 percent of P with a high probability.

3.5. (*a*) Variance is 3,247. (*b*) Variance is 17,122. (*c*) Estimate is 464, standard error being 4.7.

3.6. Variance estimates are 0.20, 0.81, 1.82, 0.16, 0.16, 0.64, 0.49, 0.12, 0.12, 0.81, 0.36, and 0.09 for the 12 samples. Variance of the estimate is 0.365.

3.7. Estimate is 941, its standard error being 70 when sampling with replacement is assumed. True value is 931.

3.8. Number of parcels is 127,120 with a standard error of 2,485.

3.9. Area under wheat is 19.790 acres, the standard error being 1,271 acres.

3.10. Variance with systematic sampling is 16.81. For random sampling variance is 12.10. Intrasample correlation coefficient is 0.037.

Chapter 4

4.1. (*a*) Sample sizes are 125, 195, 847, 924, 710, and 199, variance being 0.009129. (*b*) Same. (*c*) Variance is 0.09556 with proportionate sample.

4.2. (*a*) Average is 6.96 and standard error is 0.098. (*b*) Standard error is 0.142 with no stratification.

4.3. (*a*) Sample size is 280. (*b*) Relative error is 11.2 percent with unstratified sample and 9.2 percent with optimum allocation.

4.4. Sample numbers are 1,537, 601, 519, 587, and 312.

4.5. (*a*) Allocation is 2, 7, 8, 17, relative errors being 8.27 percent and 12.46 percent. For minimizing variance of the mean, allocation is 3, 17, 7, 7, relative errors being 6.29 and 13.77 percent. (*b*) Sample size is 60, its allocation being 3, 19, 20, 18. (*c*) Allocation is 2, 10, 11, and 11.

4.6. (*a*) y_1 is 1.68, variance being $0.3413/n$ for strata of equal size. For optimum stratification, y_1 is 1.30 and variance is $0.2860/n$. Variance is $1/n$ with no stratification. (*b*) y_2 is 0.50, variance being $0.0243/n$ for strata of equal size. For optimum stratification y_2 is 0.35 and variance is $0.0152/n$. Variance is $0.0555/n$ with no stratification.

4.7. Ratio estimate is 12.76, standard error being 0.1588. Regression estimate is 12.72 and standard error is 0.1508. Difference estimate for $k = 1$ is 12.75, standard error being 0.152. Standard error is 0.2548 if census data are not used. About 120 farms are to be selected if difference estimate is used.

4.8. Estimate of ratio is 1.236. Rate of growth is 23.6 percent, standard error being 0.0269.

4.9. Using difference estimate with $k = 1$, estimate is 1,108, standard error being 15.2. About the same results with regression estimate.

4.10. Proportion is 0.5, standard error being 0.03. For sampling by individuals 278 persons are needed. Sampling in clusters needed 108 persons only.

Chapter 5

5.1. (*a*) Average number of cattle per farm is 12.25 with a standard error of 0.77. (*b*) Estimate is 13.51 and standard error is 1.35. (*c*) Ratio estimate is 11.68, standard error being 0.93.

5.2. Average is 13.51 and its standard error is 0.76.

5.3. Area under the crop is 11,000,615 acres and its relative error is 12.62 percent. About 83 villages should be selected for a relative standard error of 5 percent.

5.4. Average age is 48.4 years, standard error being 2.46 when sampling with replacement is assumed.

5.5. About 50 villages with 5 households per village is the best choice. Standard deviation is 2.92.

5.6. (*a*) Relative error is 10 percent. (*b*) Relative error is 13 percent.

5.7. Average expenditure is 11.36, its standard error being 2.2.

5.8. Average yield per plot is 167.4 with a standard error of 15.43.

5.9. Number of persons is 109,336 and its relative standard error is 6.15 percent.

5.10. Average number of persons per room is 1.7, its standard error being 0.0717.

Chapter 6

6.1. (*a*) Variances are 29.26, 13.89, and 6.16 for the three sample sizes. (*b*) Mean squares are 48.62, 33.25, and 25.52. (*c*) Probabilities are 0.12, 0.21, and 0.42.

6.2. Initial sample size is 1,870 and subsampling rate is 1 in 2.74. Cost is $2,095.

6.3. Interviewers do not differ as $s_b^2/s_w^2 = 54.535/45.397 = 1.201$ is not significant at 5 percent level. Average size is 10.46 and its standard error is 0.738. Correlation coefficient is 0.01 and interviewer errors account for 19 percent of the sampling variance.

6.4. Proportion is 0.66, standard error being 0.046. Estimate is 0.475 when respondents alone are considered.

6.5. Proportion is 1/6 and standard error is 0.076.

6.6. About 744 plots for eye estimation and 132 for measurement. Variance is 0.00159 S_y^2. For a direct sample, 225 plots can be measured; variance is 0.00444 S_y^2. *Hint:* See D. Raj, "Sampling Theory," Sec. 7.2.

6.7. Estimate is $19,688 and standard error is 1,547. *Hint:* See D. Raj, "Sampling Theory," Sec. 7.12.

6.8. Proportion is 0.64 and standard error is 0.049.

6.9. In the optimum scheme of selection $\Pr(a, B) = 0.15$, $\Pr(b, C) = 0.10$, $\Pr(b, F) = 0.20$, $\Pr(c, A) = 0.10$, $\Pr(e, D) = 0.20$, $\Pr(e, E) = 0.05$, and $\Pr(d, E) = 0.20$.

6.10. Regression estimate is 12.7426, its standard error being 0.1891.

Index